A History Of Microbiology In Philadelphia: 1880 To 2010

A HISTORY OF MICROBIOLOGY IN PHILADELPHIA: 1880 TO 2010

INCLUDING A DETAILED HISTORY OF THE
EASTERN PENNSYLVANIA BRANCH OF THE
AMERICAN SOCIETY
FOR MICROBIOLOGY FROM 1920 TO 2010

JAMES A. POUPARD, PHD

Copyright © 2010 by James A. Poupard, PhD.

Library of Congress Control Number: 2010907122
ISBN: Hardcover   978-1-4535-0392-8
              Softcover   978-1-4535-0391-1
              Ebook   978-1-4535-0393-5

All rights reserved. No part of this book may be reproduced or transmitted in any form or by any means, electronic or mechanical, including photocopying, recording, or by any information storage and retrieval system, without permission in writing from the copyright owner.

This book was printed in the United States of America.

The cover photograph was supplied by the University of Pennsylvania Archives.

To order additional copies of this book, contact:
Xlibris Corporation
1-888-795-4274
www.Xlibris.com
Orders@Xlibris.com
77277

Contents

Foreword xv
Preface and Acknowledgments xvii
Introduction 1

PART I

PHILADELPHIA AND THE EARLY PHILADELPHIA BACTERIOLOGISTS: 1880 T0 1920

1. **BACKGROUND INFORMATION** 9
    Philadelphia: Eighteenth—and Nineteenth-century
       Medical and Scientific Capital 9
    Early Establishment of Bacteriology in America 11

2. **EARLY PHILADELPHIA BACTERIOLOGY** 13
    Bacteriology Comes to Philadelphia: 1880-1900 13
    Biographical Sketches of Seven Nineteenth-Century
       First-Generation Philadelphia Bacteriologists: 1880-1900 15
        Joseph Leidy
        Edward Shakespeare
        Henry Formad
        Lawrence Flick
        Ernest Laplace
        Samuel G. Dixon
        Lydia Rabinowitsch
    Conclusions Drawn from the Biographical Sketches of
       Seven Nineteenth-century Philadelphia First-Generation
       Bacteriologists: 1880-1900 23
    Philadelphia Institutions Associated with Bacteriology
       at the End of the Nineteenth Century 26
        The University of Pennsylvania 27
            Bacteriology in the Medical Department
            Laboratory of Hygiene
            William Pepper Laboratory of Clinical Medicine

        School of Dental Medicine
        School of Veterinary Medicine
        Jefferson Medical College 38
        Woman's Medical College of Pennsylvania 39
        Hahnemann Medical College 40
        Medico-Chirurgical College 40
        Philadelphia Polyclinic and College for Graduates ....
            in Medicine 41
        Blockley Hospital (Blockley-Almshouse) 41
        Pennsylvania Hospital 42
        Wistar Institute of Anatomy and Biology 43
        Philadelphia Health Department 43
        H. K. Mulford Company 44
    Bacteriology Books By Philadelphians at the End of the
        Nineteenth Century. 45
    Founding of the Society of American Bacteriologists 46

3. **A TRANSITIONAL OR SECOND-GENERATION GROUP OF EARLY PHILADELPHIA BACTERIOLOGISTS DURING THE LATE NINETEENTH AND EARLY TWENTIETH CENTURY** 48
        Alexander Abbott
        Michael Ball
        David Bergey
        Albert Ghriskey
        Samuel Kneass
        Joseph McFarland
        Adelaide Peckham
        Robert Pitfield
        Mazyck Ravenel
        Randle Rosenberger
        Ernest Sangree
        Allen Smith
        Alonzo Stewart
    Conclusions Drawn from the Biographical Sketches of Thirteen
        Transitional or Second-Generation Philadelphia Bacteriologists 55
    Comparison of the First and the Transitional or
        Second-Generation Philadelphia Bacteriologists 58

4. **FOUR PERSPECTIVES ON ENTERING THE NEW CENTURY: PHILADELPHIA, THE SOCIETY OF AMERICAN BACTERIOLOGISTS, EARLY BACTERIOLOGISTS, AND PHILADELPHIA INSTITUTIONS UPDATE** 59
Philadelphia at the Turn of the Twentieth Century 59
Update on the Society of American Bacteriologists: 1900 to 1945 60
The Continuation of the First-Generation Philadelphia
  Bacteriologists: 1900-1938 62
The Continuation of the Transitional Generation of
  Bacteriologists: 1900-1955 62
Philadelphia Institution Update 66
  Henry Phipps Institute for the Study,
    Treatment and Prevention of Tuberculosis 66
  Jefferson Medical College 67
  Temple College-University School of Medicine................................ 67
Books by Philadelphia Microbiologists 68

PART 2

THE EASTERN PENNSYLVANIA BRANCH OF THE AMERICAN SOCIETY FOR MICROBIOLOGY 1920 TO 2010

5. **THE MICROBIOLOGICAL CLUB AND THE FOUNDING OF THE EASTERN PENNSYLVANIA CHAPTER OF THE SOCIETY OF AMERICAN BACTERIOLOGISTS: 1912 TO 1939** 71
The Microbiological Club 1912-1923: Forerunner of
  the Eastern Pennsylvania Chapter 71
The Eastern Pennsylvania Chapter: Origin and
  Early Years (1920-1936) 73
The Eastern Pennsylvania Chapter: 1936-1939
  and the Branch Archives 77
Notes on Branch Meetings: 1936 to 1939 80
Summary of the Branch in the 1930s 81

6. **THE EASTERN PENNSYLVANIA BRANCH: 1940-1949 THE DECADE OF WORLD WAR II AND THE 1947 ANNUAL SAB MEETING** 83
   Overview of the Branch in the 1940s 83
   Summary of Branch Meetings 85
   Branch Officers 86
   Branch Events by Year 86
   Summary of the Branch in the 1940s 93

7. **THE EASTERN PENNSYLVANIA BRANCH: 1950-1959 A DECADE OF INNOVATIVE PROGRAMS** 94
   Overview of the Branch in the 1950s 94
   Summary of Branch Meetings 94
   Branch Officers 97
   Branch Events by Year 97
   Summary of the Branch in the 1950s 100

8. **THE EASTERN PENNSYLVANIA BRANCH: 1960-1969 A TIME OF TRANSITION AND TURMOIL: ANNUAL MEETING, FROM SAB TO ASM, FROM CHAPTER TO BRANCH, AND A SPLINTER GROUP WITHIN THE BRANCH THE CLINICAL MICROBIOLOGY SECTION** 101
   Overview of the Branch in the 1960s 101
      Summary of Branch Hosted Sixtieth Annual SAB Meeting
      The Clinical Microbiology Section: 1967–1969
   Summary of Branch Meetings 105
   Branch Officers 107
   Branch Events by Year 108
   Summary of the Branch in the 1960s 113

9. **THE EASTERN PENNSYLVANIA BRANCH STRUGGLING TO CREATE A UNIFIED BRANCH: 1970-1979** 115
   Overview of the Branch in the 1970s 115
      The Clinical Microbiology Section
      Topics in Clinical Microbiology

Summary of Branch Meetings 118
    Hot Topics of the 1970s
    Clinical Microbiology Oriented November Symposia
    Basic Science Symposia
Branch Officers 121
Branch Events by Year 122
Summary of the Branch in the 1970s 132

10. THE EASTERN PENNSYLVANIA BRANCH: 1980-1989 THE PEAK ACTIVITY YEARS, EXPANDING ORGANIZATION, FACING ISSUES, AND CELEBRATING FIVE HUNDRED MEETINGS 133

Overview of the Branch in the 1980s 133
Summary of Branch Meetings 135
    Hot Topics of the 1980s
    Symposia Held in the 1980s
    The Five Hundredth Branch Meeting Celebration (15 December 1987)
Branch Officers 138
Other Branch Officers 138
Branch Events by Year 139
Summary of the Branch in the 1980s 144

11. THE EASTERN PENNSYLVANIA BRANCH: 1990-1999 NEW LECTURE SERIES, SPECIAL EVENTS, AND A CHANGING ENVIRONMENT 145

Overview of the Branch in the 1990s 145
Summary of Branch Meetings 146
    Hot Topics of the 1990s
    Symposia Held in the 1990s
    Special Meeting in Honor of Smith Hall and Dr. Harry Morton 1992
    Infection and Immunity Forum Initiated 1992
    The Distinguished Branch Member Lecture Series Initiated 1996
    The Six Hundredth Branch Meeting Celebration (14 September 1998)
Branch Officers 152
Other Branch Officers 152
Branch Events by Year 153
Summary of the Branch in the 1990s 160

12. THE EASTERN PENNSYLVANIA BRANCH: 2000-2009 ADJUSTMENTS TO A NEW CENTURY, ASM GENERAL MEETING, AND THE LABORATORY OF HYGIENE...... 162
Overview of the Branch in the 2000s 162
    The 109th ASM General Meeting
    Milestones in Microbiology Dedication
Summary of Branch Meetings 164
    Hot Topics of the 2000s
    Symposia Held in the 2000s
Branch Officers 166
Other Branch Officers 167
Branch Events by Year 167
Summary of the Branch in the 2000s 174

13. MICROBIOLOGY IN PHILADELPHIA: 2010 AND BEYOND 176
The Eastern Pennsylvania Branch 176
The Pharmaceutical Industry 177
Medical and Clinical Microbiology 178
Research Institutions and Health Departments 179
Beyond 2010 180

14. SUMMARY AND CONCLUSIONS 181
Meeting Topics and Philadelphia Institutions 184
Branch Presidential Institution Summary: 1920 through 2010 185
Philadelphia Institutions that Provided Speakers
    for the Regular Branch Meetings: 1940 through 2009 185
Regular Scheduled Branch Meeting Summary 186

A Chronology of the Eastern Pennsylvania Branch Related History 188

References 227

List of Appendices 235

I.   Branch Presidents: 1920 To 2011 237
II.   Branch Nonpresident Officers: 1920 TO 2009 241

III.	A Listing of Branch Meetings: 1936 Through 2009	245
	Name Index	318
	Subject Index	331
IV.	The Clinical Microbiology Section of The Eastern Pennsylvania Branch: 1967 To 1973	337
V.	Eastern Pennsylvania Branch Sponsored Symposia 1969 To 2009	342
VI.	Eastern Pennsylvania Branch Sponsored Workshops: 1969 to 2009	359
VII.	Philadelphia Infection and Immunity Forum 1992 to 2009	362
VIII.	The Stuart Mudd Memorial Lecture Series 1976 to 1995	371
IX.	The Distinguished Branch Member Lectureship 1996 To 2006	374
X.	The Annual Industry Sponsored Guest Lectureship Series: 1962-1971	377
XI.	Celebration of the 500th and 600th Branch Meetings: 1987 and 1998	379
XII.	History of The Branch Newsletter: 1937 To 2009	382

Index 385

This book is dedicated to two members of the Eastern Pennsylvania Branch of the American Society for Microbiology, Anna Feldman-Rosen and Josephine Bartola. They both epitomize the spirit of dedication and volunteerism by serving on whatever committee-needed assistance over many years and asked nothing in return. It is this spirit that contributed to the success of the Branch and is this type of volunteerism that is needed for the continued success of Philadelphia as a center for microbiology.

Foreword

IT IS WELL TO KNOW WHERE WE HAVE BEEN, SAITH THE HISTORIAN, FOR IT PROVIDES LIGHT TO MARK THE PATH AHEAD.

In 1643, the history of Europe was being determined by the Black Death (plague), a disease caused by a rodent-transmitted bacterium. It was during this year that a small group of immigrants settled at a site where two rivers joined in their passage to the sea. This place would become the city of Philadelphia.

During the following century, diseases were recognized and described as medical entities, and physicians speculated about etiologies. Organizations of such practitioners were established as early as 1743 as indicated in the text. But microbiology remained as an undeveloped child of biomedicine.

Philadelphia has always been a city of beginnings. It was, of course, the place where the great experiment in human freedom and self-government began. It also was where the importance of clinical laboratories was conceived by men like William Pepper and Sir William Osler. The concept of computer science rises from the basement of an old engineering building and the technology of thoracic surgery received worldwide response by people who traveled to the round building behind the German Hospital on Girard Avenue.

Organization of persons with common interests permits the opportunity to compare and debate. The history of science in Philadelphia suggests that this has been a productive aspect of microbiology throughout the early years and particularly as various opinions from foreign laboratories were brought forth. Since Leeuwenhoek described his "little animals" progress has been the child of honest debate. In Philadelphia, with its complex of microbiology facilities, there is bound to be unique sessions of productive argument.

Dr. Poupard is a scientist of multiple skill and experience. In addition, he is a lifelong resident of Philadelphia. He describes the old buildings of long ago as if he paused at the front door. The impression is that he has walked to the site and heard the sounds of a time long past. It is, perhaps, the mind of the historian who murmurs of that which is long gone, "I can see it now." And his discussions of individuals who were the microbiologists of generations now past seem to suggest a bit of personal interview.

The biographies in the period of World War II are of particular interest because they were our colleagues and, most important, our friends. They were the contributors to science and education in difficult times. Some went to battle and did not return. And I remember the great visitors of our time who came to meetings we hosted in Philadelphia.

The most important accomplishments of the people of our science are that microbiologists tend to be good teachers. This may be because they frequently deal with the combating forces of living systems, which seem to defy definition and increase the depth of the unknown.

Poupard has taken us to the present in organized microbiology. I hope that the stewards of the future will remember to nurture the tale that may have no end.

JAMES E. PRIER, DVM, PhD, JD
Emeritus Director
Pennsylvania Public Health Laboratory

Preface and Acknowledgments

This book is an attempt to document some elements of the evolution of microbiology in the Philadelphia area starting in 1880. Since the Eastern Pennsylvania Branch of the American Society for Microbiology—which was founded in 1920—played such an important role in characterizing Philadelphia microbiology, the accent from 1920 to the present focuses on the activities of the Branch and how it changed over time. The accent of the period prior to 1920 is on the people and institutions that played a role in establishing bacteriology as a separate specialty of science. As bacteriology and microbiology developed, it branched out in many directions. Although subspecialties within the field—such as food, veterinary, agricultural, and industrial microbiology—are important aspects of microbiology, this book, especially in the early years, focuses on the medical aspects of bacteriology and microbiology. The author apologizes for neglecting some of these subjects but, to do justice to all these subjects, would require a much larger final product. The focus of this current work is very much limited to developments occurring in Philadelphia and the role of the Eastern Pennsylvania Branch in documenting this rich history.

The author would like to acknowledge the following people who contributed to this work: Barbara Poupard, Alice Poupard, and Dr. Linda Miller for proofreading and correcting each chapter as it developed; to Alice Poupard and Sharon Mellor for typing assistance with the numerous appendices; to Dr. James Prier for writing the foreword; to Jean Buchenhorst for supplying information on medical technology programs; to Jeff Karr, archivist, American Society for Microbiology Archives and Center for the History of Microbiology, for confirming facts and supplying material from the archives; Dr. Joseph McFarland who made significant contributions in documenting the early years of Philadelphia Microbiology. Dr. Harry Morton deserves special recognition for collecting a great deal of material relating to the Branch and realizing the importance of protecting the material for future generations. Much of his collected material now resides in the archives of the Eastern Pennsylvania Branch.

INTRODUCTION

In the eighteenth and nineteenth centuries, Philadelphia became the medical and scientific capital of the British Colonies and then the United States. Philadelphia also became the center of the developing pharmaceutical industry in the United States. Therefore, it is not surprising that in the later part of this period, as pathology and bacteriology began to develop into defined sciences in Europe, several Philadelphians traveled to France, Germany, and England to learn more about these subjects and brought that knowledge back to their city. Philadelphians traveling to Europe with the intent to study bacteriology began in the 1880s. The establishment of this new specialty of bacteriology, and its expansion in scope to become microbiology, occurred at an ever-increasing pace from that time into the next century. The first part of this book documents the development of bacteriology in Philadelphia from 1880 through the early years of the new century. From 1920 to the present, the focus is on the establishment and developments in microbiology as they relate to the Philadelphia Chapter of the Society of American Bacteriologists (SAB), which later changed its name to the Eastern Pennsylvania Branch of the American Society for Microbiology (ASM).

The Eastern Pennsylvania Branch of the ASM became one of the most active branches within the national ASM organization. During its ninety-year history, a significant amount of material, representing almost every aspect of microbiology in the Philadelphia area, was collected and stored at various locations. This collection was designated as the Branch Archives. This valuable collection of material was stored in boxes that moved from one location to another. As the number of boxes grew exponentially, the task of organizing this material became daunting. This book could not be written until the material was organized to form a true archive collection. The desire to write this book provided the incentive to first organize the collection into material from individual decades to make the ninety years of accumulated material manageable. The two-year project to organize the material finally enabled the contents of the collection to be analyzed. It became apparent that a history of the Eastern Pennsylvania Branch was, in reality, an important record of microbiology as it unfolded in the Philadelphia area.

An early issue to be addressed was to determine what date to use to start the history of bacteriology in Philadelphia. Like all urban areas of North America, infectious diseases in the form of numerous epidemics plagued Philadelphia from colonial times. These outbreaks always stimulated intense discussion on control measures and certainly played an important role in the history of this city. The epidemic of yellow fever in 1793 was a devastating event and resulted in eliminating any hopes of Philadelphia remaining as the capital of the United States. The significant epidemics prior to 1880 are well documented in the scientific and medical literature, but these all occurred prior to the establishment of bacteriology or virology as separate disciplines of either medicine or science in the United States. Since the intent of this study is to focus on bacteriology (and microbiology) as a defined specialty, epidemics prior to 1880, although quite significant, are not relevant to the current project.

Starting in colonial times, there were several Philadelphia physicians who took an interest in improving sanitary conditions in the city and addressing issues of public health. By the mid-nineteenth century, there were Philadelphians who were following the new developments in medicine and science emanating from European medical schools and universities. Also by mid-century, Philadelphia boasted prominent naturalists like Joseph Leidy. Leidy was this country's leading parasitologist and was performing comprehensive studies and classifying single-celled organisms as early as 1850. However, prior to 1880, no Philadelphian made any claim to being a bacteriologist. It was in the 1880s that Philadelphians first began identifying themselves as bacteriologists. Efforts made by later bacteriologists like Joseph McFarland to document these early years, combined with a wealth of collections in various Philadelphia Institutional Archives, resulted in a significant amount of material on the education, travels, and professional careers of these early bacteriologists. For these reasons, as well as others discussed in chapter 2, it seemed most appropriate to start the History of Philadelphia Bacteriology in 1880. I have designated this early group of Philadelphians as the "first generation" of Philadelphia bacteriologists. A group designated as the "second generation" of Philadelphia microbiologists (chapter 3) staffed the key bacteriology related positions at medical schools, pharmaceutical companies, and other Philadelphia institutions as the nineteenth century came to a close and the new century began. Several members of this second generation became members of the Microbiology Club (chapter 5), a forerunner of the Branch, and a few of this generation—like Alexander Abbott, Randle Rosenberger, and David Bergey—became early Branch presidents. Philadelphia is indebted to members of this second generation for establishing Philadelphia as a major center for teaching and research relating to the new science of bacteriology and microbiology.

The Eastern Pennsylvania Chapter/Branch has gone through several stages since its founding in 1920. From a relatively slow start, with about forty members (chapter 5), it grew into a highly organized Philadelphia institution with over one thousand active members. One of the goals of the Branch has always been to serve the needs of bacteriologists, and later microbiologists, residing or working in the Philadelphia area. Throughout its ninety-year history, it has held early evening meetings, anywhere from four to ten months each year. By the close of 2009, the number of these monthly Branch meetings exceeded seven hundred, most of which have been carefully documented. Although some of the content of the early meetings of the first sixteen years have been lost, the titles and speakers of all meetings from 1936 to the present (with rare exception) are well documented. An analysis of the content of these meetings through 2009, as presented in this book, is a demonstration of the subjects that were of interest to Philadelphia microbiologists. Since a large number of the presenters were local, it also is a reflection of the research being conducted at various Philadelphia institutions.

The monthly meetings are just one source of valuable information on the evolution of microbiology. In addition to the monthly meetings, starting in late 1969, the activities of the Branch were greatly expanded with the introduction of a comprehensive series of clinical and basic science symposia, workshops for clinical microbiologists, and a comprehensive plan for the production of innovative educational material for microbiologists, technologists, educators, and students. Although many of these activities are now common among the scientific and medical communities, these projects were considered quite innovative at the time and received national attention. This became a reflection of the talent and creativity of Branch members, as well as a demonstration of how an extremely organized Branch was able to serve the needs of a very diverse microbiology community. This spirit of innovation was expressed in the 1990s with the initiation of a new series, the "Annual Infection and Immunity Forum," which was designed to serve the needs of graduate students and to get these students more involved with the Branch. Once launched as an annual event, the responsibility of organizing this annual event was placed in the hands of the Branch Student Chapter with minimal input from senior Branch advisors. An analysis of the changing scientific content of all these activities over the years creates a wealth of material for use in the current study. An important goal of this book is to bring attention to this valuable collection of material that is now deposited in the Branch archives for future study and additional analysis.

For the current study, the material in this book from 1920 through 2009 is organized by decade as chapters 5 through 12. Over the last ninety years, the field of microbiology has experienced many changes, and this is reflected in the history of the Branch. A look at the early meetings, even in the early decades, indicates that the term *bacteriology* was too limiting to describe the field. The emphasis, particularly on virology and immunology, is evident from the early Branch records. The need to replace bacteriology with the term *microbiology* was apparent prior to the official name change by the national society, as well as the Branch, in the early 1960s. The most significant and radical change in the Branch occurred in the late 1960s. There had always been discussions in the Branch concerning the need to balance an emphasis on both the basic science and clinical aspects of microbiology. This issue came to a head in the sixties when a large number of clinical microbiologists from the Branch decided to form a separate group (chapter 8). Through the efforts of many Branch members, a compromise was reached that enabled the clinical microbiologists to remain within the Branch, but with a separate executive committee to plan meetings on an alternative evening from the regular monthly meetings and to plan other educational activities, such as workshops and symposia. What was initially perceived as a crisis eventually resulted in a period where the Branch reached its highest point if number of members and a vast expansion of activities and organizational structure are used as a gauge. This was accomplished by once again unifying the Branch under one executive committee in the 1970s (chapter 9).

Another dramatic shift in the Branch occurred in the 1990s (chapter 11). This was brought about by many factors; however, two factors were significant: a decrease in the number of clinical microbiology members and an executive committee that failed to replace older executive committee members with new members. This would not have been an issue if the goal was just to focus on the monthly meeting program. However, a critical mass of active members is needed to staff the large number of committees and maintain the diverse educational activities that Branch members had come to expect. These problems were carried over to the next decade (chapter 12), and they remain a challenge to the current Branch executive committee. Several of these issues are related to some basic changes in the concept of microbiology since the field is now divided among an increasing number of scientific specialties, such as molecular biology, genetics, and biochemistry. Many researchers in this new generation are working strictly on components of a particular microorganism; therefore, they do not view microbiology as their prime or unifying interest.

As the material in this book proves, the complex road from bacteriology to microbiology, and the development of a significant number of subspecialties and disciplines, simply reflect the continuing evolution of the field. It demonstrates the need for a new generation to step forward, as it always has in the past ninety years, to sustain the development of microbiology in Philadelphia. Therefore, the Eastern Pennsylvania Branch will continue to adapt by drawing on the more unifying aspects of contemporary microbiology, like public health, microbial ecology, and environmental studies. Microbiology in Philadelphia, over the last ninety years, has demonstrated a definite resilience. Currently, the challenges may appear daunting, but not more so than in the 1930s or during the troublesome years of the Second World War. There were always those Branch members who stepped forward to "take charge" and found innovative ways to serve the needs of microbiologists in the Philadelphia community. Microbiology remains a basic concept of contemporary biology and infectious diseases. It makes little difference if the focus is on cellular membranes, new energy sources, nucleic acid replication, microbial ecology, or the discovery of antimicrobial agents; microorganisms hold a key position for future generations. We need specialists who can work on the various components of life, as well as those who understand how a microorganism exemplifies a unified body of information that brings all these components together. Whether it is within an academic, clinical, ecological, or environmental context, the history of microbiology in Philadelphia demonstrates that the new century needs specialists who understand the broader field of microbiology. Philadelphia has the framework to stimulate innovative students who will occupy that new generation of specialists and generalists that recognize the unifying principles of microbiology. In the 1960s, the clinical microbiologists brought about change. In the second decade of this century, a new generation of Philadelphians must come forward and bring about the next stage in the evolution of microbiology. The history described in this book confirms that Philadelphia microbiologists can and will live up to these many new challenges of a changing science.

PART I

PHILADELPHIA AND THE EARLY PHILADELPHIA BACTERIOLOGISTS: 1880 TO 1920

Background information on eighteenth—and nineteenth-century Philadelphia as the primary North American city for both science and medicine. Bacteriology as a separate scientific discipline. Identification and biographical sketches of the first—and second-generation bacteriologists and prominent Philadelphia institutions relating to bacteriology. The role of the second generation bacteriologists at the transition from the nineteenth to the twentieth century.

CHAPTER ONE

Background Information

The foundation for the pharmaceutical industry in the United States was laid in Philadelphia between 1818 and 1822 with the establishment of half a dozen enduring and fine chemical manufacturers.

—J. Liebenau (1)

In the period from 1876 to 1882, before Koch's discovery of the tubercle bacillus, his early work had a limited audience in America, where there were no established public or private laboratories.

—R. Maulitz (2)

Philadelphia: Eighteenth—and Nineteenth-century Medical and Scientific Capital

In the eighteenth century, Philadelphia was not only the largest city in British North America, it also became the capital of the newly formed United States. It is not surprising that this led to the establishment of a medical community and institutions to serve the needs of a rapidly growing population. The medical heritage of Philadelphia evolved in the eighteenth century with a proposal for an almshouse in 1712, although money to build the institution was not approved until 1728. The first public almshouse finally opened in 1732, occupying the entire block between Third and Fourth Streets, as well as Spruce and Pine Streets. Regarded as a model institution, it had separate facilities for the indigent and the insane and also an infirmary. When it later moved to West Philadelphia, it became known as Blockley Hospital, and eventually became the Philadelphia

General Hospital. Pennsylvania Hospital, founded much later in 1751, is recognized as the nation's first hospital. It was founded by Benjamin Franklin and Dr. Thomas Bond. Pennsylvania Hospital not only holds the distinction as the first hospital in the colonies, but it is still in existence at the same location. The first medical school in the colonies was established in 1765 at the University of Pennsylvania. The College of Physicians of Philadelphia, established in 1787, was modeled after the Royal College of Physicians in London.

Philadelphia was destined to become the medical capital of the United States. No other North American city can claim such a rich medical heritage. It is significant that the nineteenth century ushered in the golden age of Philadelphia Medicine. With the opening of Philadelphia's second medical school, Jefferson Medical College, in 1824, Philadelphia made medical history since neither Paris nor London had two medical schools at that time. (3) In 1848, the Homeopathic College of Pennsylvania was founded; in 1868, the name was changed to the Hahnemann Medical College. The Female Medical College of Pennsylvania was founded in 1850, and the name changed to the Woman's Medical College in 1867. As the nineteenth century came to a close, the Philadelphia College of Osteopathic Medicine was founded in 1899, and a sixth Philadelphia Medical School was about to open. Temple University School of Medicine opened in September of 1901.

The oldest scientific society in America, the American Philosophical Society, was founded by Benjamin Franklin in 1743. Institutions like the the Academy of Natural Sciences of Philadelphia (1812) and the Wistar Institute of Anatomy and Biology (1892), the oldest independent biomedical research institute in the United States, encouraged studies in the biological sciences. This period also witnessed the founding of several prominent science departments in colleges and universities with well-attended courses in scientific subjects. Unique scientific institutions, such as the Philadelphia College of Pharmacy and Science, established in 1821, also gained prominence during this period.

With the presence of six medical schools, many prominent scientific institutions, and with a growing interest in public health, it was natural that Philadelphia played a significant role in embracing the new science of bacteriology that began to emerge during the close of the nineteenth century.

Early Establishment of Bacteriology in America

There were no bacteriologists, so physicians, zoologists, botanists, engineers, and old-fashioned naturalists enthusiastically took over the new tools and the new thinking.

—Paul F. Clark (4)

The above quote by Paul Clark summarizes the situation that resulted in the birth of bacteriology as a separate scientific discipline. In the early days of its development, bacteriology evolved from other disciplines of medicine and science. Therefore, the history of bacteriology is associated with many related fields. Five general fields stand out as critical forerunners: medicine, pathology (especially the subspecialty of histology), biology, microscopy, and public health (especially the subspecialty of hygiene). It is difficult to assign specific events within these general disciplines that were critical to the development of the new discipline of bacteriology since they all played a role in forming the individual specialty of bacteriology. This is particularly true if one tries to generalize to all of America. It will be left to people like Paul Clark with his book *Pioneer Microbiologists of America* to set the broader stage of early American microbiology. The task becomes easier if the focus is placed on one particular geographic area since the people and institutions within that area can be studied in detail to help draw conclusions relevant to the evolution of the concept of bacteriology in that location. Separating bacteriology as an independent discipline from public health and hygiene is particularly difficult. Obviously, factors like public health, sanitary conditions, and water quality were very compelling issues in major cities like New York and Philadelphia, with large seaports and growing immigrant populations. Interest in public health and sanitation always increased during periods of major outbreaks of communicable diseases. There is a significant amount of descriptive literature documenting various outbreaks or epidemics dating back to colonial time.

It was not until the latter part of the nineteenth century that the medical establishment came to the full realization that infectious diseases were caused by specific organisms that could be seen with the aid of a good microscope and that these organisms could be grown in pure cultures. This created the need for specialists to study, characterize, and teach others about these organisms. This need was responsible for the science of bacteriology coming to fruition during the last two decades of the nineteenth century.

As the last half of the nineteenth century progressed, there is little doubt that the early work of Louis Pasteur, Robert Koch, and their coworkers stimulated interest among several individuals in America. In the 1880s, the work performed by Pasteur on rabies aroused significant interest among both the medical and general population. The same could be said about the work of Koch on tuberculosis. However, at the start of the 1890s, the announcements by Koch, concerning the use of tuberculin as a treatment for tuberculosis, and the work of Emil von Behring on diphtheria antitoxin aroused great interest in the United States. These two discoveries were significant factors in the continuing process of separating bacteriology from its related fields of study. Therefore, the focus of this current work starts in the 1880s, when it became obvious that bacteria, and other organisms, deserved special attention. It was during this period that one can see how bacteriology starts to separate from pathology and how a different, more sound scientific element enters the fields of public health, hygiene, and infection control. It is also during the last twenty years of the nineteenth century that bacteriology starts to be incorporated into the medical school curriculum, and one can start to identify those people and institutions that began to focus on this new branch of science and medicine.

CHAPTER TWO

Early Philadelphia Bacteriology

As in Europe, many drug manufacturers in America arose from pharmacies, especially those in Philadelphia in the first half of the nineteenth century.
—J. P. Swann (1)

Philadelphia's position as an outstanding medical center played an important part in bringing medical science to the notice of the city's pharmaceutical firms at an early date.
—J. Liebenau (2)

In microbiology and related medical sciences, the transition from descriptive research to hypothesis-driven research has generally reflected the maturation of these fields. In the early stages of a field, descriptive studies may "represent the first scientific toe in the water."
—D. A. Grimes and F. F. Schulz (3)

Bacteriology Comes to Philadelphia: 1880-1900

The period of 1880 to 1900, at first, seems to be a difficult time to document a field like bacteriology. One reason is that it was not a period when Philadelphians were making significant theoretical or experimental contributions to the field. A second reason has been noted earlier. Especially in the early part of the period, no one identified themselves as a bacteriologist. Most people interested in this field were physicians, and most of these physicians would have identified themselves as pathologists or microscopists. In the early years of this period, many physicians were still adamant in their belief that the role of microorganisms in causing specific diseases remained unproven. In 1876, Joseph Lister made a visit to Philadelphia to speak on his ideas relating to antisepsis. As noted by Maulitz, Lister's visit to Philadelphia was a success and he "gained a limited degree of acceptance for the 'bacterian' thesis among the medical community." (4) By the 1880s, mainly due to the work of Robert Koch, more Philadelphians were being converted to accepting the significant role of bacteria as causes of diseases like tuberculosis.

Although it may not be the most academic approach for this period, the focus of necessity will be almost exclusively on certain people and institutions. Therefore, some disqualifiers are in order at this time. Some people who have studied this period have included those physicians who held the position of chair or professor of pathology or medicine. Men like James Tyson, first professor of pathology at the University of Pennsylvania, and others have not been included in the analysis. Although Tyson and others did give lectures on the subject of bacteriology, they identified themselves as pathologists, not bacteriologists. The professors mentioned below, all eventually became specialists in bacteriology. Therefore, what follows is a collection of Philadelphia men and women who played a direct role in separating bacteriology as a specialized field of study, focusing on concepts, which we eventually recognized as bacteriology.

Each of these physicians, and one PhD, play a role in this evolving story. As can be seen by following the careers of these people, the concept of a causative organism becomes a critical factor. This is especially significant when it comes to separating bacteriology from the broader field of public health. Another important factor that becomes apparent during this period is a sense, among the general public, that a cure for a communicable disease was a real possibility. This concept placed increasing emphasis on making a sound diagnosis prior to treatment. Starting in 1884, textbooks on various aspects of bacteriology reached Philadelphia. (5) It was during this period that increasing numbers of Philadelphia physicians started to travel to Berlin and Paris to learn more about these exciting new developments. It becomes very apparent to anyone who studies the period 1880-1900, that Philadelphia has a rich history in bacteriology. The paper written by Joseph McFarland, entitled "The Beginning of Bacteriology in Philadelphia," is an excellent account of those early years. (5) He has created a roster of the key people and institutions, as he interpreted the field in 1936. McFarland was a Philadelphian and was familiar with the institutions and most of the early Philadelphia pioneer bacteriologists personally or through conversations with their students. Anyone who is interested in studying early Philadelphia bacteriology, or the significant institutions in Philadelphia that played a role in the development of bacteriology, should start with his valuable contribution.

Biographical Sketches of Seven Nineteenth-Century First-Generation Philadelphia Bacteriologists: 1880-1900

There are five men who characterize this early period in the evolution of bacteriology in Philadelphia. These five men are Edward Orem Shakespeare, Henry F. Formad, Lawrence Flick, Ernest Laplace, and Samuel Gibson Dixon. The careers of these five men will be used to draw conclusions and attempt to make some generalizations concerning the events occurring during this period. However, like all endeavors in the history of science, there are outliers who must be included. One person is Joseph Leidy, whose career started earlier than the five men listed above. The other person is Lydia Rabinowitsch, who entered the Philadelphia scene later than the others and is unique, not just because of her gender, but because she arrived in Philadelphia with a PhD from a European University.

What follows are biological sketches of each of these seven people. The purpose of these summaries is to provide some facts relating to their relationship to Philadelphia, with an accent on how their careers evolved, in order to provide information on how their careers shifted from areas like pathology or microscopy to bacteriology. Some generalizations on the careers of these seven individuals will be provided following the last sketch on Lydia Rabinowitsch.

Joseph Leidy (1823-1891) was born 9 September 1823 in Philadelphia. He attended the Reverend William Mann's Classical Academy and received his MD in 1844 from the University of Pennsylvania. He practiced medicine for a very short time but was more interested in pursuing an academic career. Starting in 1844, he became an assistant microscopist to Paul Goddard at the University of Pennsylvania and then dissecting assistant to the professor of anatomy, William Horner. His initial work in natural science in 1851 brought him immediate recognition along with membership in the Boston Society of Natural History, as well as the Philadelphia Academy of Natural Sciences. In 1845, he became librarian, and a year later, he became curator at the academy, a position he held for the rest of his life. He became the foremost microscopist in America. He published over eight hundred scientific reports and was the founder of American Vertebrate Paleontology. He published a theory of evolution prior to the publication of *The Origin of Species*. His expertise was in taxonomy, but he did perform some experimental work on the transplantation of human cancer tissue into frogs.

Leidy's primary academic career is associated with the University of Pennsylvania. In 1853, at the age of thirty, he became professor of anatomy and taught in the medical school for thirty-eight years. His 1861 textbook on human anatomy became the standard text for medical students for decades following its release. In 1885, he founded the Department of Biology and became the first professor of biology at the university. From that time forward, he taught at both the medical school and at the college. He pioneered the need for preclinical as well as clinical aspects of a medical education. He taught both biology and the natural sciences to most of the early bacteriologists who attended the University of Pennsylvania.

In addition to the University of Pennsylvania, he was associated with several other Philadelphia institutions such as the Franklin Medical College, Saint Joseph's Hospital, and during the Civil War, the Satterlee Military Hospital in West Philadelphia. In 1871, he established the Department of Natural History at Swarthmore College. He was also president of the Wagner Free Institute of Science. He was president, fellow, and recipient of many awards from prestigious societies both in the United States and Europe.

His most significant contribution to bacteriology (really microbiology) was in the area of parasitology and protozology. He was America's first and most prominent parasitologist, a field he worked in during his entire career. His many contributions to parasitology include his 1846 identification of trichina larvae in pork, and he reported on the medical implication of the proper cooking of meat. In 1856, he discovered and described canine heartworm. In 1851, he wrote *A Flora and Fauna within Living Animals*, which was published in 1854. This massive and well-illustrated work describes, in remarkable detail, the plants, animals, protozoa, and worms that live in and on various forms of life. In this work, as well as in other shorter reports, he identified many new species of both parasitic and free-living life forms. In this publication, Leidy reports that he swallowed large doses of vibrio and many species of protozoa to prove that they were not harmful to man. This publication was well received in both the United States and Europe and brought much world attention to Philadelphia. His later 1879 publication, *Fresh-Water Rhizopods of North America*, was described as a masterpiece and is still considered a reference for this aspect of protozology. Leidy published several short works in bacteriology or infusoria. In 1882, he wrote a detailed description of his work on *Bacillus anthracis* from a cow. (11) This report included measurements of various forms of the organism and a complete description of morphological changes in the bacillus over time. Although this report was written in 1882, the language and content of his description seems as if it were extracted from a contemporary medical journal.

It is difficult to summarize Leidy's complex career in such a short space. One phrase that epitomizes his career is the title of Leonard Warren's 1998 biography *Joseph Leidy: The Last Man Who Knew Everything*. Leidy stayed in Philadelphia his entire life, except for a few excursions to Europe and out West to collect material. He died at his home on Spruce Street on 30 April 1891.

Edward Orem Shakespeare (1846-1900) is credited as the first and most distinguished early Philadelphia bacteriologist. He was born 19 May 1846 in New Castle County, Delaware. An 1867 graduate of Dickinson College, Carlisle, Pennsylvania, he received his MD in 1869 from the University of Pennsylvania. He distinguished himself early in his career as an ophthalmologist, lecturing at the University of Pennsylvania and practicing eye surgery and refraction at Blockley Hospital. He became interested in pathology and made significant translations of both French and German works into English. This gave him a definite advantage in following the latest developments in bacteriology. It is also of interest that he most likely possessed and made good use of, one of the rare oil-immersion microscopes in Philadelphia. In 1880, he became curator of the museum at Blockley Hospital. In 1882, he was appointed pathologist, and in 1889, he became bacteriologist at Blockley. In 1885, he was commissioned by the mayor of Philadelphia to study the cause of typhoid fever in Plymouth, Pennsylvania. He characterized the lesions found at autopsy. The results of his work were summarized in "The Lesson Taught by the Epidemic at Plymouth Concerning Typhoid Fever," which emphasized the importance of a pure water supply and the need for proper sanitation. Later that year, Shakespeare was appointed by Grover Cleveland to travel to Spain, and elsewhere, to study a cholera epidemic. His five years of travel and study resulted in his work, *Report on Cholera in Europe and India*. This thousand-page report was regarded as an encyclopedic history of the disease and was presented to Congress in 1890. The American Medical Association hailed it as a "great work," and Shakespeare was regarded as the leading expert on cholera in the United States. Because of this background, he served as port physician in Philadelphia during the cholera scare of 1892. Together with Walter Reed and Victor Vaughan, Shakespeare, as major and brigade surgeon of the U.S. Volunteer Army, investigated the health conditions in army camps during the Spanish American War. Their conclusions discounted the results of clinical examinations when they were contradicted by bacteriological findings. Unfortunately for the development of bacteriology in Philadelphia, Shakespeare died at the age of fifty-four at his home in Rosemont, Pennsylvania, on 1 June 1900.

Henry F. Formad (1847-1892) was born 10 March 1847 in Eastern Europe in Scheleznowodsk, a province of the Caucasus, and studied medicine in Odessa and at the University of Heidelberg. He came to America and graduated from the University of Pennsylvania as an MD in 1877. That same year, he became demonstrator of pathology at the University of Pennsylvania. In 1880, he was appointed microscopist at Blockley, then pathologist in 1887, and a Philadelphia coroner's physician in 1884. He made several trips to Europe and returned to Philadelphia with many new concepts. He was a contemporary of Edward Shakespeare, but the two men disagreed on many things. Formad was skeptical of the role a bacillus played as a cause of tuberculosis and often debated this point with Shakespeare. His negative opinion on this subject was partially due to a lack of a microscope with oil-immersion capabilities. He did have a laboratory in the Medical Building of the University of Pennsylvania and made some claims of conducting animal studies on tuberculosis, but these were apparently of no real significance, and Shakespeare challenged his results in print. Based on a report in 1881, he went to a town on Lake Michigan to study a severe diphtheria outbreak and brought back "specimens of the diphtheric virus, several of the false membranes which are invariably formed in the throats of affected persons, and portions of their viscera." (13) Formad used his laboratory as a teaching tool and had his students stain patient specimens. Although his laboratory had been described as minimal by McFarland, in February of 1981, George M. Sternberg used Formad's laboratory while he was a visiting professor at the University of Pennsylvania. It was in this laboratory that Sternberg isolated and identified an organism, which he eventually called *Micrococcus pasteuri* as an organism in human saliva that caused septicemia in rabbits. From the detailed description Sternberg gave of these experiments, the laboratory was well equipped for bacteriology research. (14) Sternberg's work involved the preparation of various culture media and quite extensive rabbit inoculations. He published this work in 1881. (15) However, Pasteur performed the same experiment and drew the same conclusion in a report he gave to the Academy of Sciences of Paris on 24 January 1881. (16) This episode leads one to the conclusion that the

Formad laboratory was certainly adequate for both research and teaching purposes. Formad died in Philadelphia on June 6, 1892.

An interesting article, which demonstrates the scope of Formad's expertise as a Philadelphia coroner's physician, appeared in the *Philadelphia Ledger*, July 14, 1885, and was reprinted in the *New York Times*, July 15, 1885. The headline was NOTHING BUT WORMS, and reported the following:

> A mass of repulsive, wriggling, threadlike worms, which came out of a down-town hydrant and was sent to the *Public Ledger* office, was turned over to Dr. Henry F. Formad, of the University of Pennsylvania, who is also one of the Coroner's physicians. "While these little creatures," said Dr. Formad last evening, "may not be very pretty to look at, they are entirely harmless. They are of the worm species, and are entirely common in Philadelphia hydrant water. In fact they are a diminutive species of earth worms, and are allied to the species called *Lumbricoides*. If a man should get a lot of these down his throat, he can rest perfectly assured that no trouble will result, for the juices of the stomach will digest them, and that is the end of the worms, and the man is so much the better off for so much additional nutriment. In other words, they would act as so much food. They are about from one-quarter to one-half an inch in length. The knotted mass which you sent me is a large collection of these worms coiled together. This coiling is characteristic of this worm tribe." (17)

Lawrence Francis Flick (1856-1938) was born to German immigrants on 10 August 1856 in Carrollton, Pennsylvania. He attended St. Vincent's College, Latrobe, Pennsylvania, and later received his MD in 1879 from Jefferson Medical College, followed by an internship at Blockley Hospital. After a bout of consumption, he adopted the eradication of TB as his cause. He championed case registration and education about the contagious nature of the disease, which provoked opposition within the medical profession. His belief in the scientific foundation of medicine and his avid interest in all research results led him to found or organize a number of institutions for the care of TB patients and the study of the disease. In 1892, he founded the Pennsylvania Society for the Prevention of Tuberculosis. Rosen notes that this is the first tuberculosis society in the United States, and it was unique in that it combined lay and professional membership to concentrate activities against a single disease. "In these respects, it set a pattern that was to be widely followed up to the present." (9) In 1902, he traveled to Europe to meet with Albert Calmette and other European leaders working in the field of antituberculosis. He was a promoter of the National Association for the Study and Prevention of TB (1904). He helped organize the 1908 International Congress on Tuberculosis, which was held in Washington DC as well as in Philadelphia. Flick was probably influential in getting Robert Koch, Albert Calmette, and others to give brief talks in Philadelphia on 25 September 1908. (4, 21) Between 1892 and 1910, he founded the Rush Hospital for Consumption and Related Diseases, the Free Hospital Association for Poor Consumptives, the Henry Phipps Institute for the Study, Prevention and Treatment of Tuberculosis, and a modest sanatorium at White Haven, Pennsylvania, which he directed until 1935. He practiced medicine until the final year of his life in 1938. He made significant contributions to the control of tuberculosis in Philadelphia and in the United States. By 1904, there were twenty-three state or local societies associated with tuberculosis based on the model he established in 1892. Flick was the author of numerous papers and published five books on tuberculosis.

One of his most important and long-lasting contributions to bacteriology was his role in the founding of the Henry Phipps Institute for the Study, Prevention and Treatment of Tuberculosis. A steel industry millionaire named Henry Phipps became very interested and impressed with the extensive work Flick was doing in the area of tuberculosis. In 1903, Phipps provided funds to establish a clinic in an area of Philadelphia with a very high incidence of tuberculosis. This was located at 236 Pine Street. Phipps wanted to ensure its permanence. In 1910, he transferred ownership to the University of Pennsylvania and provided funds to build a significant building at Seventh and Lombard Streets. Flick assembled a very significant staff and served as director until 1913. This institute earned a worldwide reputation for research on tuberculosis. In 1960, the institute was moved to a site at Forty-third and Chester Ave. As tuberculosis became more under control, the institute expanded its research areas to include other infectious diseases. (6)

Ernest Laplace (1861-1924) was of French descent, was born in New Orleans on 9 July 1861. He received his education at Georgetown College and received his MD from the University of Louisiana in 1884 and did a year of residency at the Charity Hospital in New Orleans. He then traveled to Paris, Berlin, Vienna, and other European cities to enhance his education. McFarland claims that he is possibly the only American student ever taken as a pupil and assistant by Louis Pasteur. He also worked with Robert Koch in Berlin on an acid solution of bichloride of mercury that became known as Laplace's solution. He returned to America in 1888 to become demonstrator of pathology at Tulane University. He came to Philadelphia in 1889 to the Medico-Chirurgical College of Philadelphia as chair of pathology (1889-92), professor of surgical pathology, and clinical surgery (1892-96) and became professor of surgery in 1896, a post which he held until 1924. In 1891, he was commissioned by the governor of Pennsylvania to travel to Berlin to study with Koch and learn more about the treatment of tuberculosis. He died May 15, 1924.

Samuel Gibson Dixon (1851-1918) was born in Philadelphia on 21 March 1851. He attended the Mantua Academy and the Mercantile College. He enrolled as a law student at the University of Pennsylvania and, for six years, was a practicing attorney. He then decided to pursue a medical career and received his MD from the University of Pennsylvania in 1886. In 1887, he traveled to Europe to study bacteriology at Kings College in London with Emanuel Klein and at the Pettenkoffer's Institute of Hygiene in Munich. Returning to Philadelphia in 1888, he was appointed professor of hygiene and proceeded to set up a laboratory in the medical department of the University of Pennsylvania to do research in bacteriology. Corner states that "his little laboratory, which seems to have been the first in America devoted to the teaching of hygiene, opened a year earlier than that of the University of Michigan." (6) However, due to a dispute with the authorities, he moved to the Academy of Natural Sciences with all his equipment and became professor of bacteriology and microscopial technology and eventually became president of that institution. He served as commissioner of health of Pennsylvania from 1906 until his death in 1918. His laboratory work focused on tuberculosis. While working in London, he seems to be the first to observe branched or forked tubercle bacteria. Under his tenure as health commissioner, the Division of Laboratories was established at the University of Pennsylvania in the Laboratory of Hygiene. He was influential in establishing a distribution system for diphtheria antitoxin, and he is credited with putting the state's public health activities on a firm scientific basis. (22)

Lydia Rabinowitsch (1871-1935) was born 22 August 1871 in Kovno, Lithuania. Her early education was at the gymnasium in Kovno, and she received a degree in natural science from the University of Zurich and her

PhD in medical science in 1894 from the University of Berne. She pursued postgraduate studies on tuberculosis and thermophilic bacteria with Robert Koch until she left for Philadelphia in 1895. She came to the Laboratory of Hygiene at the University of Pennsylvania and let it be known that she would accept the position as the first director of bacteriology at the Woman's Medical College of Philadelphia. The November 1895 decision to award this position to her, a PhD, rather than to the two male physicians who were after this position, was attributed to the new interest in the work of Koch and Pasteur at that time. Her initial course was offered to the third-year students, amid concerns that the practical nature of the bacteriology course would take valuable time away from needed clinical experience. However, by 1896, the laboratory was renovated and she was teaching second-, third-, and fourth-year students. In February of 1898, when she was teaching there for almost two years, the board recommended that she be made full professor of bacteriology. This raised several issues since she was not a physician, and it was felt that bacteriology was receiving undue recognition. The decision was postponed until April, when she was appointed associate professor of bacteriology. In June, only two months later, she was appointed professor of bacteriology. On 18 May 1898, at the end of the teaching session, she went to Germany to work with Koch. While there, she married Dr. Walter Kemper, another Koch assistant, with whom she had coauthored a paper on the study of bovine tuberculosis in milk in Berlin and Philadelphia. In August, she asked the faculty for permission to delay her return until January 1899. Understandably, the faculty rejected this request and appointed Adelaide Ward Peckham to the professorship.

Rabinowitsch remained at the Institute of Infectious Diseases in Berlin, became the first female assistant to Koch, and is credited with the discovery of nonpathogenic tuberculosis-like organisms in 60 percent of butter samples obtained from Berlin and Philadelphia. She traveled with Koch to Africa to study sleeping sickness and to Odessa to study plague. She remained with Koch until his death in 1910. Rabinowitsch made many additional discoveries in the area of tuberculosis and held several positions as professor in Germany; many were significant since she was the first woman professor appointed by the Prussian government. After gaining international recognition for her work, she was forced to give up her position in 1933 when the Nazi revolution began to effect widespread political change. She died in Berlin on 3 August 1935.

Conclusions Drawn from the Biographical Sketches of Seven Nineteenth-century Philadelphia First-Generation Bacteriologists: 1880-1900

I will take the liberty to call these six men and one woman "the first generation" of Philadelphia bacteriologists. A review of the biographical sketches of these seven very early bacteriologists demonstrates how the specialty of bacteriology came into being in Philadelphia during the last two decades of the nineteenth century. Each career described here is unique; however, they all were either born in the Philadelphia area or moved here for a key period of their careers.

Due to the scope of Joseph Leidy's work, it seems unfair to just consider his contributions in any one division of biology. He certainly was a well-known naturalist and one of the true American "scientists" of the nineteenth century, with an international reputation in several fields, including paleontology. He was included here, more for his work in parasitology, than his work in bacteriology. His work is also important since he realized the importance of a preclinical education of medical students and for accomplishing the formation of a separate biology department at the University of Pennsylvania, which opened the door to new scientific specialties like bacteriology. Because of these contributions, he cannot be ignored. However, for purposes of the following analysis, I would like to focus on the following five men who more closely fit together as a group that could be labeled as the first generation of Philadelphia bacteriologists. These five men are Edward Orem Shakespeare, Henry F. Formad, Lawrence Flick, Ernest Laplace, and Samuel Gibson Dixon.

1. Edward Shakespeare   (1846-1900) MD 1869   University of Pennsylvania
2. Henry Formad   (1847-1892) MD 1877   University of Pennsylvania
3. Lawrence Flick   (1856-1938) MD 1879   Jefferson Medical College
4. Ernest Laplace   (1861-1924) MD 1884   University of Louisiana
5. Samuel G. Dixon   (1851-1918) MD 1886   University of Pennsylvania

All five were born between the years 1846 and 1861; the average year being 1852, and 1851 the median year. Four were born in the United States. Three were born in the Philadelphia area, one in New Orleans, and Formad was born in the Caucasus. Four of the five received their MD in Philadelphia, three from the University of Pennsylvania, and one from

Jefferson Medical College. Laplace received his MD from the University of Louisiana. All received their MD between 1869 and 1886, the average (and median) year being 1879. The age when receiving their MD ranged from twenty-three to thirty-five, with an average of twenty-seven and median of twenty-three years. The number of potentially active career years, based on the years between obtaining their MD and their death, ranged from fifteen to fifty-nine with an average of thirty-five and median of thirty-two years. Life spans ranged from forty-five to eighty-two years with an average of sixty-two and median of sixty-three years. Year of death ranged from 1892 to 1938 with the average year of 1914 and median of 1918.

All five could read German, and at least three could read German and French. Three of the five either did an internship at Blockley Hospital or were associated with Blockley during their career. Three of the five traveled to Europe shortly or immediately after getting their MD, and four of the five visited scientific institutions in Europe during their career.

The following are their first professional and intermediate positions:

1. E. Shakespeare - curator of Blockley museum (1880), pathologist (1882)
2. H. Formad - demonstrator of pathology (1877), microscopist (1880), coroner (1884)
3. L. Flick - institute administrator (1880)
4. E. Laplace - demonstrator of pathology (1888), pathologist (1889)
5. S. Dixon - professor of hygiene (1888)

The following is when they obtained their key bacteriology-related position:

1. E. Shakespeare - bacteriologist, Blockley Hospital (1889)
2. Henry Formad - pathologist, Blockley and University of Pennsylvania (1887)
3. Lawrence Flick - tuberculosis institute administrator (1882)
4. Ernest Laplace - chair of pathology, Medico-Chirurgical College (1892)
5. Samuel Dixon - professor of bacteriology, Academy of Natural Sciences (1889)

As can be seen from the above, the most common position held prior to their more formal recognition as a bacteriologist was related to pathology. Microscopist was a position that was often part of the pathology department due to its association with histology. Dixon's career, preceding his bacteriology title, was in the hygiene department, a division of the medical department. It should be noted that Flick had an unusual career that was more administrative than laboratory oriented. He has been kept with this group since he spent considerable time in laboratories and science institutions in Europe, and his career path included a sound bacteriology base. He decided to focus exclusively on all aspects of tuberculosis, including the bacteriology of the disease, and established laboratories at several institutions in the Philadelphia area. He was responsible for bringing about, and served as the first director, of the Henry Phipps Institute. He is unusual among the group since he did not obtain a formal academic appointment.

Of the five men only two—Shakespeare and Dixon—actually had the academic title of bacteriologist. Formad and Laplace were recognized as bacteriologists but remained in pathology departments, and Flick never held the title.

From this brief survey, a generalized picture can be drawn of the typical (but theoretical) member of this first-generation group. The generalized Philadelphia bacteriologist of this period was male, born in the early 1850s, in the Philadelphia area, received an MD degree from the University of Pennsylvania in 1879 at the age of twenty-three, followed by a year of internship at Blockley Hospital, and an initial position within the pathology department. This person could read German, and probably French, and visited scientific institutions in Europe, mainly Germany. Recognition as a bacteriologist was awarded by 1889, followed by an active teaching career until about 1914.

This leaves only one additional person to discuss, Lydia Rabinowitsch. She was a contemporary of the above group but had a very different career. Like many of the others, she was initially associated with the University of Pennsylvania but was different since she was a nonphysician female, had a PhD from a European University, and had worked with Robert Koch prior to arriving in Philadelphia. Her time in Philadelphia, and her position as the first bacteriologist at the Woman's Medical College, also characterized her as different from the others. She, without a doubt, went on to make many significant contributions to bacteriology, more than all the other Philadelphia bacteriologists described above. However, most of these contributions were made while back in Germany. She faced and overcame two significant issues. She was a PhD in an MD world, and she was a woman in a field exclusively male. Her appointment as professor of bacteriology was not only short-lived, but filled with controversy. Her appointment as professor raised the question of whether too much emphasis was being placed on the new field of bacteriology at the medical college. It would have been very interesting to see how her career would have unfolded had she remained longer in Philadelphia. However, she would never have had the opportunities in Philadelphia that she obtained by spending so much of her career in Germany working with Robert Koch for so many years. The short review of her biography does not do justice to her career, but it does illustrate that she established bacteriology as a popular subject at a Philadelphia medical school and set the stage for a PhD holding such a position. Because of their contributions, Lydia Rabinowitsch, like Joseph Leidy, must be included in this group of first-generation Philadelphia bacteriologists, although they cannot be classified as typical of the new generation that entered the field of bacteriology before the close of the nineteenth century.

PHILADELPHIA INSTITUTIONS ASSOCIATED WITH BACTERIOLOGY AT THE END OF THE NINETEENTH CENTURY

By the end of the nineteenth century, bacteriology was becoming recognized as a separate scientific discipline. What follows is a partial list of those institutions in Philadelphia that were associated with these developments. This includes the following:

1. University of Pennsylvania
   A. Bacteriology in the Medical Department
   B. Laboratory of Hygiene
   C. William Pepper Laboratory of Clinical Medicine
   D. School of Dental Medicine
   E. School of Veterinary Medicine
2. Jefferson Medical College
3. Woman's Medical College of Pennsylvania
4. Hahnemann Medical College
5. Medico-Chirurgical College
6. Philadelphia Polyclinic and College for Graduates in Medicine
7. Blockley Hospital (Blockley-Almshouse)
8. Pennsylvania Hospital
9. Wistar Institute of Anatomy and Biology
10. Philadelphia Health Department
11. H. K. Mulford Company

1. THE UNIVERSITY OF PENNSYLVANIA

Of the various institutions in Philadelphia that had an influence on the development of bacteriology during the last two decades of the nineteenth century, the University of Pennsylvania played a dominant role. A majority of the first-generation (as well as the second-generation) bacteriologists were associated with the university, especially the medical department, which included pathology and microscopy. A key university department involved with the formation of bacteriology centered on hygiene and resulted in the opening of the new Laboratory of Hygiene in 1892. This resulted in the inclusion of bacteriology as a required subject in the medical student curriculum. Because events and institutions at the University of Pennsylvania are so important in the separation of bacteriology from other well-established scientific disciplines, this section will give particular attention to the University of Pennsylvania.

Its medical school, the oldest in the United States, was founded in 1765. In 1802, the medical school moved from the old site at Fourth and Arch Streets to an elegant mansion on Ninth Street between Market and Chestnut Streets. This mansion was designed to be occupied by the president of the United States but put up for auction in 1800 when the federal government was moved to the District of Columbia. A room was set aside in the mansion for chemistry apparatus. However, due to issues relating to human dissection, the anatomy faculty was not invited to move to the new facility. They were forced to remain at the old location in a building known as Surgeons' Hall, with the name changed to Anatomical Hall, but was most commonly referred to as the Laboratory. In 1806, it was agreed to put an extension on the mansion to accommodate moving the Laboratory to the new location.

In 1816, there was a desire to have chemistry and biological subjects taught somewhere in the university rather than by the medical faculty. This resulted in the formation of a faculty of natural sciences. However, the whole faculty of natural sciences consisted exclusively of physicians. In 1830, William Edmonds Horner was awarded the chair of anatomy. He was one of the first physicians to use the microscope in serious professional studies and published a classic account on the pathology of Asiatic cholera. In the late 1830s and 1840s, the importance of the lecture hall versus practical bedside clinical experience for medical students became an issue for discussion. This was precipitated by the return to Philadelphia of men like William Wood Gerhard, who spent two years studying in Paris with Pierre A. C. Louis. Louis was described as the greatest of French medical pathologists and a proponent of studying the whole disease process. He was a proponent of the need for the comparison of signs and symptoms of illness, which were noted at the bedside, with organic lesions found at postmortem examination. An outbreak of typhoid fever in Philadelphia, shortly after Gerhard returned from Paris, enabled him to apply bedside observations with his observations made in the postmortem room at Pennsylvania Hospital to clearly distinguish typhoid fever from typhus fever. He also made the first accurate clinical study of tubercular meningitis in children. During this period, emphasis was placed on the need for medical students to visit the wards of the two nearby hospitals, Pennsylvania Hospital and the Philadelphia Almshouse.

In 1870, the university bought ten acres of the almshouse property in West Philadelphia from the city. The first building, College Hall, was completed in 1872, and Medical Hall (which later became Logan Hall) at Thirty-sixth and Locust Streets, was ready to receive students for the 1874-1875 academic year. In 1874, the first university-controlled general hospital in the nation opened across from Medical Hall, the Hospital of the University of Pennsylvania, which greatly encouraged the marriage of clinical observation with new principles of hygiene—including bacteriology and other academic subjects.

As the century progressed, the new departments and institutions associated with the university, and especially the Medical School, demonstrated how various factors came together to assist in the separation of bacteriology as a new field of study for both the medical student, as well as students of natural science and biology. The following is an overview of some significant dates and events occurring at the University of Pennsylvania that relate to the eventual formation of bacteriology as a separate scientific entity:

1851

Joseph Leidy wrote *A Flora and Fauna within Living Animals*, which was published in 1854. This massive and well-illustrated work described, in remarkable detail, the plants, animals, protozoa, and worms that live in and on various forms of life. In this work, as well as in other shorter reports, he identified many new species of both parasitic and free living life forms.

1853

Joseph Leidy was appointed professor of anatomy at the age of thirty. This marked the trend of opening significant academic appointments to more youthful professors, who replaced the old more indoctrinated professors. This also opened the door for the inclusion of new ideas, like the importance of evolving developments in pathology, the biological sciences, as well as hygiene and bacteriology.

1865

Following the Civil War, hygiene was added to the medical school curriculum, which eventually opened the way for bacteriology to gain in importance. Henry Hartshorne became the first professor of hygiene.

1866

The natural sciences reentered the medical school curriculum. These courses were taught by an auxiliary medical faculty, and a certificate was granted to the students who completed the course work and passed an examination in all five subjects designated as the natural sciences, including biology. Many students demanded more formal recognition for this extra work, and in 1874, the trustees decided to employ the high European distinction of doctor of philosophy to those students who fulfilled all the requirements. This title was not previously employed in the United States. This was a distinction that was open to medical graduates of the University of Pennsylvania, or other acceptable schools, who attended all the lectures and passed the appropriate examination. However, in 1879, when Johns Hopkins University began to confer the PhD according to German standards, the university abandoned this ill-conceived PhD degree for this purpose. Therefore, starting in 1881, a bachelor of science was granted for this program in place of the PhD. The old criteria was used, but with added requirements for students to propose an original thesis in a specialty area.

1876

In recognition of the importance of pathology, James Tyson was awarded the first chair of pathology, and pathology took on the role for teaching bacteriology.

1878

The School of Dental Medicine was founded. This later resulted in an expanded need for faculty to teach bacteriology to dental students. Formal courses in bacteriology for dental students were initiated in 1896.

1879

Joseph Leidy published *Fresh-Water Rhizopods of North America*, which is still considered a reference for this aspect of protozology.

1884-1889

William Osler joined the faculty and set up a very small clinical laboratory in the University Hospital, where he conducted clinical tests and microscopic examinations on patient specimens. He established a clinical laboratory in a small brick building where postmortem examinations were held on the Blockley property. There he studied seventy cases of malaria. He was one of the first investigators in America to confirm the presence of the malaria parasite in the blood of patients with the disease. He also taught medical students how to look for the tubercle bacillus shortly after Koch published his findings on this bacillus.

1884

The School of Veterinary Medicine was founded. Formal courses in bacteriology were initiated for veterinary students in 1886.

1885

The department of biology was created, with Joseph Leidy as the first professor of biology, again stressing the need for medical students to take preclinical studies in the biological sciences.

1889

Juan Guiteras replaced James Tyson as professor of pathology. His reputation was built on his work on yellow fever and the pathology of infectious diseases.

1892

The Laboratory of Hygiene opens its doors and ushered in the real beginning of bacteriology as a separate discipline. Bacteriology became a required course of study with a dedicated faculty.

A. Bacteriology in the Medical Department

1881

Henry F. Formad had a laboratory in the Medical Building where he conducted animal studies on tuberculosis. George M. Sternberg used Formad's laboratory to isolate and identify an organism, which he eventually called *Micrococcus pasteuri*, as an organism in human saliva that caused septicemia in rabbits.

1883-1884

The small laboratory established by Henry Formad is described as "supplied with a complete outfit of materials and apparatus for the investigation of bacteria in their relation to infectious diseases, and the study of lower fungi in general." (From the 1883-1884 School Catalogue.)

1888-1891

Samuel G. Dixon, according to his assistant, Dr. Seneca Egbert, fitted and improved the medical laboratory to make it better suited for bacteriology research.

1889

Simon Flexner became chair of pathology and strengthened the need for bacteriology at the university.

1889-1890

Juan Guiteras succeeded James Tyson as professor of pathology and secured leave in 1890 to go abroad to study and select new apparatus for equipping a future laboratory of bacteriology for instruction of students.

1889-1890

While Guiteras was away, the instruction in pathology and bacteriology was carried out by Henry Formad, assisted by Allen J. Smith. Smith made several improvements in the culturing and staining of bacteria. Another Formad assistant, J. Leffingwell, improved techniques for isolating bacteria from blood.

1890-1891

After finishing their medical studies at the university, Samuel Kneass took courses at the Pasteur Institute, and Joseph McFarland studied at Heidelberg with Paul Ernst. They then returned to the university to do laboratory studies in bacteriology.

1891-1892

Guiteras makes the newly enhanced bacteriology laboratory available to J. McFarland and S. Kneass upon their return from Europe.

1892-1893

Bacteriology appears for the first time in the announcement of the medical department as a subject to be pursued during the third year. The catalogue states that there is one lecture a week for six weeks. The pathology department, under Guiteras, was responsible for teaching the course. J. McFarland assisted with the course through 1895.

1892-1897

The only link between the Laboratory of Hygiene, where the medical students took their bacteriology course, and the medical department was that J. S. Billings, director of the Laboratory of Hygiene, held his title of professor of hygiene in the medical department. With the resignation of Billings, A. C. Abbott became professor of hygiene in the medical department.

1895

McFarland is listed as lecturer on bacteriology for the above course. The course consisted of six lectures and also included six laboratory demonstrations.

1896-1899

David Reisman replaced McFarland as lecturer for the bacteriology course.

1899

Simon Flexner succeeded Guiteras in charge of pathology and insisted that pure bacteriology that was still taught in the medical department had to be taught by a bacteriologist. (The concept was that "practical" bacteriology was taught in the Laboratory of Hygiene.) A. C. Abbott became professor of hygiene and bacteriology, and later that same year he became professor of bacteriology in the medical department.

B. Laboratory of Hygiene

As early as 1881, William Pepper was planning for an institute of hygiene to be associated with the medical school. In 1889, Henry C. Lea, the wealthy publisher and historian, offered to donate funds for this project if the university could raise additional funding. Lea also placed additional requirements, which were probably instigated at the request of Dr. Pepper. These included the need for an additional $250,000 to be raised for equipment, an endowment to be used to establish a department of hygiene, and that Dr. John Shaw Billings be recruited as director. An additional requirement was that hygiene be made a compulsory study in the medical school, and that the medical student curriculum be extended to four years of study. Pepper took little time in getting John S. Billings who, among other significant accomplishments in the medical field, had designed the Johns Hopkins Hospital. In 1891, Billings signed an agreement to design an up-to-date laboratory building at 215 South Thirty-fourth Street and to become director of the new institute as professor of hygiene. He arrived in Philadelphia in 1892, as soon as the building was completed. (25)

On 2 February 1892, the Institute of Hygiene opened. The name was almost immediately changed and became known as the Laboratory of Hygiene. It became the training ground for some of the nation's pioneering bacteriologists. The facility played an important role in supporting the teaching mission of many disciplines. Primarily medical, but also dental, veterinary, engineering, and natural science students, received instruction in public health and bacteriology within this special laboratory building. The building was designed specifically for the study of hygiene and bacteriology. The laboratory was one of only about ten such facilities in the world when it was opened, and the first of its kind in the United States. It was intended not to mimic European models but to marry those models to the U.S. ideas of practical laboratory hands-on education. The entire building operated as a scientific apparatus. The architects decided not to copy existing styles but to create a new modern style befitting the spirit of the time and place. Nothing extraneous was added to its almost stark design, with just some terra-cotta work around the door frame. The heating and air circulation system was quite sophisticated and was designed to be a model for future laboratory and hospital buildings. Its state-of-the-art heating and ventilating system were specifically designed to illustrate scientific principles and safeguard the health of those working with pathogenic bacteria. Water and drainage pipes were left exposed and painted different colors so students and visitors could understand the underlying principles of water supply and drainage. (28, 29)

1891-1892

The school catalogue announces that "through the liberality of a number of citizens of Philadelphia, the University of Pennsylvania has been enabled to establish a Laboratory of Hygiene. A large building, especially planned and fitted for this purpose, is now nearing completion and will be completely equipped and ready for use on 1 February 1892."

1892

The new Department of Hygiene is formed with John S. Billings, director; A. C. Abbott, first assistant; Albert A. Ghriskey, assistant in bacteriology; and James Homer Wright, the Thomas A. Scott Fellow in hygiene. Billings, through his assistants, focused on teaching practical hygiene, while Abbott and his staff focused on teaching responsibilities for bacteriology.

1892

Two courses were offered, one in hygiene and "an elementary course in bacteriology," which began on the day the building opened and continued for eight weeks, five days per week.

1892-1893

The first course had three bacteriology students: Samuel Stryker Kneass, Adelaide Ward Peckham, and Mazyck Porcher Ravenel.

1892-1897

The only link between the Laboratory of Hygiene and the medical department was that J. S. Billings was director of the Laboratory of Hygiene and professor of hygiene in the medical department. With the resignation of Billings, A. C. Abbott became professor of hygiene in the medical department.

1893-1894

The bacteriology course had six students, one being David Hendricks Bergey, with A. W. Peckham listed as a student in the new course in advanced bacteriology.

1893-1894

The staff consisted of John S. Billings, director; A. C. Abbott, first assistant; and M. P. Ravenel, the Thomas A. Scott Fellow in hygiene.

1894-1895

The same two courses (assumed to be bacteriology and advanced bacteriology) were offered, with the following staff consisting of John S. Billings, director; A. C. Abbott, first assistant; M. P. Ravenel, assistant in bacteriology; and D. H. Bergey, the Thomas A. Scott Fellow in hygiene. There were twenty-one students, including Robert L. Pitfield.

1895-1896

The staff consisted of John S. Billings, director; A. C. Abbott, first assistant; M. P. Ravenel, assistant in bacteriology, with D. H. Bergey listed as assistant in chemistry. There were thirteen students, including Lydia Rabinowitsch, PhD.

1896

The elective laboratory course in bacteriology for medical students, conducted by A. C. Abbott and his staff, became so popular in this year that the entire class was required to take this course.

1896-1897

The staff consisted of A. C. Abbott, director, and D. H. Bergey, first assistant. There were twenty students.

1899

A. C. Abbott became professor of hygiene and bacteriology and professor of bacteriology in the medical department.

C. William Pepper Laboratory of Clinical Medicine

In 1891, William Pepper Jr. established an endowment for the establishment of the Pepper Laboratory of Clinical Medicine, which was to be part of the University Hospital system. The laboratory was to engage in clinical studies and original research. He specified that the laboratory was to be planned and organized by John S. Billings and constructed by Cope and Stewardson. The four-story brick building was completed in 1895. On the first floor were laboratories for microscopic, chemical, and bacteriological investigations. The second floor had the director's office and a laboratory for anthropometry (the study of human body measurements). A postgraduate student laboratory was on the third floor. The top floor had an additional laboratory along with the library and conference room. The building design was unique in that the corridors on the first floor connected to nursing units in one of the hospital buildings. On 17 October 1894, the board of managers of the university hospital named William Pepper as the first director of the laboratory. The laboratory was officially opened on 4 December 1895. Dr. Billings delivered the presentation address at the opening ceremony, and Dr. William Welch, professor of pathology at Johns Hopkins University School of Medicine, delivered the closing address on the "Evolution of Modern Scientific Laboratories."

The William Pepper Laboratory has the distinction of being the first clinical laboratory in the United States to be associated directly with a medical clinic and was under the control of the Department of Medicine. When Dr. Pepper died in 1898, Alfred Stengel became director, a position he held until 1911. (25, 26, 27, 29)

1895

The building was completed and contained the most modern apparatus for chemical, bacteriological, and clinical investigation. Samuel S. Kneass was placed in charge of the laboratory of bacteriology.

D. School of Dental Medicine

The Department of Dentistry was established in 1878. It was housed in Medical Hall (later Logan Hall), and in 1880 it received laboratory space in the Robert Hare Laboratory at Thirty-sixth and Spruce Streets. The department moved into its new building, Dental Hall (now Hayden Hall), at Thirty-third and Locust Streets in 1897. This new building included a laboratory designed for hands-on work by the dental students.

1879

Among the first graduates this year was Willoughby Dayton Miller (1853-1907) who later, while working in Berlin, developed the bacterio-chemical theory of dental carries. His classic work was published in 1889 as *Die Mikroorganismen der Mundhohle* (Microorganisms of the Human Mouth.)

1896

From the minutes of the Dental School Faculty, dated 3 March:

> Believing that a sufficient knowledge of Bacteriology is essential to the proper education of a dentist, it is recommended that the Trustees provide for a semi-popular course of about six lectures to be given to the senior class of the Department of Dentistry as soon as possible during the present session, which course should not be obligatory or include a final examination. But the ensuing year and thereafter, it is recommended and requested that instructions both in didactic and in laboratory work shall be given to third year dental students in

Bacteriology so far as it is adapted to their needs, and that the study of it be made an obligatory part of the third year curriculum.

1896

Arrangements were made with the Laboratory of Hygiene for A. C. Abbott to teach the bacteriology course to third-year students. In later years, D. H. Bergey also taught in the dental school.

1898-1899

Third year dental students take bacteriology for two three-hour periods per week.

E. School of Veterinary Medicine

The medical faculty passed a resolution in 1877 to establish a professor of veterinary medicine and surgery, whenever a suitable endowment could be obtained. As Corner notes, "By the 1880s, however, the development of physiologic experimentation and, above all, the new science of bacteriology had emphasized the common foundations of human and veterinary medicine." In 1882 the well-known publisher, J. B. Lippincott, offered $10,000 to establish the school. By 1883, enough funds were obtained to construct a building at Thirty-sixth and Pine Streets.

1884

The School of Veterinary Medicine was founded and opened to students on 2 October of that year.

1896-1897

The first reference to bacteriology appears in the Annual Catalogue as Practical Bacteriology. The course for second-year students was scheduled to meet after February 1 on Tuesday and Wednesday afternoons, and for third-year students, before February 1 on Monday mornings. Juan Guiteras was professor in both the medical and veterinary departments at that time, but the announcement did not state where or by whom the instruction was given.

1897-1898

One-hour lectures for the third-year students were given by J. Guiteras on Mondays at 11:00 a.m., but laboratory hours were not listed.

1898-1899

M. P. Ravenel was listed as demonstrator of veterinary bacteriology. At the end of 1899, Guiteras resigned his position to return to Cuba, and Simon Flexner succeeded him as professor of pathology in both the medical and veterinary departments.

1899-1900

Bacteriology was taught to the second-year students on Thursdays at 10:00 a.m. with a notation that "the lectures are held in the medical department." McFarland proposes that Abbott probably gave the lectures since he was professor of hygiene and bacteriology in the medical department, with the laboratory instructions given by Ravenel in the veterinary department.

2. Jefferson Medical College

Founded in 1824 by a group of physicians lead by George McCellan, it graduated its first class in 1826. It was located in the old Tivoli Theater on Prune Street, with a dispensary to give students experience in dealing with patients. Jefferson later made agreements for students to gain clinical experience at Blockley and Pennsylvania Hospital. In 1877, a detached hospital was constructed. Although Jefferson started as a rather small medical school, it attracted gifted teachers and became a driving force by competing with the University of Pennsylvania for top faculty. Many University of Pennsylvania professors made the switch to Jefferson when they felt they were not getting the attention they deserved at the university. As the century unfolded, the competition and presence of two college-based medical schools benefited both institutions. (30)

1886-1889

M. V. Ball, a medical student, states that there was no bacteriology taught; but on one occasion, Jacob M. DaCosta introduced Julius Salinger who had just returned from Berlin. He remembered Salinger drawing the "comma bacillus" on the blackboard during his presentation.

1892

W. M. L. Coplin, demonstrator of pathology, began weekly lectures in bacteriology and hygiene. The subject continued as part of the pathology department until 1909. (30)

1893

J. M. DaCosta and D. Brandon Kyle conducted a private clinical laboratory in which bacteriology examinations were made, and some private instructions given, but it was not part of the college course.

1895

A catalogue mentions one lecture a week and one demonstration a week in "Bacteriology and Clinical Microscopy" for second-year students. McFarland assumes that they were given by W. M. L. Coplin and assisted by Alonzo H. Stewart, as demonstrator of clinical microscopy.

1896

Coplin became professor of pathology and bacteriology. This appears to be the year that regular instruction in bacteriology began.

1897

Bacteriology was taught two days a week for six weeks by Coplin, with David Biven and Randle C. Rosenberger as assistants. W. M. L. Coplin accepted a position in the South, and Stewart carried on instructions until Coplin returned.

3. Woman's Medical College of Pennsylvania

The Woman's Medical College of Pennsylvania was chartered and opened in 1850 as the Female Medical College of Pennsylvania. It was the first medical college in the world exclusively for women. In 1867, the name was changed to the Woman's Medical College of Pennsylvania.

1896

Lydia Rabinowitsch, PhD, became director of the bacteriological laboratory. This seems to be the first time any research or instruction was given in bacteriology.

1898

Adelaide Ward Peckham became associate professor of pathology in charge of bacteriology, and later that same year, she replaced Lydia Rabinowitsch as professor of bacteriology. She held that position until 1919.

1899

Rabinowitsch officially resigned to continue her work in Berlin on tuberculosis at the Koch Institute.

4. Hahnemann Medical College

This medical school was founded in 1848 as the Homeopathic Medical College of Pennsylvania in a small facility at 229 Arch Street. It combined with a hospital and became the Hahnemann Medical College in 1867 and moved to Broad and Vine Streets.

5. Medico-Chirurgical College

The Medico-Chirurgical College was an old medical society turned into a medical school in 1881, mainly by the efforts of William Pancoast. It had several noted physicians on the faculty, controlled its own 135-bed hospital, and had a large outpatient service. It was eventually merged with the University of Pennsylvania-Graduate School of Medicine in 1919.

1889

William Pancoast, professor of surgery and founder of the Medico-Chirurgical College, while visiting Pasteur asked his advice on the future of bacteriology and asked for his help in selecting a professor of bacteriology for his college. Ernest Laplace was working in Pasteur's laboratory at the time. Pasteur introduced Laplace to Pancoast and recommended Laplace for the position. Laplace came to Philadelphia to fill the position, which did not have a laboratory and relied purely on didactic instruction. However, he did show the students pathological lesions and stained bacteria. E. B. Sangree became assistant in pathology and bacteriology.

1896

Laplace became professor of surgery, and Joseph McFarland became the chair of pathology and bacteriology. According to McFarland, his appointment marked a new era for the college "for it was accompanied by the establishment and equipment of modern laboratories for both pathology and bacteriology, with excellent microscopes for all of the students and a complete bacteriological outfit that enabled the students to make and sterilize their own media, plant their own cultures and perform all of the usual staining methods." With the erection of the new laboratory building a few years later, Robert L. Pitfield was appointed demonstrator of bacteriology.

6. Philadelphia Polyclinic and College for Graduates in Medicine

This college was founded in 1883 to provide additional medical and surgical course training for medical graduates who wanted additional or specialty training. Abraham Flexner referred to this school as one of the best "repair shops" available to these graduates. The school was merged into the University of Pennsylvania-Graduate School of Medicine in 1919.

1894
Approximate time that Samuel S. Kneass became adjunct professor of bacteriology and Joseph McFarland became adjunct professor of pathology, but they did not have many students.

1896
H. D. Pease replaced Kneass.

7. Blockley Hospital (Blockley-Almshouse)

The almshouse opened in 1732 to serve the charitable needs of the growing population of poor Philadelphians. It was built on a site that became adjacent to Pennsylvania Hospital at Tenth and Spruce Streets. Starting in 1815, arrangements were made for medical students from the University of Pennsylvania to spend time at the hospital, and several of the faculty members from the university taught their clinical courses and conducted "rounds" there. In 1832, the almshouse moved to a tract of land the city purchased in the former Blockley Township in West Philadelphia. It was an estate known as Old Blockley or Blockley Farm. Most people referred to the new institution simply as Blockley or Blockley Hospital. Blockley was designed to house a variety of Philadelphia's indigent population and consisted of a quadrangle of four sizable buildings, including a poorhouse, a hospital, an orphanage, and an insane asylum. As the century progressed, Blockley evolved into a more conventional public hospital and was renamed Philadelphia General Hospital (PGH). A nursing school was opened at the site in 1885. A Blockley internship or residency became open to any qualified medical graduate, and several of the first-generation bacteriologists spent at least a year there after graduating from medical school or had staff appointments there. The first identifiable bacteriologist, Edward O. Shakespeare, spent his career at Blockley.

1882

Edward O. Shakespeare, recognized as Philadelphia's first identifiable bacteriologist, is appointed pathologist at Blockley.

1884-1889

William Osler established a clinical laboratory in a small brick building on the Blockley Hospital property where postmortem examinations were performed. There he studied seventy cases of malaria. He was one of the first investigators in America to confirm the presence of the malaria parasite in the blood of patients with the disease. He also taught medical students how to look for the tubercle bacillus, shortly after Koch published his findings on this bacillus.

1889

Edward O. Shakespeare is appointed bacteriologist at Blockley.

1890

Edward O. Shakespeare presents his thousand-page publication, *Report on Cholera in Europe and India*, to Congress. The American Medical Association hailed it as a "great work," and Shakespeare was regarded as the leading expert on cholera.

8. Pennsylvania Hospital

The first and oldest hospital in the United States opened in a temporary building in 1752 on High (now Market) Street. In 1755, the cornerstone was laid for the East Wing of what would become the hospital's permanent location at Eighth and Pine Streets. Patients were first admitted to the permanent hospital in 1756. The site continued to grow through the years with the addition of more wings and buildings, extra land, and further expansion. Starting in 1815, medical students from the University of Pennsylvania, and later from Jefferson Medical College, received clinical instructions at Pennsylvania Hospital as well as at the almshouse.

9. Wistar Institute of Anatomy and Biology

The Wistar Institute of Anatomy and Biology is the nation's oldest independent biomedical institute. In 1892, General Isaac Wistar secured a charter for an institute designed to house and preserve the Wistar and Horner Museum of Anatomy. This collection was previously held at the University of Pennsylvania. The institute was also to provide facilities for advanced workers in anatomy and biology. A commodious building was erected in 1893 on Thirty-sixth Street, opposite Medical Hall of the University of Pennsylvania. The building was enlarged in 1897. Although the Wistar Institute was on the campus of the University of Pennsylvania and was associated with the university in various ways, it was and still is an independent research facility. Although it was founded toward the close of the 1800s, it is listed here only because of its future significance to microbiology in the next century.

10. Philadelphia Health Department

As stated by J. McFarland, prior to 1893, Philadelphia authorities found no interest in bacteriology until the discovery of diphtheria antitoxin when there was a public demand for a bacteriology laboratory. This laboratory was needed to receive throat cultures from suspected diphtheria patients, without expense to doctor or patient. There was also a demand for free antitoxin for the treatment of the poor. New York already had such facilities run by Willian H. Park. Also, A. C. Abbott was urging more attention to the Philadelphia water supply, and George Woodward was urging action to address problems with milk contamination. However, William M. Welch, the chief diagnostician of the City Board of Health, had been violently opposed to a laboratory-oriented program.

In 1894, the Philadelphia Health Department placed notice for a director and two assistant bacteriologists for a Laboratory of Hygiene to be established in connection with the Bureau of Health of the city of Philadelphia. To be considered, the candidates had to take a civil service examination. This examination had both a written and oral component conducted by E. O. Shakespeare and W. M. L. Coplin. Seneca Egbert and M. P. Ravenel supervised the phase of the examination that called for the identification of cultures of bacteria.

1895

B. Meade Bolton was named director of bacteriology, with Herbert D. Pease as first assistant in bacteriology. W. J. Gillespie, a graduate of Jefferson

Medical College was named second assistant in bacteriology. Bolton was a graduate of the University of Virginia with subsequent training at Gottingen. He was instructor in bacteriology in the graduate department of medicine and the undergraduate medical department, both at Johns Hopkins University. Pease was also from Johns Hopkins Medical School; he was doing graduate work with Welch and was a laboratory assistant to S. Flexner.

1895

The laboratory operations began in May in rooms on the seventh floor of city hall. The initial focus was on diphtheria diagnosis and production of antitoxin. The horses employed for antitoxin production were located at the fire department station at Frankford, but after about a year, they were moved to a larger fire station in West Philadelphia.

1896

Bolton resigned over a dispute on work carried out in city hall by W. J. Gillespie concerning the commercialization of a product that was supposed to have antibacterial properties.

1897

Bolton returned and protested his treatment to no avail and was forced to vacate his position. A. C. Abbott became the new director in February. He greatly expanded the work performed in the department. Pease left his position and was succeeded by Gillespie. Alonzo H. Stewart was put in charge of the Widal testing laboratory and later took W. J. Gillespie's position.

11. H. K. Mulford Company

The H. K. Mulford Company, pharmaceutical chemists and manufacturer of compressed tablets, (which later became Sharp and Dohme Pharmaceuticals) had its origins in Philadelphia during this period. It also has a strong connection with the development of bacteriology during the 1890s. The following account by Joseph McFarland demonstrates the entrance of bacteriology into the pharmaceutical industry at that time. McFarland was a faculty member at the University of Pennsylvania and adjunct professor of pathology at the Philadelphia Polyclinic. He relates events as they occurred one evening in 1894, although he is not positive of the exact date. Milton Campbell and H. K. Mulford called upon him at his home one evening and made the following proposition. They wanted him to organize and become the director for their company, in his spare time, a "laboratory of biology" for the manufacture of products, such as diphtheria and tetanus antitoxins, tuberculin, and various vaccines. He accepted the offer and noted that it was strictly an agreement between gentlemen with no signed contract.

They rented a stable at Thirty-ninth and Egglesfield (now Cambridge) Street in West Philadelphia. Parts of the stable were walled off for a laboratory, incubator and media preparation room, and small animal area, in addition to the stables for the horses. The staff consisted of McFarland as director; Clarence W. Lincoln, assistant; and John Adams, professor of surgery in the veterinary department of the university as veterinarian. (In a letter to McFarland from H. K. Mulford, he points out that Dr. Leonard Pearson was the first veterinarian.) (31) Their first task was to produce diphtheria antitoxin. They had no idea how to proceed until consulting with William H. Park, of the New York City Health Department, and with Paul Gibier of the Pasteur Institute of New York. They eventually resolved a large sequence of problems. After two years, E. V. Ranck replaced Adams and Robert L. Pitfield was added to the staff. In addition to producing diphtheria antitoxin, they also worked on tetanus antitoxin and tuberculin. They also made attempts to develop antistreptococcal and antipneumococcal sera. Clinical studies with the antipneumococcal serum were conducted first at the German Hospital and later at Pennsylvania Hospital, but the product proved to be more harmful than beneficial for the patient.

The Mulford team housed rattlesnakes for their antivenom work in the same room with hundreds of guinea pig cages, which sat over the stables containing over fifty horses; thus creating some interesting situations, as well as quite a variety of strong odors. These conditions continued through 1898 when the company decided to move to a more rural location in Glenolden, Pennsylvania. The new facility was completed in 1900. J. J. Kinyoun, of the U.S. Public Health Service, replaced McFarland as the new director when the move to Glenolden was completed.

BACTERIOLOGY BOOKS BY PHILADELPHIANS AT THE END OF THE NINETEENTH CENTURY.

Between 1891 and 1899, three Philadelphians wrote significant books covering the subject of bacteriology: Michael V. Ball, Alexander C. Abbott, and Joseph McFarland.

1891

Michael V. Ball, *Essentials of Bacteriology*, W. B. Saunders Company, Philadelphia.

1892

Alexander C. Abbott, *The Principles of Bacteriology: A Practical Manual for Students and Physicians*, Lea Brothers and Company, Philadelphia. The book sold for more than forty years.

1896

Joseph McFarland, *The Pathogenic Bacteria*, W. B. Saunders Company. This book was published in 1919.

1899

Alexander C. Abbott, *The Hygiene of Transmissible Diseases*, W. B. Saunders, Philadelphia. There was a second edition published in 1901.

FOUNDING OF THE SOCIETY OF AMERICAN BACTERIOLOGISTS

It is often difficult to divide historical subject matter into decades or centuries. The formation of the new single discipline society, the Society of American Bacteriologists, cooperated very nicely by coming about at the end of the year 1899. It provides a perfect ending for the story of bacteriology in the 1800s. The story culminates with the formation of a separate society dedicated to this new discipline of science and demonstrates a focal point of evidence that bacteriology was a definable specialty.

In Barnett Cohen's excellent *Chronicles of the Society*, he states the following:

> The Society of American Bacteriologists was founded through the initiative of a volunteer committee consisting of A. C. Abbott, Professor of Hygiene and Bacteriology, University of Pennsylvania, Herbert W. Conn, Professor of Biology, Wesleyan University, and Edwin O. Jordan, Assistant Professor of Bacteriology, University of Chicago. The time was ripe at the end of the nineteenth century for bringing together into a formal organization the large and growing number of investigators who were interested in the biological, agricultural, industrial, as well as the hygiene and pathological aspects of the flourishing young science. (32)

A letter of invitation, dated 16 October 1899, to join the new society was written by Jordan; signed by Abbott, Conn, and Jordan; and sent to about forty bacteriologists in the United States and Canada. One section of the letter is an excellent summary of the state of bacteriology at the time. "It is thought that such an association will conduce to unification of methods and aims, will emphasize the position of bacteriology as one of the biological sciences and will bring together workers interested in various branches into which bacteriology is now ramifying."

Responses to this letter resulted in the first scientific meeting of the society taking place on December 27, 28, and 29, 1899, in New Haven, Connecticut. This three-day scientific meeting consisted of twenty-six presentations that reflected the state of bacteriology at the close of the 1800s.

The society was destined to grow and become a dominant factor in tracing the developments in bacteriology in the twentieth century.

CHAPTER THREE

A Transitional Or Second-Generation Group Of Early Philadelphia Bacteriologists During The Late Nineteenth And Early Twentieth Century

The previous chapter focused on developments in the evolution of bacteriology during the most critical period at the end of the 1800s. The accent was on the bacteriologists and institutions that played a significant role in that important period of transition when bacteriology became a recognized separate discipline, culminating in the formation of the Society of American Bacteriologists. However, before entering into the next logical period in the evolution of bacteriology in Philadelphia, 1900 and beyond, there is a problem: What to do with a group of transitional people who were not included in the "first generation" of bacteriologists. This is a group of bacteriologists that will be designated as "the transitional or second-generation group" of early Philadelphia bacteriologists." They formed an identifiable group with qualifications almost the same as the previous grouping, but not quite. They share one quality that helps place them in a unique classification. What they all share is a birth date or a professional qualification that could almost place them in the special unique first group, but they did not face the same issues that this first group encountered. They were exposed to bacteriology in the United States through a teacher, mentor, or a course in the subject. Also their career developed during a time when bacteriology was still in a transitional or early specialty state. This transitional group faced the decision to enter a new semiformed, but basically identifiable, specialty. Therefore, this chapter will consist exclusively of short biographical sketches of these people whose careers spanned the last part of the nineteenth century and into the 1900s.

Once again relying heavily on the 1935 Joseph McFarland paper "The Beginning of Bacteriology in Philadelphia." (1) This group consists of the following thirteen bacteriologists: Alexander C. Abbott, Michael V. Ball, David H. Bergey, Albert A. Ghriskey, Samuel S. Kneass, Joseph McFarland, Adelaide Ward Peckham, Robert L. Pitfield, Mazyck P. Ravenel, Randle C. Rosenberger, Ernest B. Sangree, Allen J. Smith, and Alonzo H. Stewart.

A summary of several aspects common to their careers is presented following the last biographical sketch. A comparison of this group with the first-generation bacteriologists is also provided.

1. Alexander Crever Abbott (1860-1935) was born in Baltimore, 26 February 1860. He attended the Baltimore City College and received his MD from the University of Maryland in 1884. Prior to the opening of the Johns Hopkins Medical School, William M. Welch announced that he would be conducting a course in bacteriology and John Shaw Billings would teach a course in public hygiene. Abbott became the assistant for both courses. He had been working with Welch as a special student studying the bacteriology of drinking water as well as making observations on malaria parasites found in human blood samples. (3) Welch and Abbott have been credited with being the first to isolate the diphtheria bacillus in the United States. (2) In 1886, Welch sent Abbott to study with Max Pettenkoffer in Munich and Robert Koch in Berlin. When Abbott returned in 1889, he was put in charge of a small laboratory for research in hygiene, modeled after laboratories at the Munich Institute. (2) Billings brought him from Johns Hopkins Medical School to the University of Pennsylvania in 1891 to be the assistant in hygiene of the newly organized Laboratory of Hygiene and Bacteriology. In 1896, Abbott succeeded Billings as director and as professor of hygiene in the medical department. In 1899, Abbott became professor of hygiene and bacteriology, and later that same year, he became professor of bacteriology in the medical department. He also accepted bacteriology teaching responsibilities in the dental and veterinary schools. In 1899, Abbott, along with H. W. Conn and E. O. Jordan, was responsible for initiating the Society of American Bacteriologists. In 1903, Simon Flexner, as chair of pathology, decided that bacteriology should be separated from pathology. He placed Abbott in charge of bacteriology, and the medical student classes were transferred to the Laboratory of Hygiene for their instructions. In 1903, Abbott took a partial leave of absence while retaining his directorship of the laboratory, when he was appointed chief of the city's bureau of health, a position he held until 1909. He remained at the university until his retirement in 1928 as emeritus professor. He was author of two books, *Principles of Bacteriology*, (4) which was published in 1892 and went through many editions, and in 1899,

The Hygiene of Transmissible Diseases. (5) He died at the age of seventy-five on 11 September 1935.

2. Michael V. Ball (1868-1945) was born in Warren, Pennsylvania, 14 February 1868. He received his MD in 1889 from the Jefferson Medical College. After graduating, he visited London, Paris, Copenhagen, and Berlin. He spoke fluent German and spent most of that time in Berlin. He took a course given by Robert Koch, assisted by Carl Fraenkel and Emil Behring. The four-week course began in January 1890, and classes were held ten hours each day. He returned to Philadelphia in April 1890 and interned at the German Hospital. While still an intern, he signed an agreement with W. B. Saunders and Company on 16 January 1891 to write a book on bacteriology for $100 to be completed by July 1. He made the deadline and the book, *Essentials of Bacteriology*, (6) was very popular and went through many editions. He left Philadelphia after his time at the German Hospital to take a position as assistant in microscopy and histology at Niagara University in Buffalo, New York. In 1892, he returned to Philadelphia to become a resident physician at the Eastern State Penitentiary so he could work on another book. He died at the age of seventy-seven in 1945.

3. David Hendricks Bergey (1860-1937) was born in Skippack, Pennsylvania, in 1860. He received both his BS and MD in 1884 from the University of Pennsylvania. After some time in private practice, in 1894, he enrolled as a student in the Laboratory of Hygiene and, after a year, became the Thomas Scott Fellow in Hygiene, and then the assistant in chemistry. In 1896, when Abbott succeeded Billings as drector of the Laboratory of Hygiene, Bergey was made first assistant, a position he kept until 1903, when he became assistant professor of bacteriology. In 1916, his title became assistant professor of hygiene and bacteriology, and in 1926 he became full professor of hygiene and bacteriology. He was a charter member of the Society of American Bacteriologists and, in 1915, served as president of this organization. When Abbott retired in 1928, Bergey became professor of hygiene and bacteriology, a position he held until his retirement in 1931. He then joined the staff of the National Drug Company. In 1901, he published *The Principles of Hygiene*, which went through seven editions. His *Bergey's Manual of Determinative Bacteriology*, (7) first published in 1923, has remained a classic in taxonomy to the present time. He died at the age of seventy-seven in 1937.

4. Albert A. Ghriskey (1859-1935) was born in Philadelphia on 28 May 1859. He attended Professor Hasting's West Philadelphia Academy and received his MD in 1880 from the University of Pennsylvania. He then interned at the Mercer Memorial and Children's Seashore Home in Atlantic City, New Jersey. He was interested in bacteriology and went to the Johns Hopkins Hospital to work with William Osler. He published a paper with Hunter Robb, "The Bacteria of Wounds and Skin Stitches," that appeared in the Bulletin of Johns Hopkins Hospital (3:37:1892). In 1892, he returned to Philadelphia to become the assistant in bacteriology on the original faculty of the Laboratory of Hygiene, but only remained one year. He died at the age of seventy-six on 25 December 1835. (8) His obituary notice, published in the Pennsylvania Medical Journal, states that he joined the bacteriological staff of the Pennsylvania and Episcopal Hospitals of Philadelphia and became a fellow of the College of Physicians of Philadelphia in 1892.

5. Samuel Stryker Kneass (1865-1928) was born in Philadelphia on 16 January 1865. His early education was at the Faries Academy, and he spent a summer at a school in Vevey, Switzerland. In 1886, he received his AB degree and in 1889 an MD from the University of Pennsylvania, followed by a year internship at the university hospital. He spoke both French and German. Following his internship, he went to Paris for further course work, including a course in bacteriology at the Pasteur Institute. He had positions at the medical dpartment of the University of Pennsylvania, adjunct professor of bacteriology at the polyclinic and College for Graduates in Medicine, and bacteriologist at the "German" (later Lankenau) Hospital. In 1895, he left his previous positions to join the staff at the William Pepper Laboratory, University of Pennsylvania, where he remained for thirty-three years, until his death at age sixty-three on 8 December 1928.

6. Joseph McFarland (1868-1945) was born in Philadelphia 9 February 1868. His initial education was with the public school system and the Lauderbach Academy. He received his MD from the University of Pennsylvania in 1889. After a year residency at Blockley, he spent a year (1890-1891) in Heidelberg and Vienna studying bacteriology and pathology. Upon his return to Philadelphia, he became assistant to the professor of pathology in 1891, then assistant demonstrator of pathological histology in 1892. In 1895, he became demonstrator of pathological histology and lecturer on bacteriology. He was also adjunct professor of pathology at the Philadelphia Polyclinic and College for Graduates in Medicine from 1893-4. In 1894, he became the director of the biological laboratories of the H. K. Mulford Company in Philadelphia. He spent the summer of 1895 at Halle, working on bacteriology problems in Carl Fraenkel's laboratory under George Sobernheim. In 1896, he resigned from the University of Pennsylvania to accept the chair of pathology and bacteriology in the Medico-Chirurgical College. Also in 1896, his book, *The Pathogenic Bacteria*, was published. (9) It was a popular textbook that went to nine editions. The last edition was published in 1919. In 1900, he resigned from the H. K. Mulford Company and became consulting bacteriologist to Parke, Davis and Company until 1910. From 1907 to 1910, he was also directing laboratories at the Henry Phipps Institute. In 1903, he spent time working with streptococci and pneumococci at the Pasteur Institute in Paris under Metchnikoff. In 1910, he became professor of pathology at the Woman's Medical College of Pennsylvania and held that position until 1914. He was also a major in the medical corps of the U.S. Army during World War I, serving as chief of the laboratory service at several base hospitals in the United States, as well as director of laboratory instruction in the Medical Officers Training Corps at Camp Greenleaf. He became emeritus professor of pathology at the University of Pennsylvania in 1936 and, in 1940, became professor of pathology at the dental school of Temple University. He died at the age of seventy-seven on September 22, 1945.

7. Adelaide Ward Peckham (1848-1944) was born in Brooklyn, Connecticut, on 31 March 1848. She received her preliminary education in the public and private schools of Connecticut and received her MD in 1886 from the Woman's Medical College of the New York Infirmary for Woman and Children. (10) She was in the first class admitted to the Laboratory of Hygiene of the University of Pennsylvania and remained at that institution for several years as a student and researcher. In 1897, she published a paper on the results of her studies entitled "The Influence of Environment upon the Biological Processes of the Various Members of the Colon Group of Bacilli: An Experimental Study." (11) In 1902, she received a second MD from the Woman's Medical College of Pennsylvania. She became associate professor of pathology in charge of bacteriology in 1898 and professor of bacteriology in 1899 at the Woman's Medical College of Pennsylvania. She held this position until 1919 when she became emeritus professor until her death at age ninety-six in 1944.

8. Robert Lucas Pitfield (1870-1942) was born in the Germantown section of Philadelphia on 28 February 1870. He attended the Germantown

Friends School and the Westtown School in West Chester, Pennsylvania. He received his MD in 1892 from the University of Pennsylvania. He did his residency at the Germantown Hospital. He became a special assistant to Joseph McFarland at the H. K. Mulford Company, and then became a demonstrator of bacteriology in the Medico-Chirurgical College from 1902 to 1904. In 1907, he wrote a book entitled *A Compend on Bacteriology: Including Animal Parasites*, which went through many editions. (12) His other significant contribution to bacteriology is the method he introduced for staining flagella. He died at the age of seventy-two on 2 October 1942.

9. Mazyck Porcher Ravenel (1861-1946) was born in Pendleton, South Carolina, received his initial education at the University of the South, and his MD from the Medical College of the State of South Carolina in 1884. He spent several years in private practice and came to Philadelphia in 1892 to matriculate in the first class taught in the Laboratory of Hygiene. In 1893, he was the Thomas A. Scott Fellow in Hygiene and, in 1894, became assistant in bacteriology. He spent the summer of 1895 at the Pasteur Institute in Paris and the Hygienisches Institut at Halle in Saxony, Germany, and then went to the Laboratory of Hygiene at Princeton to become its director. In 1897, he became bacteriologist to the State Live Stock Sanitary Bureau of Pennsylvania, which was housed at the Veterinary School of the University of Pennsylvania. In 1888, he became lecturer in bacteriology in the veterinary department of the University of Pennsylvania until 1903. He then became the assistant medical director and chief of the laboratories of the newly formed Henry Phipps Institute for the Study and Prevention of Tuberculosis. During this period, he went to Genoa to spend time at the Maragliano Institute. In 1907, he became professor of bacteriology at the University of Wisconsin until 1914 when he accepted the position of professor of preventive medicine, medical bacteriology, and director of the public health laboratory at the University of Missouri. He remained in Missouri after his retirement from the university. He considered his most important contribution to be the work he began with Leonard Pearson at the University of Pennsylvania on the transmissibility of bovine tuberculosis to man. It was the basis of work presented at the 1901 London Congress on Tuberculosis, where he took issue with Robert Koch, who felt the disease was not transferred by animals. One of his more interesting contributions to bacteriology was the introduction of a platinum culture transfer wire loop fastened into an aluminum handle and covered with a glass tube protector so it could be carried in the pocket. He died at the age of eighty-five in 1946.

10. Randle C. Rosenberger (1873-1944) was born in Philadelphia on 4 March 1873. After graduating from Central High School, he entered Jefferson Medical College in 1891 and received his MD in 1894. He remained at Jefferson for the rest of his productive career. After graduation, he had an interest in both pathology and bacteriology and became assistant demonstrator of morbid anatomy and bacteriology in 1897. Two years later, he became assistant pathologist at Thomas Jefferson Hospital. In 1904, he became assistant professor of bacteriology, and in 1909, he was made professor and the first head of the new department of bacteriology and hygiene. In addition to his teaching duties, he served on several Philadelphia Public Health Commissions. He also accepted a professorship at the Woman's Medical College, served as a pathologist at Philadelphia General Hospital (1903-1919) and as a bacteriologist at the Henry Phipps Institution (1904-1908). In 1923, he became the second president of the Pennsylvania chapter of the Society of American Bacteriologists, succeeding Dr. Bergey. In 1941, he served as acting dean of Jefferson until a replacement dean could be appointed. His long and very active association with bacteriology at Jefferson ended with his death on 21 February 1944.

11. Ernest Brewster Sangree (1862-1900) was born 28 May 1862 in Huntington County, Pennsylvania, and pursued his education in the public schools. He attended Mercersburg College and then Franklin and Marshall College. His MD was received from the Medico-Chirurgical College of Philadelphia in 1889. He was assistant visiting pathologist to Blockley Hospital from 1892 to 1895. His main position, starting in 1889, was demonstrator of pathology and bacteriology at the Medico-Chirurgical College. In this position, he received personal instructions in bacteriology from Ernest Laplace. He left this position in 1899 to become professor of pathology and bacteriology at Vanderbilt University, but he had to resign this position and move to Chicago when be became ill with pulmonary tuberculosis. He died there at the age of thirty-eight on 22 February 1900.

12. Allen John Smith (1863-1926) was born 8 December 1863 in York, Pennsylvania. He attended the York Academy and Gettysburg College and received his MD from the University of Pennsylvania in 1886, followed by a year residency at Blockley. He then joined the pathology department at the University of Pennsylvania and became an assistant to Henry Formad. He greatly improved bacteriology laboratory instruction, especially for staining bacteria, and the use of culture methodology. In 1891, he left to become chair of pathology at the University of Texas. While there, his focus was on parasitology, especially on hookworm, and he wrote a book *Lessons and Laboratory Exercises in Bacteriology*. He returned to the University of Pennsylvania in 1903 as professor of pathology to replace Simon Flexner, a position which he held until his death. He continued his work in parasitology and, in 1910, he initiated a Department of Comparative Pathology and Parasitology. He died at the age of sixty-three on 18 August 1926 at his home in St. David's, outside Philadelphia.

13. Alonzo H. Stewart (1867-1955) was born in Greensburg, Pennsylvania, on 29 June 1867. He attended Greensburg Seminary and the Indiana Normal College and received his MD from Jefferson Medical College in 1892, followed by a residency at Blockley, and was then associated with the Laboratory of Hygiene. From 1894 to 1896, he was demonstrator of microscopy at Jefferson Medical College and helped establish the bacteriology laboratory at Jefferson. In 1895, he became bacteriologist to the Bureau of Health of Philadelphia and adjunct professor of bacteriology at the Philadelphia Polyclinic and College for Graduate Medicine. He invented the Stewart "cover-glass forceps." McFarland gives the following description of this invention: "This instrument was invented in 1895, in the days when bacteria were always stained upon cover-glasses, instead of the slides as at present. Only those who have experienced the facility with which capillary attraction can draw the staining solution between the blades of the ordinary forceps from the cover-glass upon which it has been dropped to the fingers, can appreciate this boon." He also invented several instruments associated with milk and water analysis and new techniques for producing antitoxins. (14) He died at the age of eighty-eight in 1955.

Conclusions Drawn from the Biographical Sketches of Thirteen Transitional or Second-Generation Philadelphia Bacteriologists

As noted earlier, these twelve men and one woman are placed into the arbitrary category of transitional or second-generation Philadelphia bacteriologists. A review of the biographical sketches of these thirteen bacteriologists demonstrates how the specialty of bacteriology started to mature during their careers, which began at the close of the nineteenth century and continued into the twentieth century. Like the previous grouping of first-generation Philadelphia bacteriologists, each career described here is unique; however, they all were either born in the Philadelphia area or moved here for a key period of their careers.

The following summarizes the dates of their birth, death, year of getting their MD along with the MD-granting institutions.

1. Alexander Abbott   (1860-1935)   MD 1884 University of Maryland
2. Michael Ball   (1868-1945)   MD 1889 Jefferson Medical College
3. David Bergey   (1860-1937)   MD 1884 University of Pennsylvania
4. Albert Ghriskey   (1859-1935)   MD 1880 University of Pennsylvania
5. Samuel Kneass   (1865-1928)   MD 1889 University of Pennsylvania
6. Joseph McFarland   (1868-1945)   MD 1889 University of Pennsylvania
7. Adelaide Peckham   (1848-1944)   MD 1886 Woman Med Col of NY Infirmary
8. Robert Pitfield   (1870-1942)   MD 1892 University of Pennsylvania
9. Mazyck Ravenel   (1861-1946)   MD 1884 Medical College of S. Carolina
10. Randle Rosenberger (1873-1944)   MD 1894 Jefferson Medical College
11 Ernest Sangree   (1862-1900)   MD 1889 Medico-Chirurgical Col. of Phila
12. Allen Smith   (1863-1926)   MD 1886 University of Pennsylvania
13 Alonzo Stewart   (1867-1955)   MD 1892 Jefferson Medical College

All thirteen were born between the years 1848 and 1873, the average and median year of 1863. All were born in the United States. Ten were born in Pennsylvania, with eight of the ten from Philadelphia or its surrounding areas. The states of birth of the three out-of-state individuals were Maryland, Connecticut, and South Carolina. Ten of the thirteen received their MD in Philadelphia, with six of the ten from the University of Pennsylvania, three from Jefferson Medical College, and one from the Medico-Chirurgical College of Philadelphia. Of the remaining three, Abbott received his MD from the University of Maryland, Peckham from the Woman's Medical College of the New York Infirmary, and Ravenel from the Medical College of the State of South Carolina. All received their MD between 1880 and 1894, with the average year being 1887 and median of 1889.

The age when receiving their MD ranged from twenty-one to thirty-eight with twenty-four as the median age and twenty-three as the average age. The number of potentially active career years, based on the years between obtaining their MD and their death, range from eleven to sixty-three with an average of fifty and median of fifty-three years. Life spans ranged from thirty-eight to ninety-six years, with an average and a median of seventy-five years. Year of death ranged from 1900 to 1955 with an average and median of 1937.

The following are their first professional positions in Philadelphia:

1. A. Abbott 1891 Assistant in Hygiene, Lab. of Hygiene and Bacteriology, U of PA
2. M. Ball 1890 The German Hospital of Philadelphia.
3. D. Bergey 1894 Laboratory of Hygiene, University of Pennsylvania.
4. A. Ghriskey 1892 Assistant in Bacteriology, Lab. of Hygiene, U of Pennsylvania.
5. S. Kneass 1890 Medical Department of the University of Pennsylvania.
6. J. McFarland 1891 Assistant to the professor of Pathology, U of Pennsylvania.
7. A. Peckham 1892 Laboratory of Hygiene of the University of Pennsylvania.
8. R. Pitfield 1893 H. K. Mulford Company.
9. M. Ravenel 1892 Laboratory of Hygiene, University of Pennsylvania.
10. R. Rosenberger 1897 Jefferson Medical College, Pathology Department
11. E. Sangree 1889 Assistant in Pathology and Bact., Medico-Chirurgical Col of Phila
12. A. Smith 1887 Pathology Department at the University of Pennsylvania
13. A. Stewart 1894 Demonstrator of Microscopy at Jefferson Medical College.

As can be seen from the above, the University of Pennsylvania provided the opportunity for eight of the thirteen individuals to start their careers in bacteriology. Of those eight, five were associated with the Laboratory of Hygiene. Of the remaining three individuals associated with this university, two were associated with the pathology division and one directly with the medical department. Of the five remaining individuals who did not initiate their careers at the university, Rosenberger and Stewart started their careers at Jefferson Medical College; Pitfield at a pharmaceutical company; Ball at the German Hospital; and Sangree at the Medico-Chirurgical College of Philadelphia. Regardless of their department affiliation, they all pursued bacteriology. This demonstrates why this group is considered a transitional group. All started their professional careers with an interest in bacteriology and in positions bearing titles that included hygiene, pathology, and microscopy; but almost all ended their careers with a title that included the word *bacteriology*.

From this brief survey, a generalized picture can be drawn of the typical (but theoretical) member of this second or transitional group of Philadelphia bacteriologists. The generalized bacteriologist of this period was male, born in 1863, in the Philadelphia area, received an MD degree from the University of Pennsylvania in 1887 at the age of twenty-three, followed by an internship in a variety of locations and showed an early interest in bacteriology as a career choice. This theoretical person had a professional career of almost fifty years and died at age seventy-five in the late 1930s.

Comparison of the First and the Transitional or Second-Generation Philadelphia Bacteriologists

What follows is a comparison of five individuals of the first generation with the thirteen individuals in the transitional group. The average birth year increased by eleven years from 1852 to 1863. The Philadelphia area remained the most common place of birth and most common location for receiving their medical training. The University of Pennsylvania was the dominant source of medical training in both groups. The year for obtaining their MD increased by eight years, from 1879 to 1887, while the average and median age of obtaining an MD remained almost the same at about twenty-five years of age. The average number of potential career years, based on the years between obtaining their MD and their death, increased by fifteen years from thirty-five to fifty years, and average life span increased thirteen years from sixty-two to seventy-five years.

Although some of the individuals in the second generation did travel to Europe after graduating from medical school, this did not seem as consistent as in the first group of bacteriologists. A major difference between the two groups was the presence of either a formal course in bacteriology, mainly at the Laboratory of Hygiene, or access to a mentor who provided them with a bacteriology laboratory to pursue their interests. This is a significant difference between the two groups and accounts for why they started in bacteriology very early in their careers and accounted for them obtaining the designation of bacteriologists early in their career. However, it should be noted that pathology, hygiene, and microscopy were still terms used to identify several of the individuals in the second generation groups, especially in the early stages of their careers. This points out that bacteriology was still in transition in spite of the major advances in identifying bacteriology as a separate discipline in a group of individuals during the end of one century and the start of the twentieth century.

CHAPTER FOUR

Four Perspectives on Entering The New Century: Philadelphia, The Society of American Bacteriologists, Early Bacteriologists, And Philadelphia Institutions Update

Philadelphia at the Turn of the Twentieth Century

The period from the Civil War into the new century saw the transformation of Philadelphia into an industrial giant.

Philadelphia presented itself as Technology—Queen of the Engine. Science, medicine, engineering, the more practical professions, were enormously respected and dominated the mind of the city.

It was during this period that the drug companies like Smith Kline and French began to flourish.

—Burt and Davis (1)

By the close of the 1800s, Philadelphia had prospered mainly due to the presence of its key industries, namely iron and steel, railroads, and coal. But at the end of the century, textiles ranked first among the city's enterprises. By 1904, Philadelphia was the world's largest and most diversified textile center. This industry accounted for 19 percent of the city's 7,100 manufacturers and employed 35 percent of the city's 229,000 workers. Factories of all sorts cropped up throughout the city, providing jobs for the growing immigrant population. This fact is microbiologically significant since it accounted for a significant increase in industry related cases of anthrax. In area, Philadelphia was the largest city in the United States and, by 1904, had three hundred thousand families living near the factories that supplied the jobs. Not unexpectedly, like all other big American cities, high immigration rates resulted in overcrowding and areas that only could be described as slums. However, as stated by Burt and Davis: "Perhaps the most important single factor in creating the differences [from other big cities] remained architecture. That is, instead of being a city of jammed-together multifamily tenements packed with renters, Philadelphia was still a city of endlessly repeated small row houses, most of them one family dwellings, many of them owner occupied." (1)

The most sophisticated water handling system in the world was showcased in Philadelphia during most of the nineteenth century. However, by the start of the new century, sewage disposal, and the water supply could only be described as inadequate. The typhoid fever death rate in Philadelphia was the highest among all the major American cities. It was three times the rate of New York City at the turn of the century. Cholera and typhoid epidemics in 1891 and 1899 caused agitation for a new water filtration system. (1) In 1908, the opening of the giant Torresdale water filtration plant had a significant effect in addressing the typhoid fever problem in Philadelphia. In 1906, there were 1,063 typhoid fever victims; by 1915, typhoid deaths had been reduced to 109.

The early years of the new century saw a significant influx of immigrants. Fortunately, the great number and variety of manufacturing jobs made it possible to absorb the new arrivals. The outbreak of World War I resulted in an expansion of Philadelphia industries, especially shipbuilding, locomotive manufacturing, and the textiles. In the last months of the war, the 1918 influenza pandemic hit Philadelphia particularly hard. Decisions made by city officials to go ahead with a parade, despite warnings from public health advocates, are thought to have contributed to the intensity of the outbreak. All measures to contain the epidemic failed. At its peak, the death toll in Philadelphia reached more than seven hundred victims a day, which required a policy of mass internment. The epidemic ran its course by early November. By that time, it had claimed many more lives in Philadelphia than had been lost due to the war. (1)

Update on the Society of American Bacteriologists: 1900 to 1945

As described in the previous chapter, as the year 1899 came to a close the Society of American Bacteriologists was formed and the first official scientific meeting took place in New Haven, Connecticut, on December 27, 28, and 29 of that year. On the last day of that meeting, 29 December 1899, a constitution prepared by H. W. Conn, A. C. Abbott, E. O. Jordan, Theobald Smith, and Wyatt Johnson was adopted and "an organization was effected by the election of the following officers for the ensuing year." (2) Those first officers serving for the year 1900 were W. T. Sedgwick as president, A. C. Abbott as vice president, H. W. Conn as secretary and treasurer, and a council consisting of H. C. Ernst, E. O. Jordan, A. E. deSchweinitz, and Theobald Smith. As stated by Cohen, this marked the birth of the new society, the first independent organization dedicated specifically to the promotion and service of bacteriology in the United States and perhaps the world. There were sixty charter members, which included Philadelphians, A. C. Abbott, D. H. Bergey, and J. McFarland. Eligibility for membership was limited to any person who had conducted and published original research in bacteriology. This requirement was maintained for the first seventeen years. The qualifications were modified in 1916 when eligibility was extended to any person interested in the objectives of the society. Initially, active membership was limited to 60 people; this was increased to 75 in 1900, 100 in 1902, 125 in 1905, 150 in 1909 and 200 in 1912. In 1913, and thereafter, the limit was fixed only by eligibility of the candidates. The actual number of active members by select years was as follows: 1899 (59), 1905 (90), 1910 (approx. 150), 1915 (283), 1920 (783), 1925 (1086), 1930 (859), 1935 (923), 1940 (1440), 1945 (2,311), and 1949 (3,643). Cohen offered some reasons for the variation in membership numbers. He noted that in 1916, the *Journal of Bacteriology* was founded and the qualifications for membership were relaxed. In 1917, *Abstracts of Bacteriology* began publication and local branches were authorized. The end of World War I marked a sharp increase in membership. The Society *News Letter* was founded in 1936, and in 1937, *Bacteriological Reviews* was founded. The end of World War II in 1945 marked the start of a further increase in membership. Cohen also notes that a major factor in the growth of the society was associated with the increase in the number of local branches. Although the constitution was amended in 1917 to authorize the establishment of branches, a provision was made for representation of branches on council in 1934. This was a significant factor in increasing the number of local branches as well as in general membership. (2)

The second annual meeting was held in Baltimore, at the Pathological Laboratory of the Johns Hopkins Hospital on December 27 and 28, 1900. It should be noted that prior to the formation of the Eastern Pennsylvania Branch in 1920, annual meetings were held in Philadelphia in 1903, 1904, and 1914. This reflects the active role Philadelphia played in the evolution of microbiology during the early years of the new century. (2)

The Continuation of the First-Generation Philadelphia Bacteriologists: 1900-1938

Of the seven bacteriologists considered previously under the designation of first generation Philadelphia bacteriologists, only three survived or continued working in Philadelphia into the 1900s. These three were Samuel Dixon, Ernest Laplace, and Lawrence Flick.

Samuel Dixon continued his work in bacteriology at the Academy of Natural Sciences and served as commissioner of health of Pennsylvania from 1906 until his death in 1918. Under his tenure as health commissioner, the division of laboratories was established at the medical school of the University of Pennsylvania. He was influential in establishing a distribution system for diphtheria antitoxin, and he is credited with stressing the need for increasing the role of the bacteriology laboratory in addressing issues relating to public health.

Ernest Laplace held his position as professor of surgery at the Medico-Chirurgical College of Philadelphia until his death in 1924.

Lawrence Flick continued his unending crusade to better understand and treat tuberculosis. In 1902, he traveled to Europe to meet with Albert Calmette and other European leaders working in the field of antituberculosis. Flick continued to promote the National Association for the Study and Prevention of TB. In 1908, he promoted the International Congress on Tuberculosis and was influential in having Robert Koch attend and give a brief lecture in the Philadelphia portion of this important International Congress. In 1903, Flick's work resulted in the formation of the Henry Phipps Institute for the Study, Prevention and Treatment of Tuberculosis. In 1910, the Phipps Institute ownership was transferred to the University of Pennsylvania. Flick practiced medicine until the final year of his life in 1938 and was the last survivor of that first important generation of Philadelphia bacteriologists.

The Continuation of the Transitional Generation of Bacteriologists: 1900-1955

Of the thirteen transitional or second-generation Philadelphia bacteriologists, only two left Philadelphia either before the new century began, or shortly after 1900. Ernest Sangree left in 1899 to become professor of pathology and bacteriology at Vanderbilt University. Mazyck Ravenel continued as lecturer in bacteriology in the veterinary department of the University of Pennsylvania until 1903. He then became the assistant medical director and chief of the laboratories of the newly formed Henry Phipps Institute for the Study and Prevention of Tuberculosis. However, during this period, he spent time in Genoa, Italy. He returned to Philadelphia, but left in 1907 to become professor of bacteriology at the University of Wisconsin.

Of the remaining eleven bacteriologists, seven of this generation continued their careers in Philadelphia and helped to define some basic areas of bacteriology that would eventually become "subspecialties" of the field.

Michael Ball left Philadelphia for a short time to take a position as assistant in microscopy and histology at Niagara University in Buffalo, New York, but returned in 1892 to become a resident physician at the Eastern State Penitentiary so he could work on another book. He died at the age of seventy-seven in 1945.

Albert A. Ghriskey worked as a staff bacteriologist at Pennsylvania and Episcopal Hospitals in Philadelphia. He died in 1935.

Samuel Kneass remained in his bacteriology position at the William Pepper Laboratory of the University of Pennsylvania until his death in 1928.

Adelaide Ward Peckham retained her position as professor of bacteriology at the Woman's Medical College of Pennsylvania until her retirement in 1919. She died in 1944.

Robert Pitfield became a special assistant to Joseph McFarland at the H. K. Mulford Company, and then became a demonstrator of bacteriology in the Medico-Chirurgical College from 1902 to 1904. In 1907, he wrote a book entitled *A Compendium on Bacteriology Including Animal Parasites* and continued to make contributions to both parasitology and bacteriology. He died in 1942.

Allen Smith left Philadelphia in 1891 to become chair of pathology at the University of Texas, but returned to the University of Pennsylvania in 1903 as professor of pathology to replace Simon Flexner. In 1910, he established the Department of Comparative Pathology and Parasitology at the University of Pennsylvania and remained at the university until his death in 1926.

Alonzo H. Stewart was bacteriologist at the Bureau of Health of Philadelphia and adjunct professor of bacteriology at the Philadelphia Polyclinic and College for Graduate Medicine. He made significant contributions by incorporating bacteriological principles into several public health studies. He died in 1955. His death in that year is significant since it marked the end of the period influenced by the transitional group of Philadelphia bacteriologists.

Four of the transition generation made significant contributions well into the 1900s. These four were A. Abbott, H. Bergey, J. McFarland, and R. Rosenberger.

Alexander Abbott continued as director of the Laboratory of Hygiene and starting in 1899, he held the position of professor of hygiene and bacteriology at the Laboratory of Hygiene, as well as professor of bacteriology in the medical department. In this capacity, as the new century began, he became the bridge between the Laboratory of Hygiene, with its focus on the more practical aspects of bacteriology in the education of both medical as well as other University of Pennsylvania students, and the teaching of medical or "theoretical" bacteriology in the medical school curriculum. He was also associated with teaching bacteriology to dental and veterinary students. In 1903, it was decided that bacteriology should be a separate discipline from pathology in the medical school structure. Therefore, as the new century began, he was charged with formulating a coherent way of teaching bacteriology to students with medical, public health, and a basic science interest. Also in 1903, he was appointed chief of the city's Bureau of Health, a position he held until 1909. And as noted previously, he played a significant role in founding the Society of American Bacteriologists in 1899 and helped to establish this new society as a driving force in defining bacteriology in the United States. He became an emeritus professor in 1928. It is apparent that Abbott played a significant role in defining the scope of bacteriology in Philadelphia for the first three decades of the twentieth century. He died at the age of seventy-five in 1935.

David Bergey started the new century as first assistant to Abbott at the Laboratory of Hygiene and did much of the bacteriology teaching. In 1903, he became assistant professor of bacteriology, and in 1916, he became assistant professor of hygiene and bacteriology. In 1926, he became full professor of hygiene and bacteriology. He was a charter member of the Society of American Bacteriologists and, in 1915, served as president of this organization. When Abbott retired in 1928, Bergey became professor of hygiene and bacteriology, a position he held until his retirement in 1931. In addition to the National Society, he was active in organizing the Philadelphia community of bacteriologists and a founding member and president of the new Philadelphia chapter of the Society of American Bacteriologists. He wrote the *Bergey's Manual of Determinative Bacteriology* that was first published in 1923, and which continues to bring honor not only to Bergey but to Philadelphia bacteriology. He retired from the university in 1931 and decided to join the staff of the National Drug Company. A review of his career demonstrates how rapidly bacteriology was changing. His entire professional career was dedicated to bacteriology in the form of teaching, research, taxonomy, and even in his later years, expanded into the new field of pharmaceutical science. He certainly was one of Philadelphia's most influential early bacteriologists. He died at the age of seventy-seven in 1937.

At the start of the new century, Joseph McFarland was still chair of pathology and bacteriology in the Medico-Chirurgical College. In 1900, he resigned his position as the director of the biological laboratories of the H. K. Mulford Company and became consulting bacteriologist to Parke Davis and Company. In 1903, he spent time working with streptococci and pneumococci at the Pasteur Institute in Paris under Metchnikoff, and from 1907 to 1910, he was also a director of laboratories at the Henry Phipps Institute. In 1910, he resigned his position at Parke Davis and Company and became professor of pathology at the Woman's Medical College of Pennsylvania, a position he held until 1914. He was a major in the Medical Corps of the U.S. Army during World War I, serving as chief of the laboratory service at several base hospitals in the United States, as well as director of laboratory instruction in the Medical Officers Training Corps at Camp Greenleaf. He became emeritus professor of pathology at the University of Pennsylvania in 1936 and, in 1940, became professor of pathology at the Dental School of Temple University. Needless to say, McFarland took advantage of many new aspects of bacteriology that were unfolding as the new century progressed. It was not unusual at the time to hold several positions at once, since many of these positions paid very little and it was common, especially early in the 1900s, to teach in a number of settings in addition to having a primary teaching appointment at a different institution. McFarland was not typical, but his career certainly demonstrates the array of opportunities available to a bacteriologist in Philadelphia. He was one of the first to work for two pharmaceutical companies, a precedent that would become more common later in the century when so many pharmaceutical companies were established in the Philadelphia area. His other contribution was his interest in documenting the history of Philadelphia bacteriology up to 1935. He certainly knew both the people and the institutions that played a part in the new discipline of bacteriology. McFarland died at the age of seventy-seven in 1945.

Randle C. Rosenberger remained at the Jefferson Medical College for the rest of his career. In 1909, he became the first head of the new department of bacteriology and hygiene at Jefferson. He also held joint positions at the Woman's Medical College, Philadelphia General Hospital, and the Henry Phipps Institution. He served on several Philadelphia Commissions relating to public health and designed several innovative pieces or equipment used for analysis in the bacteriology laboratory. In 1923, he became the second president of the Pennsylvania chapter of the Society of American Bacteriologists. He died in 1944.

Philadelphia Institution Update

Three institutions, two new and one undergoing significant changes, must be mentioned as significant at the turn of the century. These are the Henry Phipps Institute for the Study, Treatment and Prevention of Tuberculosis, Jefferson Medical College, and Temple School of Medicine.

Henry Phipps Institute for the Study, Treatment and Prevention of Tuberculosis

A steel industry millionaire named Henry Phipps became very interested in the extensive work Lawrence Flick was doing in the area of tuberculosis. In 1903, Phipps provided funds to establish a clinic in an area of Philadelphia with a very high incidence of tuberculosis. It was called the Henry Phipps Institute for the Study, Treatment and Prevention of Tuberculosis with L. Flick as director of the institute. Mazyck Ravenel left his bacteriology position at the veterinary department of the University of Pennsylvania to become the assistant medical director and chief of the laboratories of the newly formed institute. Early research work focused on experimental methods for treating tuberculosis patients as well as conducting both animal studies and human trials of various antisera.

In 1910, Phipps transferred ownership of the institute to the University of Pennsylvania and provided funds to build a large building at Seventh and Lombard Streets. Flick, as director of the institute, assembled a significant staff and served as director until 1913. The new institute rapidly earned a worldwide reputation for research on tuberculosis. The initial staff consisted of young investigators which included Joseph Walsh, H. R. M. Landis, George W. Norris, and Daniel McCarthy. In 1910, Paul A. Lewis, a noted experimental pathologist, became director of the laboratory and initiated work on the use of dyes as possible therapeutic agents for treating tuberculosis. Dr. Flick retired as director of the institute in 1913, and Charles J. Hatfield became director. When Hatfield arrived, all the staff became university faculty, and Landis was made director of the institute's clinics. As the staff grew, almost every aspect of contemporary research associated with the pathogenesis, immunity, and the chemotherapy of tuberculosis was included in the institute's research programs. Investigators like Eugene Opie and Esmond R. Long held significant positions at the institute.

Jefferson Medical College

As presented previously, Jefferson Medical College was founded in 1824, and as the century unfolded, the competition and presence of two college-based medical schools benefited both institutions. However, instruction in bacteriology got off to a slow start. It was not until 1892 that W. Coplin began weekly lectures in bacteriology and hygiene with a formal course offered on a routine basis starting in 1896. Randle C. Rosenberger was made associate in bacteriology in 1903 and assistant professor of bacteriology the next year. In 1909, bacteriology was separated from pathology. W. Coplin remained in the pathology department and Randle C. Rosenberger became professor of bacteriology. Rosenberger was first to hold the chair of bacteriology and hygiene, a position he retained until his death in 1944, when he was succeeded temporarily by William A. Kreidler. Kenneth Goodner became the second chairman of bacteriology (1946-1967). Therefore, in the new century, bacteriology at Jefferson under Rosenberger became a significant institution in the development of bacteriology in Philadelphia.

Temple College-University School of Medicine

As noted previously, in 1901, the sixth Philadelphia Medical School was established at Temple College, and in 1907, the Philadelphia Dental College, Garreston Hospital, and Samaritan Hospital merged with the former Temple College to form Temple University. As the century progressed, bacteriology studies at Temple University would join the other medical schools in Philadelphia to strengthen the city's prominence in medical microbiology.

Books by Philadelphia Microbiologists

1907

Robert Lucas Pitfield, *A Compendium on Bacteriology Including Animal Parasites*, P. Blakiston's Sons and Company, Philadelphia.

1915

John A. Kolmer, *Practical Text-book of Infection, Immunity and Specific Therapy*, W. B. Saunders Company, Philadelphia.

1921

Joseph McFarland, *Fighting Foes Too Small to See*, F. A. Davis Company, Philadelphia.

1923

David H. Bergey, *Bergey's Manual of Determinative Bacteriology: A Key for the Identification of Organisms of the Class Schizomycetes*, Williams and Wilkins Company, Baltimore.

PART 2

THE EASTERN PENNSYLVANIA BRANCH OF THE AMERICAN SOCIETY FOR MICROBIOLOGY 1920 TO 2010

The Microbiological Club, forerunner to the Eastern PA Branch; the formation of the Eastern PA chapter of the Society of America Bacteriologists; early years of the Branch including World War II; early years with focus on regular meetings; transition from Society of American Bacteriologists to American Society for Microbiology and chapter to Branch; the split to a Clinical Microbiology Section and reformation of one unified Branch; a new energy of workshops, symposia, forums and educational activities for all students, undergraduate through graduate and postdoctorates, as well as initiation of special lecture series, and issues faced by the Branch as it enters the twenty-first century, and 2010 and beyond.

CHAPTER FIVE

The Microbiological Club and The Founding of The Eastern Pennsylvania Chapter of The Society of American Bacteriologists: 1912 To 1939

In 1912, a group of Philadelphia bacteriologists formed a club that met on an irregular basis to discuss their common interests. No minutes were taken, and there was no formal charter. According to several accounts, there was considerable discussion on the name of the group with no real consensus on whether it should be called the Microbiological Club or the Bug Club. Regardless of the name, the Microbiological Club was a definite forerunner of the Eastern Pennsylvania Branch. Dr. Harry Morton in 1984 assembled all the available material on the Microbiological or Bug Club as well as the early years of the Eastern Pennsylvania Branch. What follows is a summary of Dr. Morton's research (1), comments made by C. J. Bucher at a roundtable on the history of Philadelphia bacteriology held at the Society of American Bacteriologists General Meeting in Philadelphia on 13 May 1947 (2), and additional material from other sources. Starting in 1937, material from the Branch archives collection is used to supplement the material collected by Drs. Morton and Bucher.

The Microbiological Club 1912-1923: Forerunner of the Eastern Pennsylvania Chapter

The idea of organizing a group of individuals in the Philadelphia area that was interested in the relatively new science of bacteriology, as well as pathology, was conceived by A. Parker Hitchens, MD, most likely during a visit to Boston with his friend Dr. Benjamin White in 1911 or 1912. Following this visit, Dr. Hitchens brought together a group of twelve to fifteen individuals interested in either bacteriology or pathology. They formed an organization in 1912. The 1912 date is based on the recollection of Dr. Claude P. Brown, one of the original members of this group. Dr. Brown stated that C. Y. White served as secretary for eleven years. Since there is firm evidence that Dr. White served as secretary until 1923, it makes the beginning of his eleven years in the office of secretary as the year 1912. This was the beginning of what was to become known to many early Philadelphia bacteriologists as the Philadelphia Bug Club.

The first meeting and some of the subsequent meetings were held at Palumbo's Restaurant, Eighth and Catherine Streets, Philadelphia. It was a dinner meeting that started with antipasto and liberal amounts of Chianti, then a pasta dish, followed by a meat and vegetable course and ended with Lachrymal Christi.

Those present at the first official organization meeting were the following:

Alexander C. Abbott, MD, David H. Bergey, MD, Claude P. Brown, MD, Domaso deRivas, MD, Herbert Fox, MD, A. Parker Hitchens, MD (president), John Reichel, VMD, Randle C. Rosenberger, MD, George Robinson, MD, Otto Schoble, MD, George H. Smith, PhD, and C. Y. White, MD (secretary).

Frequently, at those early meetings, discussions took place as to whether the young organization should be known as a Microbiological Club or a Bug Club. It was noted that the amount of wine consumed at these meetings produced a state of conviviality and often stimulated the discussions, especially the discussions between A. C. Abbott and Domaso deRivas. The forces proposing the name Microbiological Club were led by Abbott, and they appear to have won, at least on paper, but the proponents for the name Bug Club led by deRivas had a popular following for a long time. Dr. Abbott thought the name Bug Club was rather undignified. Other cities, such as Boston, had a Bug Club so a similar name for the new Philadelphia club was not without precedent.

The programs were arranged by the directors of several Philadelphia hospital laboratories, the city of Philadelphia Public Health Laboratory and the Mulford Laboratories in Glenolden. This would determine where the meetings would be held. This arrangement was continued for some time, but it was finally decided that a more or less regular central meeting place was desirable since not everybody had an automobile. Most of the subsequent meetings were held at the University of Pennsylvania, Jefferson Medical

College, the Henry Phipps Institute at Seventh and Lombard Streets, and the Mulford Laboratories in Glenolden. The latter was conveniently located near a local train station.

Usually, there were three scientific papers, followed by a not altogether coherent discussion. Dr. Carl J. Bucher claimed that the essays were erudite, eloquent, not always scientifically sound, but usually witty. There was a spirit of camaraderie in those early days, possibly because the club was comparatively small. This camaraderie would not be possible in later years with so many diversified interests being represented.

Claude P. Brown noted that the existence of the Bug Club was somewhat erratic. There is no record indicating that it ever had a constitution or bylaws. There are no records of minutes, dues, or officers, except that A. Parker Hitchens served as the president during the first year of the organization, and that C. Y. White served as secretary throughout the existence of the club.

In 1914, the annual meeting of the Society of American Bacteriologists was held in Philadelphia for the third time. A. P. Hitchens served as chair of the Local Committee on Arrangements assisted by D. Bergey, J. McFarland, and J. Leidy Jr.

During World War I, the club almost became extinct due to members joining the Armed Forces or taking positions elsewhere. Drs. Abbott, Bergey, Hitchens, Brown, and Aronson joined the army, and Hitchens made the army his career.

Dr. C. J. Bucher provided the following list of bacteriologists who routinely attended the Microbiological Club meetings (2):

A. Abbott, David Bergey, Claude Brown, C. J. Bucher, Domaso deRivas, Arthur Hitchens, John Kolmer, Paul Lewis, Frank Lynch, John Reichel, Randle Rosenberger, S. W. Sappington, and C. Y. White.

On 24 February 1920, Dr. Bergey called on many of the above members, along with several new recruits, to revive the Microbiological Club as an official branch of the Society of American Bacteriologists. The Eastern Pennsylvania Branch is indebted to Drs. Morton and Bucher for recording the activities of the Microbiological Club, a definite forerunner to the official formation of the Eastern Pennsylvania Branch.

The Eastern Pennsylvania Chapter: Origin and Early Years (1920-1936)

At the 1916 Society of American Bacteriologists meeting, a committee was appointed to look into the feasibility of adding local branches. The committee reported to the general business meeting that year, offering a constitutional amendment establishing the guidelines for branches. The amendment passed the business meeting and was sent to the complete membership in January 1917. A month later, the amendment was declared approved. On 24 February 1920, Dr. David H. Bergey invited a group of persons interested in bacteriology and pathology to meet in the Laboratory of Hygiene, University of Pennsylvania (Thirty-fourth Street, between Spruce and Walnut Streets), for the purpose of organizing a group to be known as the Eastern Pennsylvania chapter of the Society of American Bacteriologists. Most of the individuals present were former members of the Microbiological Club, and it was proposed to form a local chapter around that nucleus.

At the March 1920 meeting, Dr. C. P. Brown presented a constitution and a set of bylaws. The name of the local group was to be the Eastern Pennsylvania chapter. The term chapter was most likely proposed by Bergey. The Society of American Bacteriologists looked with disfavor on the name chapter and the name was eventually changed to the Eastern Pennsylvania Branch in 1961. The bylaws noted that the meetings were to be held on the second Tuesday of each month, October to May, inclusive. Beginning in 1925, the meetings were scheduled for the fourth Tuesday instead of the second Tuesday. The secretary-treasurer, by direction of the executive committee, was to select a meeting place and arrange for a dinner, not to exceed $1.50 per plate. On special occasions, the price was not to exceed $3. The dues were $1.50 per year.

Not much is recorded on the actual meetings, but some early meeting problems were noted. There is a note about how a threatened railroad strike interfered with the April 1920 meeting scheduled to be held at the Mulford Laboratories in Glenolden, since most members took the train to Glenolden. Also it is noted that at the October meeting the secretary-treasurer petitioned for an assistant.

During the years 1920, 1921, and 1922, Dr. Bergey served as president, Dr. C. Y. White served as secretary-treasurer, and Miss Lola S. Hitch served as assistant secretary-treasurer. Dr. Claude P. Brown was chairman of the executive committee. There were approximately forty active members each year.

At the meeting on 11 January 1921, the first annual dinner was held at Bookbinder's restaurant. The 13 December 1921 meeting was devoted to pneumococci and pneumonia, and on December 26, 27, 28, Philadelphia was host to the National Society Meeting for the fourth time. Dr. C. Y. White was chair of the Local Committee on Arrangements, with a committee consisting of the following members: A. Abbott, C. P. Brown, R. Rosenberger, J. Reichel, and F. M. Huntoon.

In 1923, Dr. A. Parker Hitchens was listed as the first guest speaker. His topic was "Emergency Production of Potable Water."

After the first three years, Dr. Claude P. Brown served as secretary-treasurer from 1923 through 1936. Most of the presidents of the chapter served for two years. The ten presidents for the period 1920 to 1939 were as follows:

Dr. David H. Bergey, University of Pennsylvania (1920-1922)
Dr. Randle C. Rosenberger, Jefferson Medical College (1923-1924)
Dr. A. C. Abbott, University of Pennsylvania (1925-1926)
Dr. Eugene L. Opie, Henry Phipps Institute (1927-1928)
Dr. Joseph L. T. Appleton, Jr., Univ. of PA, Dental School (1928-1929)
Dr. Joseph D. Aronson, Henry Phipps Institute (1930-Oct. 1931)
Dr. Jefferson H. Clark, Philadelphia General Hospital (1931-1934)
Dr. Evan L. Stubbs, University of PA, Veterinary School (1934-1935)
Dr. Carl Bucher, Jefferson Hospital (1936-1937)
Mr. Christopher G. Roos, Sharp and Dohme Laboratories (1938-1939)

The chapter met at various places, usually on the premises of the institution that was putting on the program. There may have been about seventeen different places where the meetings were held at least once. These locations included:

Jefferson Medical College
Temple University
Woman's Medical College
Drexel Institute
H. K. Mulford Laboratories in Glenolden
Pennsylvania State Laboratories
City of Philadelphia Health Laboratories
Philadelphia General Hospital
Graduate Hospital
College of Physicians
Research Institute of Cutaneous Medicine
University of Pennsylvania: Laboratory of Hygiene, Medical School, Veterinary School, Dental School, William Pepper Clinical Laboratory, and the Medical Clinic

It is quite possible that in 1925, when the chapter began meeting on the fourth Tuesday of the month, the meeting in December came too close to the holiday season as well as to the annual meeting of the National Society which took place that month. The chapter discontinued its December meeting and began meeting only seven times each year. Dr. Bucher noted the decline in 1930-31 when membership "reached a new low ebb of 25 members but clung tenaciously to life." He gives credit to David Bergey, Randle Rosenberger, and Claude Brown for keeping things going during "these precarious times."

In 1926, the annual meeting of the Society of Bacteriologists was held in Philadelphia for the fifth time. (3) C. P. Brown served as chair of the Local Committee on Arrangements with A. Abbott serving as chairman ex-officio and a committee consisting of F. M. Huntoon, D. Bergey, R. Rosenberger, and C. Y. White. In December 1933, the Thirty-fifth Annual Meeting of the Society of American Bacteriologists was once again held in Philadelphia. This was the sixth time the meeting was held here. David Bergey was Honorary Chairman, and J. H. Clark was chairman of the Local Committee on Arrangements. Committee members were J. D. Aronson, C. P. Brown, F. Konzelmann, S. Mudd, W. L. Obold, J. Reichel, B. Rose, E. L. Stubbs, and R. C. Rosenberger.

In 1936, two people did much to rejuvenate the Branch. Drs. Carl Bucher (from the clinical laboratories at Thomas Jefferson Hospital) and Harry Morton (from the newly reorganized bacteriology department, University of Pennsylvania) were determined to bring about significant changes. Dr. Bucher was elected president in 1936, and Dr. Morton became secretary-treasurer in 1937. This ushered in a period of regrowth for the Branch. Membership became open to anyone who was interested in bacteriology, immunology, or the allied sciences. Some members, particularly some of the original active members, were not happy with this liberal admission policy. It was felt that the admission of "technicians" could destroy the nature of the organization. This apprehension did not seem to exist with the newer members.

One of the changes that Dr. Bucher initiated was to establish a permanent location for the Branch meetings starting in 1936. He was able to get the auditorium of the Philadelphia County Medical Society Building at the Southeast corner of Twenty-first and Spruce Streets. This new location apparently attracted several new members. Because of this increase in membership, the annual dues were reduced from $1.50 to $1.00. Dr. Bucher continued as president in 1937 and, once again, membership rose to 123 in that year and 166 the following year.

In February 1935, the Society of American Bacteriologists passed a constitutional amendment providing for Branch representation on council. The minutes of the December 1935 council meeting list nine Branch councilors as being present. Dr. Rosenberger represented the Pennsylvania Branch as councilor. Another initiative to Branches that was introduced in 1935 was the first publication of the Proceedings of the Local Branches in the *Journal of Bacteriology*. The first publication of the Proceedings of the Eastern Pennsylvania chapter was in a 1936 issue of the *Journal of Bacteriology* (31: 439), which reported on the 28 January 1936 chapter meeting that took place at Temple University School of Medicine.

It should be noted that during this period, the national organization held the annual meeting in Philadelphia in the following years: 1921, 1926, and 1933.

The Eastern Pennsylvania Chapter: 1936-1939 and the Branch Archives

The year 1936 marks the start of the current Branch Archives Collection. Fortunately, a few early members recorded events during the first sixteen-year history of the Branch, but the original documents, in spite of many efforts to locate them, remain lost. Since there was no central Branch location, the original documents were moved among various sites and have not been recovered. Archival material, starting in 1936, consists of meeting notices, abstracts of all meetings, two annual newsletters that summarize the activities for the respective years, detailed membership lists, and summary notes prepared by contemporary Branch members with an interest in recording events.

In September of 1937, Harry Morton started what he referred to as an *Annual Newsletter*, which he designated volume 1, no. 1. Between 1937 and 1946, eight issues were published. Some facts extracted from the 1937 newsletter provide a detailed picture of the Branch at that time. The officers were the following:

President—Dr. Carl Bucher, Jefferson Hospital.
Secretary/Treasurer—Dr. Harry Morton, Univ. of Penn. School of Medicine.
Councilor—Dr. Randle C. Rosenberger, Jefferson Medical College.
Alternate Councilor—David Bergey, National Drug Company
Program Committee—Dr. Stuart Mudd, Chairman.

The dues for 1938 were $1.00. There were 123 paid members, double the number from the previous year; 43 were also members of the national organization. Also published in this newsletter is the revised, comprehensive set of bylaws, and it notes the use of the new address-o-graph system. The names and addresses of all individuals on the mailing list were put on address-o-graph plates for more efficient mailing. The programs were printed on U.S. penny postal cards and then run through the address-o-graph. The newsletter also includes a full listing, with addresses, of all the dues-paying members.

The following is also noted in this first newsletter: "Our most prosperous year was marred by one event, that was the death of Dr. David H. Bergey, the founder of the society. We shall greatly miss him at the meetings." On 26 October 1937, at the regular Branch meeting, Dr. Randle C. Rosenberger of Jefferson Medical College delivered a talk on Dr. Bergey. Some of his comments are as follows:

> Dr. Bergey made friends because of his personality and scholarship. He was tolerant of others and understandable in all his trials, genuinely sincere in all his earthly dealings, rarely differed from any one, talked and let others talk, was at home in the presence of scholars and enjoyed the respect and confidence of his associates to an unusual degree. His greatest hobby was genealogy and his greatest tome was the compilation of the Bergey family, which was 1,047 pages in length and consisted of notes on 5,759 individuals. His family life was ideal and his devoted wife was a bulwark in aiding him in his accomplishments.

The second edition of the *Annual Newsletter* was published November 1938. The officers for that year were the following:

President—Mr. C. G. Roos, Sharp & Dohme Laboratories, Glenolden. Pa.
Secretary/Treasurer—Dr. Harry Morton, University of Pennsylvania School of Medicine.
Councilor—Dr. Carl J. Bucher, Jefferson Hospital.

Membership increased to 166, with 60 new members, and 38 were also national members. There were seven meetings that year with an average of more than 100 people in attendance for each meeting.

The 134th Branch meeting was held on 18 October 1938, at the Philadelphia County Medical Society. At this meeting, Mr. C. G. Roos from Sharp and Dohme Laboratories in Glenolden, Pennsylvania, gave a presentation entitled "The Original Minutes of the Society of American Bacteriologists." An abstract of this presentation was published in the *Journal of Bacteriology* (37:107). Mr. Roos made the announcement that the original minutes of the first meetings of the Society of American Bacteriologists had been located at Sharp and Dohme Laboratories. It was known that some records had been left at the laboratories in Glenolden in 1917 by the society's secretary, Dr. Parker Hitchens, but searches for the documents had been fruitless. In the spring of 1938, laboratories in the main building were remodeled and provisions were made for safe keeping of valuable records. This resulted in the discovery of the minutes of the early society's first meeting. A record book contained the original transcript of the society's constitution, minutes of the meeting of organization, names of the society's organizers with a list of the names of its charter members, the first to the eleventh programs presented at the annual meetings, and the minutes of transactions at those meetings. The records cover the period from 16 October 1899 to 30 December 1909.

Roos stated that "Photostats are being made of the records and will be available at the library of the Mulford Biological Laboratories of Sharp and Dohme, Glenolden, Pennsylvania. The original record book will be delivered to the archives committee of the Society of American Bacteriologists."

This incident raises hope that many of our Branch records, which are currently missing from 1920 to 1935 will some day be located.

No newsletter was issued in 1939; however, there are short minutelike notations recorded for each meeting. See below for some of the details. Election results for 1940 were announced at the November meeting. The treasurer reported a balance of $306.64 for the period ending February 1.

It was noted that the monthly meeting programs were changing with guest speakers from outside of the Philadelphia area who spoke on timely topics, as well as joint meetings with other societies. One was with the New York City and New Jersey Branches at Princeton, NJ. A cold buffet supper with beer on draft was served for $1 per person. Another joint meeting was with the Physiological Society of Philadelphia and the Pathological Society of Philadelphia. It was also noted that there were meetings which looked back into history. At the 28 March 1939 meeting, Dr. Hitchens talked about the introduction of agar into bacteriology, Dr. McFarland reported on "Hunting Tubercle Bacilli Fifty Years Ago," and Dr. Schramm presented a history of pure culture studies.

Notes on Branch Meetings: 1936 to 1939

As noted, little information exists on the details of Branch meetings prior to 1935 since these records have been lost. However, there is extensive information on Branch meetings starting with the 118th Branch meeting in 1936 and continuing to the present. All meetings since 1936 are listed in the appendix. The following are some observations on the meetings for 1936 through 1939 (meetings 118 to 142).

In 1936, there were four meetings. Some of these meetings, such as the November meeting, were probably very long, since there were nine presentations. The March meeting had five presentations, and there were four presentations at the October meeting. January of that year was the only meeting that had only one presentation.

In 1937, there were seven meetings. There was only one presentation at the November meeting and up to six presentations in several meetings that year. The 124th meeting of 23 March was listed as a "Symposium on Hemolytic Streptococci." This is the first recorded time the designation *symposium* was used to describe a Branch meeting or event. There were five presentations relating to beta hemolytic streptococci, all given by speakers connected with the University of Pennsylvania. It is assumed the designation of symposium was used because all five talks were on the same subject, since it was not unusual to have five presentations at one meeting during this period. On the first of May, the 125th meeting was held at the Rockefeller Institute for Medical Research in Princeton, New Jersey. A regular meeting was also held later that month.

Starting in 1937, Dr. Morton, in his role as secretary, initiated the practice of assigning a number to each meeting. He got this idea after discovering that the Baltimore Branch had started doing this for their meetings. By searching the past records, he determined that the 25 May meeting was the 125th Branch meeting and was the first meeting to be listed by title and number, a practice that continues to the present. It should be noted that it was later discovered that this was actually the 126th meeting.

In 1938, there were seven meetings. Only two meetings consisted of a single presentation.

In 1939, there were seven meetings with three or four presentations per meeting except for the meeting of 25 April, which was designated "A Symposium on *Lymphopathia venereum*, the Sixth Venereal Disease," which consisted of six presentations all from the proctologic department of the Graduate Hospital. Seventy-eight people attended this meeting. Attendance at the meetings ranged from a low of 59 in March to 111 in November. Average attendance for the seven meetings was 83 attendees.

The following are institutions represented at the meetings for 1936 through 1939:

PHILADELPHIA INSTITUTIONS
Claude Brown Clinical Laboratory
Henry Phipps Institute
Jefferson Medical College
Pennsylvania State Health Laboratories
Research Institute of Cutaneous Medicine, Philadelphia.
Temple University School of Medicine
University of Pennsylvania
Woman's Medical College
Hospitals: Abington Memorial, Bryn Mawr, Children's, Easton, Graduate, Jefferson, Philadelphia General, University of Pennsylvania

OTHERS
Department of the Interior, Washington, D. C.
Easton Hospital, Easton, PA
Johns Hopkins University, Baltimore, MD
Lafayette College, Easton, PA
National Drug Company, Swiftwater, PA
National Institute of Health, Washington, D. C.
Post-Graduate Medical School and Hospital, Columbia University, NY
Rockefeller Institute for Medical Research, Princeton, NJ
Rockefeller Institute for Medical Research, NY
Sharp & Dohme Laboratories, Glenolden, PA

Summary of the Branch in the 1930s

In summary, the Branch went through several changes in the 1930s, which included innovations, such as meeting at the same location, accepting all members who were interested in bacteriology, presentations by bacteriologists outside the Philadelphia area, as well as meetings at institutions outside of Philadelphia. Other changes included the assigning of numbers to meetings and publication of meeting abstracts in the *Journal of Bacteriology*. These changes resulted in a significant increase in membership and set the stage for future innovations.

CHAPTER SIX

The Eastern Pennsylvania Branch: 1940-1949 The Decade of World War II and the 1947 Annual SAB Meeting

Overview of the Branch in the 1940s

As noted in the previous chapter, there was a successful program initiated in the late 1930s, by Carl Bucher, the Bacteriologist at Jefferson Hospital and Harry Morton, from the Bacteriology Department of the Medical School at the University of Pennsylvania, to reorganize and expand Branch membership. The new decade started with a continuing sense of momentum. During this period, the official name, due to the continued use of the original bylaws, was still the Eastern Pennsylvania chapter. The officers of the chapter consisted of a president, secretary/treasurer, councilor and a councilor-alternate. The only standing committee was a Program Committee. The activities of the chapter, during the entire decade, centered on the monthly meetings, which were held on a Tuesday evening during the months of January through May, and then October and November. The meetings during this period were held at the Philadelphia County Medical Society Building at Twenty-first and Spruce Streets. Annual dues throughout this period were $1. Meeting notices were mailed to each member in the form of a "penny-postal-card" that some months contained a remarkable amount of information, often requiring the aid of a magnifying glass to read the fine print.

Dr. Harry Morton appears to be the most active member during a major part of this period, serving as secretary/treasurer through 1945 and president through 1947. In his role as Secretary, he recorded the short business meetings which occurred at the start of each regular Tuesday evening meeting. Proposed members were voted on by all members present at the meeting. He also recorded the number of attendees at each meeting. The records become sparse during the last few years of the decade.

Two major events occurred during the 1940s: World War II and the 1947 Society of American Bacteriologists Annual Meeting in Philadelphia. As noted in the Annual Newsletters, all written by Dr. Morton, during the early War Years, there was concern that members were not sufficiently prepared for emergency situations. As a result, the Branch sponsored special meetings such as the one that focused on culturing anaerobes from severe wounds. One result of these concerns was a request by Dr. Morton to Selman Waksman, then President of the Society of American Bacteriologists, to address this issue. Waksman appointed a Committee on Materials for Visual Instruction in Microbiology, with Dr. Morton as Chairman. The Committee collected materials for lantern slides, photographic prints, and motion picture films to be made available, by rental or purchase, to all interested bacteriologists.

There was also expressed concern, noted in the newsletter, for the number of Branch members who were in the Armed Services and that the level of Branch activity was going to be affected. In spite of these expressed concerns, the opposite occurred, mainly due to the presence of bacteriologists who were in the Armed Services being assigned to positions in Philadelphia, especially at the University of Pennsylvania and the Henry Phipps Institute. As a result, some meetings had over 230 attendees, and the number of meetings was not reduced during the war years.

The second major event was the 1947 SAB Annual Meeting held in Philadelphia, with Dr. Morton as chairman of the Local Committee on arrangements, and Willard Verwey serving as vice-chair. Several new Annual Meeting features were introduced at this Philadelphia meeting. The local committee introduced television for the first time as a feature in the exhibit area. Other firsts included mimeographed lists of members in attendance so that members could have a complete list of all those present to take home with them. Particularly significant was the introduction of a daily newsletter called the "Incubator" which was available each of the five mornings of the meeting. The Incubator contained announcements for the day, news from the day before and a generous sprinkling of humorous and personal notes. An official vote of appreciation was received from the National Society.

Summary of Branch Meetings

The following is an attempt to characterize the sixty-nine meetings held in the 1940s, (meeting numbers 143 to 211). It should be kept in mind that almost all meetings consisted of more than one presentation. For the sixty-six meetings that have full details available, there were 214 individual presentations. Attendance at these meetings averaged 95 attendees, with at least one meeting having over 230 attendees. Most presenters were local; however, there were presentations given by non-local speakers at approximately 20 percent of the meetings. The local speakers at these meetings were from a wide variety of Philadelphia institutions. The dominant institution was the University of Pennsylvania. Other well represented institutions were: Jefferson Medical College, Mulford Laboratories, and Temple Medical School. The following hospitals were represented: Children's Hospital of Philadelphia, Philadelphia General, Graduate, Bryn Mawr, Abington, Pennsylvania, Lankenau, Philadelphia Naval Hospital and the Jewish Hospital of Philadelphia.

The following is the percent for each category of the 214 presentations in the 1940s:

Bacteriology, 16; antimicrobials, 10; virology, 7; immunology/vaccines, 5; electron microscopy, 4; serology, 4; air-sampling, 3; methodology, 3; biochemistry, 3; mycology, 2; parasitology, 2; and disinfectants, 1.

As would be expected, the dominant subject focused on bacteriology, especially relating to three diseases: tuberculosis, syphilis, and gonorrhea. In addition, there were several presentations on staphylococci, streptococci, brucellosis, *H. pertussis*, tularemia, whooping cough and intestinal diseases. There were a significant number of presentations on virology, especially on influenza, swine influenza, bacteriophage, poliomyelitis and hepatitis. Starting in 1941, there were a significant number of presentations on sulfa drugs, streptomycin and the penicillins. The first presentation on betalactamases was in 1945. There were also several presentations on mycology, parasitology and general methodology. Three topics are worth noting. Tuberculosis related research received a lot of attention due to the research being conducted at the Henry Phipps Institute. There were many presentations on air sampling due to a department specializing in this at the University of Pennsylvania.

Also, many presentations were on electron microscopy because of the research teams working with Dr. Anderson at Radio Corporation of America (RCA) in Camden and the group working with Dr. Mudd at the University of Pennsylvania.

Branch Officers

Branch Presidents and Secretary/Treasurers
There were five presidents (the eleventh through the fifteenth) who served the Branch during this decade:

Stuart Mudd, University of Pennsylvania (1940-1941)
William A. Kreidler, Jefferson Medical College (1942-1943)
Earle H. Spaulding, Temple University School of Medicine (1944-1945)
Harry E. Morton, University of Pennsylvania (1946-1947)
William Verwey, Sharp and Dohme Laboratories, Glenolden (1948-1949)
Secretary/Treasurers for this period were: Harry E. Morton, University of Pennsylvania, (1937-45), Amedeo Bondi, Temple University (1946-47), and W. G. Hutchinson (1948-49) University of Pennsylvania.

Councilors
A. Parker Hitchens, University of Pennsylvania, (1940-1), Harry E. Morton, University of Pennsylvania, (1942-5), W. Verwey, Sharp and Dohme Laboratories, (1946-7), James A. Harrison, Temple University, (1948-9).

Councilor Alternates
Fred Boerner, Graduate Hospital (1940-1), Earle H. Spaulding, Temple University (1942-3), Ruth E. Miller, Woman's Medical College (1944-5), James A. Harrison, Temple University (1946-7), Amedeo Bondi, Hahnemann Medical College (1948-9).

Branch Events by Year

1940
There were seven meetings in 1940, with an average of four (range of two to five) presentations at each meeting, which resulted in a total of twenty-eight presentations. Average attendance at the seven meetings was eighty (range of fifty-one to ninety-eight) attendees.

Dr. Morton published Volume III of the Annual Newsletter. In this edition, he noted that the Branch continued to prosper, with 171 paid-up members and good attendance at each meeting. The Branch was asked by the Commonwealth of Pennsylvania to demonstrate some laboratory methods at the Pennsylvania Medical Society meetings held at the Bellveue-Stratford Hotel from 30 September to 3 October. Dr. Morton complained that the booth provided to the Branch was too small, but based on the list of methods that were demonstrated and the number of prominent Branch members who participated in these demonstrations, it was a major undertaking, and great use was made of the space that was provided.

1941 There were seven meetings in 1941, with an average of four (range of two to five) presentations at each meeting. This is not counting the February 25 meeting, which had four regular presentations followed by seven short presentations on techniques used to culture anaerobes. As a result, there were a total of thirty-one presentations at the seven meetings. Average attendance at the seven meetings was 106 (range of 87 to 150) attendees.

In January, Dr. Morton sent a letter to all members asking them to state what techniques or questions needed to be addressed. He used an anaerobic technique symposium that was in the planning stage as an example of how such requests would be met. Near the end of the letter he included the following: **"It is also felt that in view of emergencies which will arise from the unsettled condition of the world that it would be well to consider procedures which would be very important in such emergencies."**

The above mentioned symposium "Technics for the Cultivation of Anaerobic Microorganisms" was held in February. It was the first symposium at which a stenographer recorded the proceedings in shorthand and prepared a detailed transcription of all presentations and discussions. A summary of the transcript was distributed to all members in the form of Dr. Morton's Fourth Volume of the Annual Newsletter.

1942

In 1942, Dr. Morton produced Volume Five of the Annual Newsletter. He stated that, "The Society [meaning the Branch] is experiencing its first war period. More than ever, bacteriologists need to get together to discuss their problems and make known the results of their labors." He goes on to note that, "Many who attended our meetings in the past will not be with us this season, having entered the Armed Forces." This is followed by a list of Branch members who were in the Armed Services as of October

1942. Several years later he described one of the effects of World War II as follows:

> In 1942, many of our members, like other bacteriologists, served in the armed forces of World War II. Many were in positions where they had to train personnel in microbiology and they had very little with which to work. They urged us to make some teaching aids available to them. Dr. Morton urged Dr. Selman A. Waksman, then President of the Society of American Bacteriologists, to appoint a committee to fulfill that need. Waksman appointed the Committee on Materials for Visual Instruction in Microbiology with Dr. Morton as Chairman, a position he held until 1971. The Committee collected materials for lantern slides, photographic prints, and motion picture films. These were made available by rental or purchase. Bacteriologists in many medical schools, universities, colleges and teachers of biology in high schools used the various visual aids. It became such a voluminous business that finally the National Society placed the production and distribution of the visual aids in the hands of a commercial producer and distributor of visual aids. The Committee was disbanded, and activities in the field expanded under what became known as the Board of Education and Training (B.E.T.). The name was proposed by Dr. Earle H. Spaulding.

There were seven meetings in 1942, with an average of 3 (range of 1 to 6) presentations at each meeting, resulting in a total of 22 presentations. Average attendance at the seven meetings was 90 (range of 48 to 178) attendees. At the May meeting Dr. Selman Waksman gave a presentation entitled "The Production of Antibiotic Agents by Microorganisms and Its Significance in Natural Processes."

At the 24 November meeting, a motion was passed which stated that members serving in the Armed Forces are to be relieved from payment of dues, and that their names are to be temporarily removed from the mailing list, and upon their return they are to be invited to again become active members without going through the formalities of applying for membership.

1943

There were eight meetings in 1943, with an average of four (range of one to eight) presentations at each meeting, resulting in a total of thirty-one presentations. Average attendance at the eight meetings was seventy-seven (range of sixty-two to ninety-nine) attendees.

On a Saturday afternoon, 15 May, there was a joint meeting of the New Jersey, New York City and the Eastern Pennsylvania Branches, held at Guyot Hall, Princeton, New Jersey. A cold buffet supper with beer on draught followed the meeting. There were various subjects included in this special meeting. The accent was on developments concerning antibacterial agents, which was significant, since this was a time when the nation was at war, and the need for effective treatments for war wounds was becoming critical. In addition to this special joint meeting on 15 May, the Branch maintained its regular meeting schedule, and ten days later the normally scheduled Branch meeting was held on 25 May.

1944

There were six meetings, with an average of 2 (range of 1 to 3) presentations at each meeting, resulting in a total of 12 presentations. Average attendance at the six meetings was 129 (range of 66 to 238) attendees. For the 25 January meeting, members of the Pathological Society of Philadelphia were invited to hear Professor Paul R. Cannon, from the University of Chicago, give a presentation entitled: "Protein Metabolism and Resistance to Infection." The attendance for this meeting was 238 members and guests.

In January, Dr. Morton produced Volume Six of the Annual Newsletter. He listed additional Branch members who had joined the Armed Services. He also noted that:

> "In spite of the severe restrictions being placed upon the release of certain research data, the Program Committee has already arranged excellent programs for the first two meetings in 1944. As results of confidential research projects become available throughout the year, the Committee plans to have them reported promptly to the Branch. Because so many of our active members are serving in the Armed Forces at the present time, it will be difficult to obtain the full quota of papers by our own local members. The fields of Bacteriology and Parasitology are more important today than they ever were before. Therefore, it is appropriate, if not essential, that the Branch's activities in no way be reduced, even though individually the accelerated pace of our routine duties may leave us little time."

1945

There were seven meetings in 1945, with an average of three (range of one to three) presentations at each meeting, resulting in a total of twenty presentations. In January, Dr. Morton produced Volume Seven of the Annual Newsletter. He noted that there were 174 paid-up members with an average attendance of 128 at the seven meetings, and goes on to note, "The war has brought many new bacteriologists into our midst and has interested for the first time many other individuals in the science of bacteriology."

1946

There were seven meetings in 1946, with an average of three (range of one to four) presentations at each meeting, resulting in a total of twenty presentations.

In January, Dr. Morton produced Volume Eight of the Annual Newsletter. He noted that the Branch was growing as a possible sign of "post-war expansion" and provided membership figures for the last ten years as: 1936-9 (67, 125, 166, 155), 1940-5, (172, 170, 168, 167, 174, 195). It was also reported that an audit conducted on 7 January indicated that the balance at that time was $392.20.

1947

There were six meetings in 1947, with an average of three (range of two to four) presentations at each meeting, resulting in a total of twenty-two presentations.

The following was reported in the newsletter of the Society of American Bacteriologists:

> The Eastern Pennsylvania Chapter is one of the most careful in regard to the collection and publication of abstracts in the JOURNAL OF BACTERIOLOGY and periodically reprints of these abstracts are collected and bound. The latest, for the season of 1945-46, has just appeared and is a most attractive little reprint. A membership list as of January 1947 has just been prepared and lists 279 [the real figure for 1947 was three hundred] members. Ninety of these names are starred to indicate that they are members of the SOCIETY.

The most significant event for the Branch in 1947, as noted by Dr. Morton, was the following: "The Branch was host to the National Society for its Annual Meeting in Philadelphia. We had a very energetic membership and put on a very good annual meeting with many innovations. For one thing, we started the custom of having a public lecture and a mixer on the evening preceding the beginning of the Annual Meeting. This allowed the attendees to get settled in their rooms, meet their friends and get their visiting taken care of so the meetings could get started in full force early the next morning and make possible three full days of meetings. The two functions became known as the Opening Session and Annual Reception and became incorporated into the official program of the Annual Meetings of the Society of American Bacteriologists."

Another innovation was the establishment of a daily newsletter or bulletin issued during the Annual Meetings. This was an idea proposed by Dr. Grant O. Favorite during a meeting at the Union League. In searching for an appropriate name, Joseph Smolens suggested the *Incubator*. That seemed like a logical name as nearly every morning bacteriologists say, "Let's see what's in the incubator!" The comments about the 47th Annual Meeting as they appeared in the July 1947 issue of the S.A.B. Newsletter describe the *Incubator* and the meeting very well:

> "Several new annual meeting features were introduced in Philadelphia. The local committee not only provided television for the entertainment of the members who wished to mix in a little baseball with their roundtables, but it mimeographed the list of members in attendance so efficiently that members who desired had a complete list of all those present to take home with them. Particularly significant was the daily appearance of the "Incubator" which was available each of the five mornings neatly mimeographed on gray, deep pink, light pink, blue and yellow paper containing announcements for the day, news from the day before and a generous sprinkling of humorous and personal notes. The registration and information service was good, entertainment features were well planned and administered, and hotel accommodations and service satisfactory. The vote of appreciation from the Society to Local Committee Chairman Morton and his colleagues was as sincere and well deserved as any ever registered."

It was later noted that the publication of the *Incubator* continued for many years until the Annual Meetings became so large that other methods of communication became more convenient, such as the daily computer printout of registrants by name, hotel and home address.

As was then customary, at each Annual Meeting of the Society of American Bacteriologists, there was a session on the history of bacteriology in the region where the Annual Meeting was held. Naturally, at the 1947 meeting there was a session on the History of Bacteriology in the Philadelphia Area. Because of Dr. Bucher's interest in history and in bacteriology, he volunteered to be chairman of the session and organized the program. There were seven presentations:

Introductory Remarks by Dr. Carl J. Bucher, Chairman
History of Veterinary Bacteriology in Philadelphia, by Dr. E. L. Stubbs
History of Bacteriology of Tuberculosis in Philadelphia, by Dr. Florence B. Seibert
History of Freeze Drying of Biological Products in Philadelphia, by Dr. Earl W. Flosdorf
Ten Years' Work of the Laboratories for the Study of Airborne Infections, by Professor William F. Wells
Electron Microscope in Philadelphia Bacteriology, by Dr. Thomas F. Anderson
History of Bacteriology in Commercial Laboratories, by Dr. Christopher Roos

Drs. Bucher, Stubbs and Seibert handed in prepared manuscripts and a public stenographer prepared transcripts of the talks by the other four speakers on the program. Copies of the proceedings of the session are in the Branch Archives. They form a valuable source of information on the history of microbiology, especially as it relates to the development of research and commercial/pharmaceutical bacteriology in the early decades of the twentieth century.

1948

There were seven meetings in 1948, with an average of three (range of two to four) presentations at each meeting, resulting in a total of eighteen presentations. It must be noted that there is no information available for the April meeting. Dr. Morton produced a short newsletter dated 28 May. The newsletter noted that April marked the 200th Branch meeting and that there were 326 paid-up members in1948.

1949

There were seven meetings in 1949, with an average of three (range of one to three) presentations at each meeting, resulting in a total of twelve presentations. It must be noted that there is currently no information available for the March or April meetings.

Summary of the Branch in the 1940s

As can be seen from the information above, the main goal of the Branch during this decade was to conduct monthly meetings on a regular schedule and at one location. The decade started with an element of doubt that this goal could be met due to the state of the world political situation. The entrance of the U.S. into World War II resulted in Branch members joining the Armed Services and the belief that scheduled meetings could probably not be maintained. However, several factors became apparent as the decade progressed. There was a need for clinical bacteriologists to become fully aware of the latest techniques available in such areas as culturing anaerobic organisms, how to provide teaching aids to train new physicians and technicians, and to learn more about the new field of antibacterial drugs. There was also a need for bacteriologists to keep up-to-date on contemporary research in many areas relating to microbiology. These factors were just some of the elements that resulted in the educational activities of the Branch becoming more important than ever before. This, coupled with an influx of researchers who were assigned by the Armed Forces to research positions in Philadelphia, resulted in increased attendance and interest in the monthly meetings.

Later in the decade, the 1947 annual meeting of the Society of American Bacteriologists offered the opportunity for Branch members to serve on the Local Organizing Committee. Committee members were charged with making this meeting a great success with several new innovations that would be adopted in future meetings. At this meeting, the tradition of holding a special session on the history of the designated Annual Meeting location was an outstanding success. This gave the opportunity for several prominent members of the Branch to bring together many aspects of Philadelphia's contribution to microbiology. The resulting transcripts of this session remain a valuable source of information on the contributions Philadelphia microbiologists and institutions made to the developing field of microbiology.

A close look at the activities of the Branch in the 1940s demonstrates how, with just a few active members, working with a Program Committee that was attuned to the needs of Branch members, the Branch was able to expand membership participation during a very difficult period of American History.

CHAPTER SEVEN

The Eastern Pennsylvania Branch: 1950-1959 A Decade of Innovative Programs

Overview of the Branch in the 1950s

The decade of the 1950s was much like the previous one for the Branch. The focus was still almost exclusively on the monthly meetings. The official name was still the Eastern Pennsylvania chapter. The officers of the chapter consisted of a president, secretary/treasurer, councilor and a councilor-alternate. The only standing committee was a Program Committee. The monthly meetings continued to be held on a Tuesday evening during the months of January through May, and then October and November. For the first five years, the meetings continued to be held at the Philadelphia County Medical Society Building. Starting in October 1955, the meetings were moved to Medical Alumni Hall, Maloney Building, of the University of Pennsylvania, at Thirty-sixth and Spruce Streets. Meeting notices continued to be mailed to each member in the form of a "penny-postal-card" that now became two cents. Except for the last two years, attendance records at the monthly meetings were not kept; however, during 1958 and 1959, there was an average of 85 (range 37 to 180) attendees. This figure is approximately the same as the figure for the previous decade, which had an average of 95 attendees.

Summary of Branch Meetings

There were a total of 70 Branch meetings (meeting numbers 212 to 281) from 1950 through 1959. Details on one of the meetings during this period have been lost. Therefore, figures presented in this section are based on 69 meetings. Since almost all meetings had more than one presentation, these 69 meetings resulted in 215 individual presentations. Most presenters were local; however, there were presentations given by non-local speakers at approximately 35 percent of the meetings, an increase of 17 percent above the previous decade. The local speakers at these meetings were from a wide variety of Philadelphia institutions. As in the previous decade, the dominant institution was the University of Pennsylvania (29 percent). Other well represented institutions were the following: Temple Medical School, 7 percent; Sharpe and Dohme, 7 percent; Children's Hospital of Philadelphia, 6 percent; Jefferson Medical College, 5 percent; and Hahnemann Medical College, 4 percent. The following hospitals (in decreasing order of participation) were represented: Children's Hospital of Philadelphia, Albert Einstein Northern, Lankenau, Philadelphia General, Graduate, Pennsylvania, and Delaware Hospital. Except for Children's Hospital, there was a trend toward decreasing numbers from hospital presenters, but an increase in presenters from the local Medical Schools. It should be noted that presenters from hospitals that are part of a medical school are included as representing the medical school. There was a significant increase in presentations by researchers from the pharmaceutical companies, especially Sharp & Dohme which accounted for seven percent of the presentations. Other pharmaceutical companies represented were Smith Kline & French, Wyeth, Lederle and Squib.

The following is the percent for each category for the 215 presentations in the 1950s; numbers in brackets indicate the percent increase or decrease when compared to the previous decade results. Items in bold print indicate a new category:

- bacteriology 23 [+7]
- immunology/vaccines 14 [+9]
- virology 13 [+6]
- antimicrobials 10 [0]
- biochemistry 7 [+4]
- **PPLO 6**
- mycology 5 [+3]
- electron microscopy 4 [0]
- serology 2 [-2]
- methodology 2 [-1]
- **antimicrobial susceptibility testing 2**

All other categories are less than 1 percent.

Several points can be noted from the above percentage of presentations when compared to the percent of presentations during the previous decade. The following categories, included in the 1940s summary, were dropped due to low number (less than 1 percent) of presentations: air-sampling, parasitology, and disinfectants. Antimicrobial susceptibility testing was added. This is mainly due to presentations on disk testing. Also new is a separate category for pleuropneumonia-like organisms. There were a few presentations on this subject in the previous decade, but due to the group working on this at the Bacteriology Department of the University of Pennsylvania Medical School, there were whole meetings with several presentations dedicated to this subject. The single biggest increase was in the area of immunology/vaccines, and this subject, in addition to the antimicrobial category, partially accounts for the increase in pharmaceutical company presentations in the 1950s.

The following is a summary of the subject content of the 215 presentations in the 1950s. As expected, the dominant subject focused on bacteriology (23 percent); however, except for tuberculosis related topics, no single genus or species dominates. There were several presentations on staphylococci, streptococci, mycoplasmas (PPLO), *Bacillus sp.*, *E. coli*, *Haemophilus pertussis*, *Serratia sp.*, *Pseudomonas sp.*, mycoplasmas, enterics, and oral flora. *Serratia marcescens* appears in a presentation for the first time in 1950, and *Streptococcus faecalis* appears as a topic for the first time in 1959. There were a significant number of presentations on virology (13 percent), especially on influenza and poliomyelitis related subjects. There were also several presentations on bacteriophage and rabies. In 1950 there was a presentation on the relation of viruses to cancer with several additional presentations on this topic as the decade progressed. The topic of virus induced Rous sarcoma was presented in 1958. Ten percent of the presentations were on antimicrobial agents, and this figure does not include susceptibility testing, which is treated as a separate subject. As in the previous decade, the penicillins, sulfa drugs and streptomycin received considerable attention, but these shared the field with aureomycin, neomycin, nitrofurans, stilbamidine and terramycin. The first presentations on disk susceptibility testing appeared in 1956. Thirteen percent of the presentations focused on immunology or vaccines. Other presentations were on mycology, parasitology and general methodology. The term "Diagnostic Microbiology" first appeared at a 1952 meeting, in the form of a panel discussion dedicated to this subject. The subject of tissue culture first appeared in 1955 with a full meeting dedicated to the subject.

Branch Officers

Branch Presidents
There were five presidents (the sixteenth through the twentieth) who served the Branch during this decade:

Amedeo Bondi, Temple University School of Medicine (1950-51)
James Harrison, Department of Biology, Temple (1952-53)
Ruth Miller, Women's Medical College (1954-55)
Kenneth Goodner, Jefferson Medical College (1956-57)
Morton Klein, Temple University School of Medicine (1958-59)

Secretary/Treasurers
1950	W. G. Hutchinson, University of Pennsylvania
1951-53	Ruth E. Miller, Woman's Medical College
1954-57	Theodore G. Anderson, Temple University School of Medicine
1958-59	Elizabeth H. Fowler, Temple University School of Medicine

Councilors and Councilor-Alternates
1950-1957	Unknown
1958-59	Ruth Miller, Councilor, and Councilor-alternate, Dr. Bernard Briody

Branch Events by Year

1950
There were seven meetings in 1950, with an average of three (range of two to four) presentations at each meeting, resulting in a total of twenty-four presentations.

A special meeting was held on the afternoon of 27 May at the Institute for Cancer Research adjacent to Jeans Hospital in Fox Chase. The meeting was entitled "Some Microbiological Aspects of Cancer." There were four presentations, one each focusing on mycology, mycobacteriology, virology and immunology. The meeting was preceded by a tour of the laboratories.

1951
There were seven meetings in 1951, with an average of four (range of one to seven) presentations at each meeting, resulting in a total of twenty-six presentations.

1952

There were seven meetings in 1952; however, details for the October meeting have been lost. Of the remaining six meetings there was an average of two (range of one to four) presentations at each meeting, resulting in a total of fifteen presentations. The first meeting for this year was entitled "Some Aspects of Viral Diseases" presented by Frank M. Burnet, Director of the Walter and Eliza Institute in Melbourne, Australia. At the May meeting, Dr. Hilary Koprowski spoke on "The Present Status of Immunization against Rabies."

1953

There were seven meetings in 1953, with an average of three (range of one to three) presentations at each meeting, resulting in a total of eighteen presentations. The October meeting was a special meeting held on a Saturday afternoon at the Research Laboratories of the Sharp and Dohme division of Merck Inc. in West Point, Pennsylvania. A tour of the laboratories was held before the meeting. There were three presentations at the meeting covering topics relating to bacteriophage, anaphylaxis in mice due to *Haemophilus pertussis* and effects of chemotherapy on intestinal flora.

1954

In 1954, Dr. Ruth Miller of the Women's Medical College became the first female Branch President.

There were seven meetings in 1954, with an average of four (range of three to five) presentations at each meeting, resulting in a total of twenty-four presentations. The October meeting consisted of five presentations on pleuropneumonia-like organisms.

1955

There were seven meetings in 1955, with an average of three (range of one to five) presentations at each meeting, resulting in a total of twenty-two presentations. Starting in October, the meetings were held at the Medical Alumni Hall, Maloney Building, of the University of Pennsylvania, at Thirty-sixth and Spruce Streets. At the November meeting, Merrill W. Chase from the Rockefeller Institute for Medical Research in New York spoke on "Some Studies on Drug Allergy." The talk was preceded by a turkey dinner at Houston Hall for $2.20 (tip included).

1956

There were seven meetings in 1956, with an average of three (range of one to five) presentations at each meeting, resulting in a total of twenty-three presentations. The April meeting was held at Wyeth Laboratories in Radnor. The topic was "Current Problems in Medical Microbiology" and it consisted of three presentations on the following topics: the polio vaccine by John. H. Brown of Wyeth, immunity relating to *H. pertussis* by Willard F. Verwey of Merck Sharp & Dohme and fungal chemotherapy by Raymond C. Bard of Smith Kline & French.

1957

There were seven meetings in 1957, with an average of three (range of one to four) presentations at each meeting, resulting in a total of twenty-two presentations. The January meeting had three presentations on epidemiology, and the April meeting consisted of a panel discussion on the combined use of antibiotic agents. Although not a Branch function, Branch members were invited to attend a lecture in October by Wendell M. Stanley of the University of California. The lecture, at the University of Pennsylvania, was entitled "Recent Advances in Virus Research."

1958

There were seven meetings in 1958, with an average of three (range of one to four) presentations at each meeting, resulting in a total of twenty-one presentations. Average attendance at the seven meetings was 86 (range of 37 to 170) attendees. The January meeting was designated as a symposium on the microbiology of cancer research, and the February meeting was a panel discussion on changes occurring in medical microbiology. Other meetings with several presentations centering on a single theme were on oral microbiology, and germ-free animals.

1959

There were seven meetings in 1959, with an average of three (range of one to four) presentations at each meeting, resulting in a total of twenty presentations. Average attendance at the seven meetings was 83 (range of 40 to 180) attendees. Meetings with several presentations centering on a single theme were on viral studies, immunology, hospital infections, and pleuropneumonia-like organisms.

Summary of the Branch in the 1950s

Except for the information on the individual meetings, records are incomplete for the years 1950 through 1957. However, it is safe to conclude that the accent was on maintaining top quality presentations at the monthly meetings. There was an increase in the use of outside speakers and a tendency to designate a single topic for many of the meetings. Some of these meetings were designated as "symposia". There was also a tendency to have panel discussions at several meetings. Based on available attendance records, it appears that the programs were meeting the needs of the membership.

CHAPTER EIGHT

The Eastern Pennsylvania Branch: 1960-1969
A Time of Transition and Turmoil: Annual Meeting, From SAB to ASM, From Chapter to Branch, and a Splinter Group within the Branch The Clinical Microbiology Section

Overview of the Branch in the 1960s

This decade started much like the previous two decades; the focus was still on providing top quality monthly meetings. There was an increase in communication with the national organization. The officers of the chapter still consisted of a president, secretary/treasurer, councilor and a councilor-alternate. The only standing committee was still just a Program Committee, but there was some increase in ad hoc committees to address specific issues such as government funding for science studies, especially issues related to funding for the biological sciences. The monthly meetings continued to be held on a Tuesday evening during the months of January through May, and then October and November. Most of the meetings were held at the Medical Alumni Hall, Maloney Building, of the University of Pennsylvania, at Thirty-sixth and Spruce Streets; however, there was a trend toward experimenting with alternative sites and reliance on the pharmaceutical companies for financial support, especially for special meetings involving leading researchers from other cities. In May of 1960, the Branch was host to the Society of American Bacteriologists for the sixtieth annual meeting. (See Summary of Branch Hosted Sixtieth Annual SAB Meeting below.)

A significant event occurred on 27 December 1960 when the Society of American Bacteriologists changed its name to the American Society for Microbiology. This change demonstrated the growing complexity of the field that now was officially referred to as microbiology. One result of this name change was that it provided incentive for the Branch to take action in resolving an issue that had been ignored since 1920. On 24 October 1961, the name of the Eastern Pennsylvania chapter of the Society of American Bacteriologists was officially changed to the Eastern Pennsylvania **Branch** of the American Society for Microbiology. Until this time it was the only Branch that called itself a chapter, and this was looked upon with disfavor by the National Society.

From the start of the new decade until 1967, there was little evidence of the unrest that was brewing within the Branch. The year 1967 started as a normal year with a full schedule of meetings. The Branch minutes during this period were sparse. The only recorded business meetings were held immediately prior to the monthly Tuesday meetings and included all who were attending the lecture for that evening. It remained a normal year until 25 September 1967 when a special meeting was held that resulted in the formation of a splinter group within the Branch. (See the Clinical Microbiology Section 1967-1969 below.)

Summary of Branch Hosted Sixtieth Annual SAB Meeting. In May of 1960, the Branch was host to the Society of American Bacteriologists for the sixtieth annual meeting. The meeting was quite successful, and once again the Local Committee on arrangements was chaired by Harry Morton with Carl Clancy as vice-chair. At that meeting, teachers and students from high schools in Philadelphia were invited to attend the meeting on the last day of the exhibits. Talks by representatives of various fields of microbiology were presented in the morning, followed by a complimentary lunch, and then the students and teachers were encouraged to visit the exhibits. This was so enthusiastically received by everybody that it became a part, in various forms, of every Annual Meeting of the National Society. Also, a successful session was held on the History of Bacteriology in the Philadelphia Area which covered the period of 1947 to 1960. This encouraged Dr. Morton to prepare a document that identified the institutions in Philadelphia associated with microbiology. This included detailed information on microbiology in Philadelphia at seventeen educational institutions and departments, twelve clinical and research institutions, and two pharmaceutical companies. A copy of this report is on deposit in the Branch archive collection.

The Clinical Microbiology Section: 1967-1969. Starting in May of 1967, there were a series of letters and questionnaires written by Richard Clark, the Microbiologist at Pennsylvania Hospital, concerning the need and interest to have more clinical microbiology oriented programs in the

Philadelphia area. As later summarized by Dr. Morton, "For some time, clinical microbiology had been neglected by the Branch in favor of molecular biology and virology." A decision was made by several Branch members to bring microbiologists who were interested in Clinical Microbiology together to discuss a way forward. A meeting notice was sent to all branch Members, as well as other Clinical Microbiologists, who might be interested. On 25 September 1967, approximately seventy people attended a special meeting that was held in the auditorium of the Public Health Laboratories, 500 South Broad Street. The following are the official minutes of the meeting, first drafted by Dr. Moat, the Branch President and sent to Herman Friedman, the Branch Secretary to finalize:

Official Minutes of the 25 September 1967 Meeting

An organizational meeting of a Clinical Microbiology Section of the Eastern Pennsylvania Branch of the American Society for Microbiology (ASM) was held at 7:30 p.m. on Monday, September 25, 1967, in the auditorium, Public Health Laboratories, City of Philadelphia, 500 South Broad Street, Philadelphia, Pennsylvania. Mr. Richard Clark, Head, Dept. of Microbiology, Pennsylvania Hospital, 8th and Spruce Streets, Philadelphia, was the convener for the meeting. He was introduced by Dr. Albert Moat, Hahnemann Medical College, President of the local society, who welcomed the group and made a few statements related to desires of a number of microbiologists in the city to organize a clinical microbiology section.

Mr. Clark made a few statements relating to the need for such a section. Mr. John McKitrick, of Children's Hospital, moved that a Clinical Microbiology Section be organized as a subsidiary of the local ASM branch. This motion was seconded and passed. A steering committee was formed. Dr. Moat appointed the following persons: Richard Clark, Program Chairman, Dr. Harry Morton, Pepper Laboratory, University of Pennsylvania, Dr. Theodore Anderson, Temple University, Dr. Eileen Randall, Jefferson Hospital, Dr. James Prier, Director, Pa. State Public Health Laboratory, Dr. [Ralph] Knight, Women's Medical College Hospital, Mr. James Copeland, Director, City Public Health Laboratory. Dr. Albert Moat, Hahnemann Medical College, President of the local society, Dr. Herman Friedman, Albert Einstein Medical Center, Secretary-Treasurer, and Dr. George Warren, Wyeth Laboratories, Program Chairman, were appointed members ex officio.

The committee was directed to meet and set meeting dates and plan the programs for a clinical section. The meeting adjourned for coffee. After the break, a request was made by Dr. Moat for all those who were not members of the local branch to join so that the clinical section could help support itself. Dr. Moat then introduced Dr. Earle H. Spaulding, Chairman of the Microbiology Department at Temple University, who spoke on the topic: "Training in Clinical Microbiology."

The announcement was made to the general membership by way of the lead news story in the December 1967 Branch Newsletter written by Dr. Zubrzycki. He announced the event as follows:

Microbiology, Weaned on Pus: To conquer the infectious diseases was the first command leveled at our infant science by history. Because five medical schools and numerous pharmaceutical houses are in the Philadelphia area, we have a large group of microbiologists who have maintained a strong interest in this unfinished job. They have organized a section of clinical microbiology within our local ASM.

The first Steering Committee meeting was held on 9 October 1967 to plan the activities for the remaining months of 1967 and for the next year. The second official meeting of the new Clinical Microbiology Section was held on 4 December 1967. Several meeting dates were set for 1968, making every effort to avoid conflicting dates with the normal Branch activities. In a letter written by Dr. Friedman, dated 30 November, he noted that at the regular Branch meeting in October 1967, "the membership voted to accept this Section. It is planned that the Clinical Microbiological Section will have 5 meetings in addition to the regularly scheduled meetings of the Eastern Pennsylvania Branch." The new Section increased in member numbers and activity almost from the start. A second Steering Committee meeting was held on 18 December to discuss soliciting members from the Branch and soliciting technicians to join the Clinical Microbiology Section.

Dr. Friedman, the Branch Secretary, nicely summarized the situation at the end of 1967 as follows:

It is the expectation of those interested in Clinical Microbiology that this section will become an important forum for the dissemination of information and for the scientific discussion of clinical microbiology among those interested in public health and medical laboratory aspects of microbiology in the Philadelphia and Eastern Pennsylvania area. It is hoped that this Clinical Section will augment and not compete with the primary functions and goals of the parent branch of ASM.

Before the end of the decade, the Section was increasing the number of meetings per year, and became so well organized that they initiated a series of several day workshops. Conducting workshops became a constant feature of the Branch which continues to the present time. The Section also was influential in working with the Branch to initiate a two day symposium on Rubella in November of 1969. This became the first in a long series of November symposia that continues to the present time. The tradition of having the symposia published in book form was also initiated as part of the first symposium. The symposia and book sales eventually instituted a mechanism to generate financial stability for the Branch. Although not originally intended, the presence of the Section also created a certain tension within the Branch that eventually proved to have many positive repercussions for the future of the Branch. It should be noted that this was not always apparent as the decade was coming to a close. One aspect of historical confusion at this time was the fact that, for the remainder of this period, there were the official Tuesday evening meetings, which were assigned official Branch Meeting numbers, and separate meetings conducted by the Clinical Microbiology Section that did not get official Branch meeting numbers. In reality, this new Clinical Microbiology Section would significantly alter the history and nature of the Branch for decades to come.

Summary of Branch Meetings

The following is an attempt to characterize the 67 regular meetings held in the 1960s, (meeting numbers 282 to 348). For the 67 meetings, there were 182 individual presentations. Attendance at these meetings averaged 78 attendees. Most presenters were local; however, there were presentations given by non-local speakers at 36 percent of the meetings, essentially the same percent as the previous decade. The local speakers at these meetings were from a wide variety of Philadelphia institutions. The following institutions are represented as a percent of the 182 presentations: As in the previous decade, the dominant institution was the University of Pennsylvania, 20 percent. Other well represented institutions were the following: Temple Medical School, 14 percent; Albert Einstein Medical Center, 12 percent; Hahnemann Medical College, 11 percent; Wistar Institute, 10 percent; Bryn Mawr College, 9 percent; Children's Hospital of Philadelphia, 8 percent; Jefferson Medical College, 7 pecent; Wyeth 5 percent; Merck Sharp & Dohme, 4 percent; and Smith Kline and French, 4 percent. Other institutions represented by a small number of presentations were the following: Woman's Medical College, Drexel Institute, Fox Chase Cancer Institute, Haverford College, Philadelphia General Hospital, Valley Forge General Hospital, and Philadelphia/Pennsylvania Health Departments.

The following is the percent for each category for the 182 presentations in the 1960s, numbers in brackets indicate the percent increase or decrease when compared to the previous decade results. Items in bold print indicate a new category:

- immunology/vaccines 21 [+7]
- bacteriology 21 [-2]
- biochemistry 20 [+13]
- virology 19 [+6]
- PPLO 4 [-2]
- methodology 2 [0]
- **bacteriophage 2**
- **tissue culture 1**
- **interferon 1**

All other categories were less than 1 percent.

The following categories, included in the 1950s summary, were dropped due to a very low number (less than 1 percent) of the presentations: antimicrobials, mycology, electron microscopy, serology, and antimicrobial susceptibility testing. However, several of these topics were included in the presentations that were held separately in the Clinical Microbiology Section meetings.

Bacteriology (21 percent) was no longer the single dominant subject due to an increase in presentations in the areas of immunology/vaccines, biochemistry and virology. In the bacteriology category, several groups of organisms received considerable attention during this period. This included: mycobacteria, staphylococci, streptococci, *E. coli,* and *Haemophilus sp.* Other organisms of note were: *Salmonella sp., Clostridium sp., Bacillus sp.,* pneumococci, anaerobes, and the Mima/Herrellea group. Virology presentations focused considerable attention on adenovirus, coxsackievirus, poliovirus, the herpes viruses and cancer associated viruses. Other viruses receiving attention were bacteriophage and viruses associated with rabies, rubella and smallpox. The following viruses appeared in presentations for the first time during this period: SV40, Burkitt's lymphoma/EB virus, B virus, dengue, rubella, and leukemia associated viruses. The first presentation on antiviral agents was presented in 1962. Also, the drug actinomycin appeared for the first time during this period. The first presentation on interferon was in 1966, and this subject appeared in several presentations during the remainder of the 1960s.

In addition to the sixty-seven regular meetings, there were twelve Monday-evening meetings of the Clinical Microbiology Section, starting in September 1967. Most of the meetings were in the form of panel discussions employing local microbiologists. Topics included the following: urinary and respiratory tract infections, fluorescent microscopy, blood culturing, mycoplasmas, antimicrobial susceptibility testing, viral and rickettsial infections, mycobacteria, quality control and non-fermenting gram-negative rods.

The Clinical Microbiology Section held two workshops on fluorescent antibody techniques in 1969. The first Branch symposium was held in November of 1969 on rubella.

Branch Officers

Branch Presidents of the 1960s

There were five presidents (the sixteenth through the twentieth) who served the Branch during this decade:

Joseph S. Gots, University of Pennsylvania (1960-61)
L. Joe Berry, Bryn Mawr College (1962-63)
Harold S. Ginsberg, University of Pennsylvania (1964-65)
Albert G. Moat, Hahnemann Medical College (1966-67)
George Warren, Wyeth Laboratories, Radnor (1968-69)

Secretary/Treasurers
Elizabeth H. Fowler (1960-61)
Leonard Zubrzycki (1962-65)
Herman Friedman, (1966-69)

Councilors and Councilor-Alternates
1960-61 Carl Clancy   Alternate-George Warren
1962-63 George Warren   Alternate-Albert Moat
1964-65 Albert Moat   Alternate-Herman Friedman
1966-67 Joseph Gots   Alternate-Samuel Ajl
1968-69 Leonard Zubrzycki   Alternate-Richard Crowell

Branch Events by Year

1960

At the February meeting, the annual dues were increased from $1 to $2 due to increased postage, printing, and duplicating costs. Also at the February meeting, thirty-eight individuals were voted in as new members. It is unusual, since all thirty-eight were enrolled in a college bacteriology class at Temple University, and all wanted to join the Branch as a group. In May, the Branch was host to the Society of American Bacteriologists for its sixtieth annual meeting, and once again, the local organizing committee insured that the meeting was very successful.

There were seven meetings in 1960, with an average of three (range of one to four) presentations at each meeting, resulting in a total of twenty presentations. Average attendance at the seven meetings was seventy-eight (range of sixty-five to one hundred) attendees. The January meeting was designated a Symposium on Microbiological Problems in the Food Industry, and the March meeting was designated as a Symposium on Bacterial Endotoxins. The April meeting was held in the Klahr Auditorium at Hahnemann Medical College for a special lecture by Herman C. Lichstein of the University of Minnesota, who presented a talk on "Physiological Control Mechanisms in the Bacterial Cell." The April meeting was an afternoon meeting held at Wyeth Laboratories in Radnor, hosted by George Warren.

1961

At the 24 October meeting, Dr. Morton proposed a motion to change the name of the group to the Eastern Pennsylvania Branch of the American Society for Microbiology rather than chapter. The motion was passed and the Branch designation became official. There were six meetings in 1961, with an average of three (range of one to four) presentations at each meeting, resulting in a total of seventeen presentations. Average attendance at the six meetings was 92 (range of 42 to 182) attendees.

The May meeting was held as an afternoon meeting at Merck Sharp & Dohme Laboratories in West Point. There was a special Branch event on a Sunday in July in the form of a "Summer Picnic and Swim" held at Cappel Laboratories in West Chester. The meeting notice stated the following: "There is plenty of space to romp, pony rides for children and swimming for all. Bring supper for you family. A fire-place is available for cooking. The

Cappel's will supply beverages." There was no scientific session so this did not get a Branch meeting number. The November meeting was designated as a Symposium on Biology and Chemistry of Antibody Formation.

1962

Dr. Clancy produced an Annual Newsletter for 1962-63. It listed those several members who have been members for twenty-five years or more. The death of long time member, Mr. Roos, was noted. It was also noted that funding was received from seven industrial sponsors, each contributing $50, to initiate a guest lecturer series, the first to be held in November. (See appendix for listing of topics and speakers.) At the October meeting, it was agreed that the annual dues for students would be $1 rather than the $2 for regular membership.

There were seven meetings in 1962, with an average of three (range of one to four) presentations at each meeting, resulting in a total of eighteen presentations. Average attendance at the seven meetings was 76 (range of 48 to 125) attendees.

The February meeting was held at Barton Hall, Temple University. The March meeting was designated a Symposium on Bacterial Endotoxins, and the April meeting as a Symposium on Bacterial Cell Wall and Lysis. The May meeting was held in the Scheer Auditorium of Albert Einstein Medical Center, Northern Division, at Tabor Road at Eleventh St. The First Annual Industry Sponsored Guest Lecture was in November and was given by Merrill W. Chase, of the Rockefeller Institute for Medical Research, N.Y., who spoke on "Experiments on the Role of Heredity in Experimental Sensitization."

1963

There were seven meetings in 1963, with an average of three (range of one to four) presentations at each meeting, resulting in a total of nineteen presentations. Average attendance at the seven meetings was fifty-one (range of thirty-three to eighty) attendees. The January meeting was designated a Symposium on Biochemistry and Physiology of Protozoa. The February meeting was held at the Woman's Medical College, and the March meeting was a joint meeting with the Philadelphia chapter of the American Association of Clinical Chemists. At the May meeting, there was a Panel Discussion on Susceptibility Testing. The theme of the October meeting was Virus—Tumor Relationship, and the November meeting was the Second Annual Industry Sponsored Guest Lecture.

1964

There were seven meetings in 1964, with an average of three (range of one to six) presentations at each meeting, resulting in a total of twenty presentations. Average attendance at the seven meetings was 67 (range of 40 to 130) attendees. Most of the meetings had a theme. The following were meetings with a designated theme: January-Tuberculosis, February-The Biology and Chemistry of Endotoxin, March-Scientific Reports on Studies Performed Abroad, April—Graduate Student Presentations, and October-Metabolism and Control Mechanisms in Microbiology. The February meeting was held at Temple University, and the May meting was held at Bryn Mawr College, which was a foundation for microbiology lecture. The November meeting was the Third Annual Industry Sponsored Guest Lecture.

1965

At the start of 1965, all members were encouraged to join the national organization and it was announced that under the Branch rebate program, the Branch receives a $2 rebate for each new member that joins the National Society.

There were six meetings in 1965, with an average of three (range of one to six) presentations at each meeting, resulting in a total of seventeen presentations. Average attendance at the six meetings was 93 (range of 51 to 137) attendees. The January meeting was a foundation for microbiology lecture. The following were meetings with a designated theme: February-Staphylococcal Infections, March-Graduate Student Program, May-Immunologic Aspects of Virus-induced Tumors, October-Structure and Synthesis of Antibodies. November was the Fourth Annual Industry Sponsored Guest Lecture.

1966

A newsletter was produced by Dr. Leonard Zubrzycki and designated as the Spring Issue of 1966. Some notes from this newsletter follow: In April, a group flight delivered thirty-two Branch members to the ASM Annual meeting in Los Angeles, and Dr. Carl Abramson was the master of ceremonies at the annual banquet. Members were reminded that the Sixth Interscience Conference on Antimicrobial Agents and Chemotherapy was going to be held in Philadelphia at the Sheraton Hotel on 26-28 October.

There were seven meetings in 1967, with an average of two (range of one to six) presentations at each meeting, resulting in a total of twenty presentations. Average attendance at the seven meetings was 96 (range of 42 to 195) attendees. In January, Dr. Salvador Luria, of MIT, gave a talk entitled

"Sweet and Sour DNA." The following were meetings with a designated theme: February-Advances in the Biology and Chemistry of Mycoplasmas and L-forms, March-Graduate Student Presentation, April-Immunogenetics and Transplantation, May-Microbiology of Leukemias. The November meeting was the Fifth Annual Industry Sponsored Guest Lecture.

1967

In 1967, Dr. Zubrzycki produced a Branch newsletter in December designated as volume 14, number 1. There must have been some rationale for the designation of volume 14, which currently remains a mystery. In this newsletter, in addition to describing the formation of the new clinical microbiology section, the following newsworthy items were presented:

It was noted that all other Branches had a vice-president, and it was explained that the Eastern Pennsylvania Branch did not have one by raising the question "Well, what would he do? . . . Right now there does not seem to be a good reason to change our minds about this." However, in November of 1969, Dr. James Prier was elected as the first Branch vice-president. There was also an appeal for members to take an active role in obtaining more funding for biology research in the government budget. This was precipitated by the launch of the Russian *Sputnik*, and it was felt that this was the time to support all sciences, not just physics. There was a request that interested Branch members join the new Immunology Club of Philadelphia that was forming due to the efforts of Dr. Francis Haven of Temple University. It was also noted that the Branch lost two very active members due to the deaths of Dr. Kenneth Goodner, former Chairman of Microbiology at Jefferson Medical College, on 30 August, and Dr. Manasseh Sevag of the University of Pennsylvania, on November 23.

There were seven official (numbered) meetings in 1967, with an average of two (range of one to four) presentations at each meeting, resulting in a total of eighteen presentations. Average attendance at the seven meetings was 128 (range of 50 to 250) attendees. The January meeting was a Foundation for Microbiology Lecture. The following were meetings with a designated theme: February—The Role of Nucleic Acids in Antibody Production, March—Graduate Student Program, April—New Look at Infectious Diseases, May-Rabies, October—Recent Developments in Immunology. November was the Sixth Annual Industry Sponsored Guest Lecture.

There were two meetings of the newly formed Clinical Microbiology Section. These meetings dealt with training needs in Clinical Microbiology and diagnosis and management of urinary tract infections.

1968

In June of 1968, the Branch initiated a program to attract sustaining members. This was precipitated by the presence of the Clinical Microbiology Section, which increased mailing and printing costs. Since one of the goals of the section was to attract new members, the decision was made to keep the dues at $2 and to solicit $25 for the new membership category of sustaining members. At a clinical microbiology steering committee meeting in October, the subject of conducting workshops was raised, and Eileen Randall was charged with developing this possibility. The final Clinical Microbiology Section steering committee meeting was held on 12 December.

There were four Clinical Microbiology Section meetings in 1968, and seven official (numbered) regular meetings. There were two meetings in February; one was considered a joint meeting between the clinical microbiology section and regular Branch members. This meeting was counted as a regular meeting and designated as number 336.

For the seven regular meetings in 1968, there was an average of two (range of one to three) presentations at each meeting, resulting in a total of fourteen presentations. Average attendance at the seven meetings was 57 (range of 40 to 85) attendees. The January meeting was a foundation for microbiology lecture. The February and November meetings were held at the Public Health Laboratories, 500 South Broad Street. The March meeting was the Graduate Student Presentation meeting. The May meeting was held at the Smith Kline and French Laboratories, in Gulph Mills, and the October meeting was the Seventh Annual Industry Sponsored Guest Lecture.

There were four Clinical Microbiology Section meetings. Subjects covered in the four meetings were respiratory infections, fluorescent microscopy and blood culturing.

1969

There were six official (numbered) meetings in 1969, with an average of three (range of one to five) presentations at each meeting, resulting in a total of nineteen presentations. Average attendance at the six meetings was forty-five (range of thirty to sixty) attendees. The January meeting was a foundation for microbiology lecture, and the March meeting was the Graduate Student Presentation meeting. The May meeting was held at Temple University in the new Kresage Science Building at Broad and Tioga Streets. The November meeting was the Eighth Annual Industry Sponsored Guest Lecture.

There were six Clinical Microbiology Section meetings that covered the following subjects: mycoplasmas, antimicrobial susceptibility testing, diagnosis of viral and rickettsial disease, mycobacteria, quality control and non-fermenting gram-negative rods.

By mid year of 1969, the Clinical Microbiology Section had over 150 members and was growing steadily.

The first workshop sponsored by the section was held July 8-11 on "Fluorescent Antibody Techniques in the Clinical Laboratory." The registration fee was $25, and the workshop was held at the Pennsylvania Department of Health Laboratories, 2100 West Girard Avenue.

A second workshop, "Advanced Fluorescent Antibody," was held October 21-24 at the Temple University School of Medicine. The registration fee was $25.

The first Branch symposium was held on 6-7 November and was called the Northeast Regional Conference on Rubella. James E. Prier was the chairman with J. E. Satz and Richard Gutekunst as co-chairmen. The symposium was held at the Holiday Inn on City Line Avenue. The registration fee was $25. The book of proceedings was Published by Charles C. Thomas Inc., in 1973, with Herman Friedman as editor.

Summary of the Branch in the 1960s

There were several significant Branch events that occurred during the 1960s. The decade started with the Branch hosting a successful Society of American Bacteriologists Annual Meeting. In 1961, the chapter officially changed its name to the Eastern Pennsylvania Branch of the American Society for Microbiology. The high quality of Branch regular meetings continued through the decade.

The single most important event that occurred in the later part of the decade was the formation of the Clinical Microbiology Section. It took great effort on the part of all parties involved to keep this new Clinical Microbiology Section as a functioning part of the Branch. This effort paid off, since the new section not only served the needs of the clinical microbiologists in the Philadelphia area, but would greatly change the nature of the entire Branch in the next decade. The regular Branch had only two standing committees, a program and, starting in 1969, a symposium committee. The Clinical Microbiology Section was run by a steering committee with many standing committees to handle the greatly expanding activities, such as organizing the meetings, workshops, and a major clinical microbiology educational audio tape project, as well as several other ad hoc committees to address issues relevant to clinical microbiology.

Needless to say, the nature of the Branch had changed rapidly with the activities of the Clinical Microbiology Section. In spite of all these positive aspects, by the end of the decade it was becoming apparent that increasing numbers of members were attending the Clinical Microbiology Section meetings, often at the expense of the regular Branch meetings. This issue would become more apparent in the new decade and several Branch members came to the realization that many issues were going to have to be addressed early in the 1970s.

Chapter Nine

The Eastern Pennsylvania Branch
Struggling to Create A Unified Branch: 1970-1979

Overview of the Branch in the 1970s

In previous decades, the main activity of the Branch was to produce a top quality lecture series that was of interest to Branch members. This function was carried out by having only one standing committee, the program committee. Starting in 1969, there was need for an additional standing committee, the Symposium Committee. In the early 1970s, the gap in levels of activities between the regular Branch and the Clinical Microbiology Section was becoming wider. By 1973, it was apparent that significant changes had to occur due to the continuing presence of the Clinical Microbiology Section. The need to merge the regular Branch with the Clinical Microbiology Section became a priority. This was precipitated by a substantial decrease in attendance at the regular meetings and a significant increase in attendance at the clinical microbiology meetings.

A true merger could not be brought about without a major restructure of the governance of the Branch. For a true merger to occur, the Branch would have to take on the various functions of the Clinical Microbiology Section, which had committees for workshops, educational tapes, programs, education and a number of specialized committees to address such issues as new legislation and regulations. After much discussion, and a certain amount of reluctance on the part of many Clinical Microbiology Section members, a decision was made to form a new official executive committee to assist in unifying the Branch. Once this was in place, a merger agreement was reached in August of 1973. (See the Clinical Microbiology Section below.) There were no longer two sets of meetings, and the new executive committee met prior to each regularly scheduled Branch meeting and kept official minutes of these meetings. This was a significant step in documenting the history of the new executive committee as it attempted to coordinate all activities of the various working committees. This was also the start of written descriptions of the duties for the various officers and committee chairpersons. The difficulties in unifying the Branch overshadowed all other activities during this decade.

This decade was an active decade filled with many accomplishments. One of the most significant early accomplishments was the introduction of the extremely successful *Topics in Clinical Microbiology* audio tape set. (See the Topics in Clinical Microbiology section below.) A grant from Smith Kline and French enabled the Branch to initiate a series of special awards to Branch members for teaching and research. In April 1972, Philadelphia was host for the ninth time to the National Society for the seventy-second annual meeting. The *Topics in Clinical Microbiology* tape series was introduced to national members at this meeting. Also, the Branch set a precedent for future National ASM meetings by sponsoring a pre-convention workshop on computerization for clinical microbiology laboratories. In 1976, The First Stuart Mudd Lecture was initiated. This lecture was held every year thereafter, through 1995. This became a focal point for the Branch each spring with a prominent speaker, the awarding of the Stuart Mudd plaque, and the presence of Stuart Mudd family members. (See The Stuart Mudd Lecture Series in the Appendix Section for more information and a complete list of lecture titles and speakers.) In November 1978, a "Branch Past Presidents Night" was celebrated. All 17 past presidents were invited and nine were able to attend. Each was given a Silver Reserve Bowl inscribed with their names and dates of service.

The Branch held fourteen symposia in the 1970s. Each symposium lasted two or three days, and were all well attended by microbiologists from a broad geographical area. They also included a special banquet speaker and dinner for all attendees, and most symposia resulted in a published book. In addition to the ten November symposia, there were four basic science symposia in the 1970s. The Branch also held twenty-two workshops during this period. (See the Appendix Section for more information on the symposium and workshop series that includes a complete list of each.) Both the symposium and workshop series continue to the present time.

The combined income and royalties from the symposia, workshops, books, and tapes provided financial stability to the Branch and permitted the Branch to experiment with new educational programs.

The Clinical Microbiology Section. The Clinical Microbiology Section came about on 25 September 1967. Many people took a "wait-and-see attitude" to determine if the new section would survive. The answer came very rapidly. The initial goal was to have five meetings, independent of the regular Branch meetings, and two workshops a year. The goal was met, and the section proceeded to take on new tasks. By 1972, the number of meetings was increasing, and additional workshops were in the planning stages. Members of the Clinical Microbiology Section produced the innovative, and financially rewarding, *Topics in Clinical Microbiology* tape set, and worked with the regular Branch to initiate the clinical-microbiology-oriented November symposium series. Numbers of attendees at clinical microbiology Monday-evening meetings were increasing. At the same time, there were meetings of the regular Branch on Tuesday evenings that were attracting significant numbers of attendees; however, there were also regular Branch meetings with as little as five attendees. Regular Branch meeting attendance was on a definite decline. Talks were held on what to do and eventually, in August of 1973, the decision was announced that a compromise was reached. The solution was to form one organization with a defined percent of regular meetings associated with clinical microbiology, and all the Clinical Microbiology Section programs carried over to the unified Branch. The decision was made, and the next few years included much discussion within the Branch between clinical microbiology and basic science oriented members. Many of these same discussions and issues were also taking place within the national ASM organization. There was concern in the 1970s that clinical microbiologists would split from the national organization for many of the same reasons that caused the formation of the Clinical Microbiology Section of the Branch. It took much effort on the part of the Branch leadership to hold the Branch together during these transitional years.

It should be noted that the National ASM also made efforts to accommodate clinical microbiologists by forming a clinical microbiology division in 1972, which resulted in dedicated programs as part of the general meeting, and by creating the *Journal of Clinical Microbiology* that was first published in 1974. (1) Like the ASM on a national level, the Branch faced the same issues of how to handle a growing population of clinical microbiologists, who demanded specialized programs and continuing education to satisfy new regulations applying to directors and supervisors of clinical microbiology hospital and private laboratories. Once again, Eastern Pennsylvania Branch members and Philadelphians were at the forefront of this movement.

Topics in Clinical Microbiology. In 1972, "A Major Audio-Visual Publishing Event" occurred with the release by Williams & Wilkins

Publishing Company of *Topics in Clinical Microbiology*, edited by Richard Clark of Pennsylvania Hospital. This was a joint project between the Eastern Pennsylvania Branch and Williams & Wilkins Publishing Company. The topics were described as a complete course in clinical microbiology and an official publication of the Eastern Pennsylvania Branch of the American Society for Microbiology. This was the culmination of a project that was started in late 1970 by members of the Clinical Microbiology Section of the Branch. It was noted in a news release by Williams & Wilkins that "The editor has ably coordinated the eclectic contributions of thirteen members of the Eastern Pennsylvania Branch. Harry Morton and James Prier, officers of the Eastern Pennsylvania Branch, assisted in the preparation." The full set consisted of twenty-four compact C-30 cassettes with eight cassettes in each of three color-coded volumes, a complete explanatory manual with references above and beyond the taped material, and a set of colored slides. The complete set was initially priced at $187, or one volume of eight cassettes plus manual for $80, or a minimum of two individual cassettes plus the manual for $30. All profits from the topics went to the Branch. By current standards, in this time of iPods and educational programs on the internet with real-time feedback, an audio-tape program seems rather simplistic. However, this was essentially the first of its kind on a national level in clinical microbiology. It dominated the field for many years. In the early 1980s, an attempt was made to completely update and modernize the series, but by that time there were several competing series and the publishers were not interested in the investment. In addition to its educational and financial value to the Branch, this project was important for other reasons. It is a tribute to the Clinical Microbiology Section of the Branch and, in addition, it provided publicity for the Branch as the set was used by laboratory microbiologists as well as for teaching purposes in the United States and several other countries.

Summary of Branch Meetings

At the start of the decade, the regular meetings were held on the third or fourth Tuesday evening, January through April or May, as well as October and November, at the Medical Alumni Hall of the Hospital of the University of Pennsylvania, but in 1971, the meetings were moved to the Alumni Hall of Thomas Jefferson University on Locust Street. The Clinical Microbiology Section meetings, which continued until mid-1973, were held from five to seven times a year on Monday evenings at the Philadelphia Public Health

Laboratories on Broad Street. After the merger, all regular meetings were held on Monday evenings at Jefferson Alumni Hall.

The following is an attempt to characterize the 81 regular meetings, (meeting numbers 349 to 427). For the 81 meetings, there were 115 individual presentations, although details on two meetings have not been found. Accurate attendance records were only kept for the first three years. For these years, there was an average of forty-eight attendees per meeting. This was a decrease from the previous decade which averaged seventy-eight attendees per meeting. It should be noted that this was considerably lower than attendance at the Clinical Microbiology Section meetings, which averaged more than eighty attendees for this same period. There was a considerable increase in meetings with a single presenter and in non-local presenters. Non-local presenters accounted for 54 percent for this decade as compared to 36 percent for the previous two decades. The local speakers at these meetings were from a variety of Philadelphia institutions. The following institutions are represented as a percent of the seventy-nine documented presentations: As in the previous decade, the dominant institution was the University of Pennsylvania (18 percent). Other well-represented institutions were: Temple University, 11 percent; Fox Chase Cancer Center, 10 percent; Thomas Jefferson University, 7 percent; and Hahnemann Medical College, 6 percent. Other institutions included the following: Albert Einstein Medical Center, Wistar Institute, Children's Hospital of Philadelphia, Wyeth, Merck Sharp & Dohme, Smith Kline & French, Woman's Medical College, Philadelphia Health Department, Pennsylvania Hospital, and La Salle College.

The following is the percent for each category for the seventy-nine documented presentations in the 1970s. The number in brackets indicates the percent increase or decrease when compared to the previous decade results. Items in bold print indicate a new category:

- bacteriology 41 [+20]
- virology 27 [+8]
- immunology/vaccines 25 [+4]
- **education 15**
- biochemistry 11 [-9]
- **mycology 6**
- **parasitology 5**
- **antibacterials 2**
- **mycoplasmas 2**
- **history 2**

All other categories were less than 1 percent.

It should be noted that the number of subjects relating to clinical microbiology increased after 1974 when the Clinical Microbiology Section meetings were combined with the regular meetings.

In the area of bacteriology, there were several presentations on *E. coli, Salmonella sp.,* anaerobes, streptococci, mycoplasmas, and the venereal diseases. The first presentation on Legionnaire's Disease was in 1977. Unlike previous decades, only a few presentations dealt with tuberculosis-related topics. Virology accounted for 27 percent of the talks and included a wide range of viruses or viral diseases that included: herpes, adenovirus, hepatitis, EB virus, tumor-related viruses, coxsackie, smallpox, SV40, and slow viruses. The term *ecology* is used in these presentations for the first time in 1970 and appears often in subsequent presentations.

In addition to the sixty-seven regular meetings, there were twenty-two Monday-evening meetings of the Clinical Microbiology Section, from 1970 through April 1973. Many of the meetings were in the form of panel discussions. These meetings covered a wide range of topics relating to clinical microbiology. Topic summaries of these meetings are provided in the Branch Events by Year section below.

Hot Topics of the 1970s. What were the hot topics in microbiology as decided by the microbiology community in Philadelphia? One answer to this question can be obtained by looking at the titles for the symposia that occurred each year. These topics provide an answer to this question from two perspectives. The topics for the November symposia are from the clinical microbiology perspective, and the Basic Science symposia topics are from the more academic oriented microbiology community.

Clinical Microbiology Oriented November Symposia:
1970—Symposium on Immunoglobulins
1971—Symposium on Australia Antigen
1972—Symposium on Opportunistic Pathogens
1973—Symposium on Quality Control in Microbiology
1974—Modern Methods in Medical Microbiology: Systems and Trends
1975—Clinical Laboratory as an Aid in Chemotherapy of Infectious Disease
1976—Infection Control in Health Care Facilities: Microbiological Surveillance
1977—Immunoserology in the Diagnosis of Infectious Disease
1978—Diagnosis of Viral Infections: The Role of the Clinical Laboratory
1979—Clinical Parasitology: Current Concepts in a Changing Science

Basic Science Symposia:
1972—Virus Tumorigenesis and Immunogenesis
1975—Tumor Virus Infections and Immunity
1977—Infection, Immunity, and Genetics
1978—Microbial Infection and Immunity

BRANCH OFFICERS

Branch Presidents
There were six presidents (the twenty-sixth through the thirty-first) who served the Branch during this decade:

Herman Friedman, Albert Einstein Medical Center (1970-71)
James Prier, Pennsylvania Department of Health Laboratories (1972-73)
Richard L. Crowell, Hahnemann Medical College (1974-75**)
Norman P. Willett, Temple University School of Dentistry (1975-77*)
Robert J. Mandle, Thomas Jefferson Medical College (1977-79*)
Joseph F. Pagano, Smith Kline & French Laboratories (1979-81*)
(**Jan 74 to June 1975)   (* July to July)

Vice-Presidents
1970-71 James Prier, Pennsylvania Department of Health Laboratories
1972-73 Richard Crowell, Hahnemann Medical College
1974-75 Norman P. Willett, Temple University School Dentistry
1975-77 Robert J. Mandle, Thomas Jefferson Medical College
1977-79 Joseph F. Pagano, Smith Kline & French Laboratories
1979-81 John C. McKitrick, University of Pennsylvania

Secretary
1970-73 Eileen Randall, Jefferson Medical College (Until April 1973)
1973-75 Elizabeth Free, Hahnemann Medical College (Starting May 1973)
1975-79 Toby Eisenstein, Temple University
1979-81 Josephine Bartola, Pennsylvania Department of Health Laboratories

Treasurers
1970-73 Norman P. Willett, Temple University School Dentistry
1974-77 Joseph F. Pagano, Smith Kline Diagnostics
1977-79 Carl Abramson, Pennsylvania College of Podiatric Medicine
1979-81 Henry R. Beilstein, Philadelphia Public Health Department

Councilors
1970-71 Richard L. Crowell, Hahnemann Medical College
1972-73 Joseph F. Pagano, Smith Kline and French Laboratories
1974-75 Robert J. Mandle, Thomas Jefferson University
1975-77 Ralph A. Knight, Medical College of Pennsylvania
1977-79 John McKitrick, University of Pennsylvania
1979-81 Donald Stieritz, Hahnemann Medical College

Councilor-Alternates
1970-71 Joseph F. Pagano, Smith Kline and French Laboratories
1972-74 Ralph A. Knight, Medical College of Pennsylvania
1975-77 John McKitrick, University of Pennsylvania
1978-79 Burt Landau, Hahnemann Medical College
1979-81 James A. Poupard, Bryn Mawr Hospital

Branch Events by Year

1970

In May of 1970, the "First Annual Smith Kline and French Award for Excellence in the Teaching of Microbiology" was awarded to L. Joe Berry at the May meeting. A dinner was held in his honor. It was noted that Dr. Berry was leaving his position at Bryn Mawr College to become chairman of the microbiology department at the University of Texas.

A newsletter (volume 15, number 1) was produced in June, with Dr. Albert Moat as editor. Some items of interest in this newsletter were the following:

> An announcement that the secretary/treasurer position was now two separate positions, and that a vice-president position was added. A membership committee was formed with Dr. Zubrzycki as chairman. It was noted that there were approximately four hundred paid members, but meeting attendance was declining to an average of thirty to fifty attendees per meeting, while attendance at the Clinical Microbiology Section was increasing to more than eighty attendees per meeting. It was also noted that past Branch president, Dr. James Harrison, professor of biology at Temple University for thirty-five years, was retiring on 1 July, and that Dr. Kurt Pauker was appointed the new professor and chairman of microbiology at Women's Medical College.

Most regular meetings were held on Tuesday evening at the Medical Alumni Hall of the Hospital of the University of Pennsylvania. There were seven regular Branch meetings, with an average of three (range of one to five) presentations at each meeting, resulting in a total of twenty-three presentations. Average attendance at the seven meetings was fifty-four (range of twenty-five to eighty-nine) attendees. The January meeting was a Foundation for Microbiology Lecture and was considered a joint meeting of the Branch and the Clinical Microbiology Section. The March meeting was the Graduate Student Presentation meeting. The September meeting was a joint meeting with the Allegheny Branch and was held on a Saturday at the Milton S. Hershey Medical Center in Hershey, Pennsylvania. The meeting also included a tour of the Medical Center, a picnic lunch and was followed by a swimming outing. The November meeting was the Ninth Annual Industry Sponsored Guest Lecture given by Rene Dubos on "The Ecology of Infectious Diseases."

The six Clinical Microbiology Section meetings covered the following subjects: immune response of patients with bacterial infection, *Haemophilus* related infections, anaerobic infections, antibiotic susceptibility testing, and current problems in Clinical Microbiology. Most meetings were held on Monday evenings at the Philadelphia Department of Public Health Auditorium on South Broad Street.

A workshop entitled "Advanced Enteric Bacteriology" was held 3-6 March at Jefferson Medical College and was directed by Eileen Randall.

A four day workshop on "Systemic Mycosis" was held 17-20 November at Thomas Jefferson University.

The second November "Symposium on Immunoglobulins" was held on 12-13 November at the Marriott Hotel on City Line Avenue. Dr. H. M. Rawnsley was the Chairperson for the symposium, and Jay Satz was Chairman of the new, well organized, Branch Symposium Committee. Over three hundred people attended.

1971

A newsletter (volume 16, number 1) was produced in June 1971, with Dr. Albert Moat as editor. Some items of interest in this newsletter were the following:

> The Smith Kline and French Award for Excellence in the Teaching of Microbiology was awarded to Joe Gots, University of Pennsylvania, and two additional awards sponsored by Smith Kline and French were

awarded: the Excellence in Graduate Studies to George Mark, Temple University and the Excellence in Research was awarded to Doctors Gertrude and Werner Henle, Children's Hospital of Philadelphia. Current Branch membership was reported as 358 members.

The first three meetings continued to be held at Alumni Hall at the Hospital of the University of Pennsylvania; the May meeting was held at the Smith Kline & French Animal Health Farm near Paoli, and starting in November, the location for regular meetings was moved to the Solis-Cohen Auditorium at Jefferson Alumni Hall on Locust Street.

There were six regular Branch meetings, with an average of three (range of one to four) presentations at each meeting, resulting in a total of fifteen presentations. Average attendance at the six meetings was sixty (range of twenty-eight to eighty-three) attendees. The January meeting was a Foundation for Microbiology Lecture, and March was the Annual Graduate Student presentation meeting. The May meeting was held at Smith Kline & French Animal Health Farm near Paoli, and included the award program mentioned above as a newsletter item. The November meeting was the Tenth Annual Industry Sponsored Guest Lecture and was held at Jefferson Medical College.

The seven Clinical Microbiology Section meetings covered the following subjects: computers, non-gonococcal venereal diseases, problem gram negative bacilli, gonorrhea, anaerobes, environmental microbiology, and rapid diagnostic methods. The April meeting was a "kick-off" for the workshop. Most meetings were still held on Monday evenings at the Philadelphia Department of Public Health Auditorium on South Broad Street. The December meeting was held at Pennsylvania Hospital and included a tour of the new Clinical Microbiology facilities.

A workshop on "Non-Fermenting Gram-Negative Rods" was held April 13-16 at Thomas Jefferson University.

A workshop on "Identification of Anaerobes" was held on October 5-8 at Thomas Jefferson University

The third November Symposium was held 8-9 November at the Marriott Hotel on City Line Avenue. The "Symposium on Australia Antigen" was chaired by Baruch Blumberg and Jay Satz. James Prier and Herman Friedman were designated editors of the book that was to be published by the University Park Press. Registration fee stayed at $35 and three hundred people attended.

1972

In 1972, Philadelphia was host for the ninth time to the National Society for the seventy-second annual meeting on 23-28 April. Once again, Dr. Morton was chair of the Local Committee on Arrangements, with Herman Friedman and James Prier serving as vice-chairs. The meetings and exhibits were held at the Philadelphia Civic Center. Dr. Morton reported that while the space for the exhibits was adequate, the meeting room facilities in the auditorium, which was built in the 1920s, left much to be desired. He also noted that the "History of Microbiology in the Philadelphia Area, 1960-1972" was the most comprehensive report of its nature prepared for any of the annual meetings held in Philadelphia.

The *Topics in Clinical Microbiology* tape series was introduced to national members at this meeting. Also, the Branch set a precedent for future National ASM meetings by sponsoring a pre-convention workshop on 22 April, just prior to the ASM General Meeting in Philadelphia. The workshop conducted at Pennsylvania Hospital was entitled "Practical Techniques in Computerization of Clinical Microbiology." The workshop attracted 110 attendees from the United States as well as other countries. It is now common practice to have at least a day of premeeting workshops at the General Meeting; however, in 1972 this was truly innovative.

All regular Branch meetings were held on Tuesday evenings at the Solis-Cohen auditorium at Jefferson Alumni Hall. There were six regular Branch meetings, with an average of two (range of one to four) presentations at each meeting, resulting in a total of twelve presentations. Average attendance at the six meetings was thirty-one (range of ten to forty) attendees. The February meeting had a theme of New Approaches to Bacterial Vaccines. March was the Annual Graduate Student presentation meeting. The May meeting was the Third Smith Kline & French Award meeting. The Award for Excellence in Teaching was presented to two co-recipients, W. G. Hutchinson of the University of Pennsylvania, and E. H. Spaulding of Temple University. The Excellence in Graduate Studies was awarded to Maria Ensinger of the University of Pennsylvania, and the Excellence in Research was awarded to B. Blumberg of the Institute for Cancer Research.

The five Clinical Microbiology Section meetings covered the following subjects: identification of yeast-like fungi, automation of blood cultures, antibiotic susceptibility testing, nosocomial infections, malaria and diagnosis of bacteremia. The January meeting was held at the Philadelphia Department of Public Health Auditorium on South Broad Street; however, the remaining meetings were held at Thomas Jefferson Alumni Hall on Locust Street. All meetings were on Monday evenings.

In addition to the precedent setting pre-convention workshop on Clinical Microbiology Computerization in April, a workshop was held on 6-9 June on "The Isolation and Identification of Mycobacteria" at the biology department of the University of Delaware.

A workshop "Parasitology" was held on 14-17 November.

There were two symposia:

The first Basic Science symposium on "Virus Tumorigenesis and Immunogenesis" was chaired by Herman Friedman, with the book published by Academic Press edited by W. Ceglowski and H. Friedman.

The fourth November Symposium on "Opportunistic Pathogens" was held on November 2-3 and chaired by Earle H. Spaulding and Jay Satz. It was held at the Marriott Hotel on City Line Avenue. The book was published by the University Park Press and was edited by James Prier and Herman Friedman.

1973

In August of 1973, the decision was made to merge the Clinical Microbiology Section with the regular Branch. The decision was to establish a single Executive Committee to assist in merging the two groups. Due to the "overwhelming success" of the Clinical Microbiology Section, it was agreed that two-thirds of the meetings would "continue to carry a clinical microbiology orientation."

New standing committees were formed in addition to an Executive Committee. The following standing committees were formed: Program, Workshop, Clinical Symposium, Editorial, Basic Science, Career Opportunities, News, and Bylaws. The President was designated ex-officio of all committees. For the November election, a mail-ballot system was initiated for the first time to elect officers.

All meetings were moved to Monday evenings and held at Jefferson Alumni Hall on Locust Street. There were seven regular Branch meetings but details on two of these meetings are missing. Of the five documented meetings, there was one presentation per meeting. Average attendance at the five meetings was forty (range of twenty to seventy-five) attendees. The meetings in May, October and November were combined meetings between the Section and regular Branch. All meetings after this were considered the new unified Branch meetings.

The four Clinical Microbiology Section meetings covered the following subjects: cholera, uncommon gram-negative rods, contamination of biologicals, infections acquired in hospitals and by laboratory personnel, and laboratory improvement activities at the Center for Disease Control. The June meeting was a kick-off meeting for the workshop. All meetings were held on Monday evenings at Jefferson Alumni Hall on Locust Street.

A workshop was held 5-7 June entitled "Hospital Epidemiology; Hospital Environmental Surveillance and Quality Control in Clinical Microbiology Surveillance." It was held at Thomas Jefferson University.

A second workshop, "Bacterimia and Blood Culturing" was held in October.

The fifth November Symposium was entitled "Symposium on Quality Control in Microbiology" chaired by Richard R. Gutekunst and Jay Satz. The symposium was held at the Marriott Hotel on City Line Avenue. The book was published by the University Park Press and was edited by James Prier, Josephine Bartola and Herman Friedman.

1974

After an absence of three years, another newsletter (volume 17, number 1) was published in July of 1974, edited by Dr. Moat. Much of this newsletter was spent explaining the merger of the Clinical Microbiology Section with the regular Branch, expressing a definite goal of moving forward as one united Branch. Other items of interest in this edition were:

It was announced that Thomas Jefferson University graduated its first class of eleven Clinical Microbiology Master's Degree students. Of this first class of eleven, six took either supervisory or director positions in microbiology at Philadelphia area hospitals. They all became active in helping to make the transition of merging the Clinical Microbiology Section with the regular Branch. The class included the following: Fritz C. Blank (Wilmington Medical Center), Diane Bleckman (Daroff Division of Einstein Medical Center), Lance Ericson (Philadelphia State Hospital, Byberry), Georganne Kretschman (Jefferson University Hospital), James Poupard (Bryn Mawr Hospital), Loretta Rocco (Methodist Hospital) and Jane Stipcevich (West Jersey Hospital).

E. G. Scott provided "A Personal Memory" in the newsletter commemorating the death of W. Robert Bailey of the University of Delaware.

Largely due to the financial success of the workshops, clinical microbiology tapes, symposia and published books, there was a need to file for tax-exemption status and to hire an accountant to review the Branch books for the last three years. Also, the decision was made to now collect dues every two years. The ByLaws were voted on in May to handle all the changes that were occurring. Regular dues were changed to $3; student rate $2 per year.

In March, it was decided to have an outside auditor review the Branch records, since royalties were arriving for the published symposium books, the Clinical Microbiology tapes and excess revenue from symposia and workshops. Decision was also made to bond the treasurer and president, and to file the request for tax-exempt status of the Branch with the I.R.S. Duties of science fair judges were formulated. The new ByLaws were approved.

Meetings were held on Monday evenings at Jefferson Alumni Hall on Locust Street, with few noted exceptions. There were nine regular Branch meetings, with an average of one (range of one to three) presentations at each meeting, resulting in a total of eleven presentations. Average attendance at the six meetings was 96 (range of 45 to 145) attendees. April was a Foundation of Microbiology lecture. The March meeting was a kick-off meeting for the enterobacteriaceae workshop. The June meeting was held at the Temple University Conference Center at Sugarloaf. The September meeting was the kick-off meeting for the mycology workshop, and the November meeting was a joint meeting with the Theobald Smith Branch and was held at the E. R. Squibb Laboratories in Princeton.

There was a workshop at Thomas Jefferson University on 5-8 March entitled "The Identification of the Enterobacteriaceae," organized by Fritz Blank, that had ninety attendees. A mycology workshop was held on 15-18 October.

A second workshop entitled "Mycology" was held in October.

The sixth November Symposium entitled "Modern Methods in Medical Microbiology: Systems and Trends" was held 14-15 November at the Marriott Hotel on City Line Avenue, chaired by Vern Pidcoe and Jay Satz. The book, published by the University Park Press, was edited by James Prier, Josephine Bartola and Herman Friedman.

1975

On 30 May 1975, many Branch members attended the memorial service for Dr. Stuart Mudd who died on 6 May. Dr. Mudd was our eleventh Branch President (1940-41) and former ASM President. In memory of Dr. Mudd, the Branch decided to establish a Stuart Mudd Memorial Lectureship. This was to be a series of lectures that occurred at a Branch meeting each spring.

The lecture series was initiated in 1976 and continued each year through 1995. The lecturer was a prominent scientist often associated with Dr. Mudd. (See the Appendix Section for a complete history of this prominent lecture series.)

Meetings were held on Monday evenings at Jefferson Alumni Hall on Locust Street, with few noted exceptions. There were eight meetings, usually with one speaker per meeting. Attendance records were not recorded. January, February and September were Foundation of Microbiology Lectures. The May meeting was held at the Temple University Conference Center at Sugarloaf. The October meeting was the kick-off meeting for the Mycology workshop.

A workshop entitled "Identification of Unusual Gram Negative Rods from Clinical Specimens" was held in June.

A workshop was held in October entitled "Anaerobic Bacteriology: The VPI Method." There were fifty attendees.

The second basic science symposium, "Tumor Virus Infections and Immunity" was held on 24-25 April, at the Marriott Hotel. Herman Friedman and Richard Crowell were chairmen and the book published by University Park Press was edited by Richard Crowell, Herman Friedman and James Prier.

The seventh November Symposium, "The Clinical Laboratory as an Aid in Chemotherapy of Infectious Disease" was held on 20-21 November at the Hilton Hotel on Thirty-fourth Street. Amedeo Bondi and Jay Satz were chairmen and the book published by the University Park Press was edited by Amedeo Bondi, Josephine Bartola and James Prier. There were 225 paid attendees.

1976

The year 1976 singled out Philadelphia for special attention during America's Bicentennial Celebration. The planned celebrations were poorly attended due to the presence of an unknown disease that proved fatal to many Legionnaires attending a convention at the Bellevue-Stratford Hotel in Philadelphia. The number of discussions on this newly identified disease, which eventually became known as Legionnaires' Disease, outnumbered the talks relating to our country's bicentennial. In addition to the regular Branch meetings, there were many informal discussions on possible etiology and epidemiology of this new disease state.

Regular Branch meetings were held on Monday evenings at Jefferson Alumni Hall on Locust Street, except when noted. There were ten meetings, usually with one speaker per meeting. Attendance records were not recorded. In April, the first Stuart Mudd lecture was held. Dr. Alan W. Bernheimer spoke on "Lipid Specific Exotoxins." The May meeting was held on a Friday at the Temple University Conference Center at Sugarloaf. The June meeting was connected with the Mycobacteriology Workshop, and the December meeting was a social event featuring wine and cheese.

On 7-10 June, there was a workshop entitled "Isolation and Identification of Mycobacteria" at Thomas Jefferson University.

A second workshop entitled "General Immunology" was held in November.

The eighth November Symposium "Infection Control in Health Care Facilities: Microbiological Surveillance" was held 11-12 November at the Hilton Hotel at Thirty-fourth Street. William Ball, Kenneth Cundy and Jay Satz were co-chairpersons and the book published by University Park Press was edited by Kenneth Cundy and William Ball.

1977

After an absence of several years, a new series of newsletters began in November 1977, (volume 18, number 1.) This newsletter series continued each year through 2008. The editor of this series was Steven Specter of Albert Einstein Medical Center.

Meetings were held on Monday evenings at Jefferson Alumni Hall on Locust Street, except when noted. There were ten meetings, usually with one speaker per meeting. Attendance records were not recorded. The February meeting was a joint meeting with the Theobold Smith Branch. The second Stuart Mudd lecture was presented in April by Dr. Baruch S. Blumberg, who spoke on Australia Antigen. The May meeting was held at the Temple University Conference Center at Sugarloaf. The June meeting was a Foundation of Microbiology lecture as well as a kick-off meeting for the workshop on "Antimicrobial Susceptibility and Assay Methods." The October meeting was a kick-off lecture for the workshop on opportunistic fungi.

A June workshop was on "Antimicrobial Susceptibility and Assay Methods."

There was a one day workshop held in October entitled "Chlamydia Infections Old and New Facets of the Disease."

A third workshop was held 18-21 October at Thomas Jefferson University on "Isolation and Identification of Dermatophytes and Opportunistic Fungi." There were seventy-eight participants.

The third basic science symposium, "Infection, Immunity, and Genetics," was held 6-7 June at the Marriott Hotel on City Line Avenue. Herman Friedman and James Prier chaired the symposium, and the book published by University Park Press was edited by H. Friedman, T. J. Linna, and J. Prier.

The ninth November Symposium "Immunoserology in the Diagnosis of Infectious Disease" was held 10-11 November at the Hilton Hotel on Thirty-fourth Street. T. Juhani Linna, H. Friedman, and Jay Satz were co-chairpersons, and the book published by University Park Press was edited by Herman Friedman, T. J. Linna, and James Prier.

1978

In January 1978, it was decided that the Branch needed a logo. A competition was held to design an appropriate logo. From many submissions, one designed by Dr. Harry Stempin was chosen.

Meetings were held on Monday evenings at Jefferson Alumni Hall on Locust Street, except when noted. There were ten meetings, usually with one speaker per meeting. Attendance was not recorded. The third Stuart Mudd lecture was presented in April. The June meeting was a kick-off meeting for the workshop on infection control. The November meeting was designated as "Branch Past Presidents Night". All 17 past living presidents were invited and nine were able to attend. They were: Earle Spaulding, Harry Morton, Morton Klein, Joseph Gots, George Warren, Herman Friedman, James Prier, Richard Crowell, and Norman Willett. Prior to the start of the regular meeting, each was presented with a Silver Reserve Bowl inscribed with their names and dates of service.

There were three workshops: "An Overview of Diagnostic Virology" in March, "Hospital Infection Control-Microbiology and Epidemiology" in June, and "Sexually Transmitted Diseases" in October.

The fourth Basic Science Symposium, "Microbial Infections and Autoimmunity" was held on 17-18 April at the Marriott hotel on City Line Avenue. It was co-chaired by Herman Friedman and T. Juhani Linna.

The tenth November Symposium "Diagnosis of Viral Infections: The Role of the Clinical Laboratory" was held 9-10 November at the Hilton Hotel on Thirty-fourth Street. The symposium was co-chaired by Steven Specter, David A. Lennette and Kenneth Thompson. The book published by the University Park Press was edited by D. A. Lennette, S. Specter and K. Thompson.

1979

In April 1979, Linda Creeden replaced Steven Specter as Newsletter Editor.

Meetings were held on Monday evenings at Jefferson Alumni Hall on Locust Street. There were ten meetings, usually with one speaker per meeting. Attendance records were not recorded. The fourth Stuart Mudd lecture was presented in April by Dr. Gots on bacterial genetics. Also in April, there was a special Branch afternoon seminar consisting of three speakers on "Professional Affairs in Clinical Microbiology." The June meeting was a kick-off meeting for the workshop on parasitology.

There was a workshop on 26-28 June on "Clinical Protozoology."

The eleventh November Symposium "Clinical Parasitology: Current Concepts in a Changing Science" was held 8-9 November at the Hilton Hotel on Thirty-fourth Street. William Ball, C. Iralu and Robert Strauss co-chaired the symposium.

Summary of the Branch in the 1970s

The decade of the 1970s was a very active time for the Branch. The Branch survived the split of the Clinical Microbiology Section and worked to form a new unified Branch. It perfected a system of producing two sets of highly successful symposia: the annual clinical microbiology oriented November Symposia series and several Basic Science Symposia. Branch members worked with two national publishing companies to produce a successful series of books based on these symposia. During this period, the Branch produced an innovative and widely used audiotape program that was distributed extensively both in the United States and elsewhere. The needs of bench technologist and microbiologist were served by holding at least two workshops each year targeted to this audience, and a new annual lecture series to honor Dr. Stuart Mudd was initiated. In addition to all these activities, the Branch hosted its ninth ASM Annual meeting, and maintained a lecture series in the form of monthly meetings that some years were held twelve times a year.

CHAPTER TEN

The Eastern Pennsylvania Branch: 1980-1989
The Peak Activity Years, Expanding Organization, Facing Issues, and Celebrating Five Hundred Meetings

Overview of the Branch in the 1980s

The Branch entered the new decade of the '80s with much momentum from the previous decade. There was a new energy within the Branch that pushed it to move in several directions and tackle some difficult issues. The role of the president was very different from earlier decades. The president presided over the Executive Committee, which increased in size and complexity as new committees were formed. What follows are just some of the activities that characterized this decade.

For all ten years, there were ten meetings per year with July and August the only months without a regular Monday night meeting. Each meeting was preceded, with few exceptions, by a two hour Executive Committee Meeting in which there was never enough time to cover all the issues that needed to be discussed. Each chairperson wanted more time and the president had to set priorities at these meetings. Branch standing committees included the following: academic affairs, archives, education, finance, industrial affairs, newsletter, placement, program, publications, publicity, membership, legislative, legal, scholarship/awards, public and scientific affairs, symposium, industrial affairs and workshop. There were also several ad hoc committees that handled annual events like the Stuart Mudd lectures and special Poster Session meetings.

Issues that received attention included discussions of new regulations concerning qualifications for laboratory directors and Medicare reimbursements; both involved pending restrictions being placed on non-physician laboratory directors. New minimum requirements for laboratory technicians and technologists received considerable attention. The role branches should play in the national ASM also received considerable attention. There was a strong feeling in the Branch that the ASM was not working closely with the Branch due to a general lack of communication in sharing membership lists, mailing labels, and the scheduling of ASM workshops that conflicted with Branch workshops and the general role of branches versus divisions in the governance of the ASM. Many Eastern Pennsylvania Branch members became active on ASM committees during this decade.

In addition to conducting one hundred regular meetings, the Branch conducted ten very successful symposia, some with 225 attendees, and most resulting in a published book. There were also nineteen well-attended workshops. In addition, there were several one day mini-symposia and mini-workshops on specific subjects. Also, the Education Committee started a series of meetings and seminars designed for science teachers at all levels to increase their knowledge of microbiology. This committee also produced printed microbiology materials for teachers and awarded grants to science fair winners. There was also a renewed interest in getting graduate students interested in the Branch. The old graduate student presentation night was replaced with a Poster Session Night with awards to the best posters in several categories.

Discussions on issues of meeting content, and number of meetings relating to clinical microbiology versus basic science, were carried over from the previous decade. The clinical microbiology oriented members felt the meetings were not being geared to their needs, and there were noticeable differences in attendance depending on how clinically oriented the topic was for a particular meeting. In addition, the presence of two Clinical Masters Programs with considerable enrollments strengthened the need to capture this active group since many graduates took positions in the Philadelphia area. Some of these needs were resolved by forming groups outside the Branch that were clinically relevant and open to anyone who wanted to attend. The Jefferson Alumni Group and the Lehigh Valley Clinical Microbiology Group were just two examples. Meeting notices for these groups were routinely included in the Branch Newsletter. The Clinical Microbiology debate continued throughout the decade; however, rather than this debate being negative, it resulted in members on both sides of the issue working hard to initiate new programs. It became a very active and progressive decade for the Branch.

A definite highlight of the decade was the celebration of the 500th Branch Meeting in December 1987. This meeting brought all elements of the Branch together for a celebration of the Branch's sixty-seven interesting years of existence. (See The 500th Branch Meeting Celebration in the Summary of Branch Meetings section below.)

Summary of Branch Meetings

There were ten monthly meetings each year. Meetings regularly were held on the fourth Monday of every month, except July and August, at Alumni Hall of Thomas Jefferson University on Locust Street.

The following is an attempt to characterize the 100 meetings, (meeting numbers 428 to 529). Almost all meetings consisted of only one presentation. Attendance records were not routinely recorded. Nonlocal presenters accounted for 52 percent for this decade, which was almost the same as for the previous decade, which was 54 percent. The local speakers at these meetings were from a variety of Philadelphia institutions. The following institutions are represented as a percent of the one hundred presentations:

With 10 percent of the presenters from the University of Pennsylvania this still made it the dominant institution. Other well-represented institutions were the following: Fox Chase Cancer Center, 7 percent; Thomas Jefferson University, 4 percent; E.I. DuPont Laboratories, 4 percent; Hahnemann Medical College, 3 percent; Medical College of Pennsylvania, 3 percent; Saint Christopher's Hospital, 3 percent; Smith Kline and French, 3 percent; Temple University, 2 percent; Wistar Institute, 2 percent, and Merck Sharp & Dohme, 2 percent. Other institutions included: Bryn Mawr Hospital, Will's Eye Hospital, and the pharmaceutical companies, Pfizer and Sterling.

The following is the percent for each category for the one hundred meetings. The number in brackets indicates the percent increase or decrease when compared to the previous decade results. Items in bold print indicate a new category:

- bacteriology 25 [-16]
- virology 22 [-5]
- immunology/vaccines 13 [-12]
- parasitology 7 [+2]
- **clinical microbiology 5**
- mycology 4 [-2]

- antmicrobials and susceptibility testing 4 [+2]
- history 4 [+2]
- biochemistry 3 [-8]
- **venereal diseases 3**
- mycoplasmas 1[-1]

One previous category from the 1970s, education, was eliminated.

Presentations in bacteriology included the following: anaerobes, streptococci, staphylococci, venereal diseases, mycobacteria, *Legionella*, *Pseudomonas sp.*, *Actinobacillus sp.*, enterobacteriaceae, actinomycetes, and mycoplasms. Four bacterial diseases appeared in presentations for the first time in the 1980s: lyme disease, toxic shock syndrome, listeriosis, and pelvic inflammatory disease. Virology accounted for 22 percent of the talks and included a wide range of viruses or viral diseases. The first presentation on AIDS was in 1983. Other viruses that were in presentations for the first time in this decade were CMV, parvovirus and the retroviruses. Presentations were also made on a variety of viruses or viral diseases including, hepatitis, slow viruses and rabies. Meetings dedicated to the theme of "Grand Rounds" were introduced in the 1980s.

Hot Topics of the 1980s. The symposia titles are one measure of the "hot topics" in microbiology as decided by the microbiology community in Philadelphia. These included the following subjects:

Symposia Held in the 1980s
1980—New Horizons in Medical Mycology
1981—Host Defenses to Intracellular Pathogens
1982—Hepatitis B: The Virus, the Disease and the Vaccine
1983—Infections in the Compromised Host: Laboratory Diagnosis and Treatment
1984—Urogenital Infections: New Developments in Laboratory Diagnosis and Treatment
1985—Clinical Implications of Antimicrobial Resistance: Mechanisms, Testing Problems and Epidemiology
1986—Host Defenses and Immunomodulation to Intracellular Pathogens
1987—Rapid Methods in Clinical Microbiology: Present Status and Future Trends
1988—Infection Control: Dilemmas and Practical Solutions

1989—Emerging Developments in Infectious Diseases: Challenges for the 90's

The 500th Branch Meeting Celebration 15 December 1987

On 15 December 1987 the Branch took the occasion of the 500th Branch meeting to celebrate sixty-seven years of continuous activity. The celebration, at the University of Pennsylvania, started with an afternoon of presentations on the history of the Branch, microbiology in general and historical events that occurred during each of the past seven decades. The following presentations were included in the afternoon program:

WELCOME, James Prier, Philadelphia College of Osteopathic Medicine
INTRODUCTION, James Poupard, Medical College of Pennsylvania
THE EARLY YEARS 1920-1932, James Poupard, Medical College of
    Pennsylvania, and Harry E. Morton, University of Pennsylvania
THE SECOND DECADE 1933-1939, Henry Beilstein, Hahnemann
    University; PA College of Podiatric Medicine; Beaver College
THE WAR YEARS—INFECTIOUS DISEASE AND MICROBIOLOGY,
    Carl Abramson, Pennsylvania College of Podiatric Medicine
POST-WAR MICROBIOLOGY 1951-1970, Paul Actor, Smith Kline and
    French Laboratories
THE ROLE OF THE BRANCHES IN THE AMERICAN SOCIETY FOR
    MICROBIOLOGY, Donald Shay, American Society for Microbiology
    Archives Committee, Center for the History for Microbiology
MICROBIOLOGY TODAY AND PERSPECTIVE FOR TOMORROW,
    Nick Burdash, Philadelphia College of Osteopathic Medicine
DINNER LECTURE: PHILADELPHIA BACTERIOLOGY AT THE
    TURN OF THE CENTURY, Edward Morman, The Institute of the
    History of Medicine, Baltimore

The presentations were followed by a reception and a spectacular dinner attended by 164 guests at the faculty club of the University of Pennsylvania. In addition to Dr. Morman's lecture, following dinner, several long-term Branch members were honored. Honorees included Drs. Harry Morton, Ruth Miller, Earle Spaulding, Amedeo Bondi and Morton Kline. A publication, *A History of the Eastern Pennsylvania Branch of the ASM: 1920 to 1987*, edited by Drs. Linda A. Miller, Harry E. Morton and James A. Poupard, was introduced and distributed to those in attendance. This publication became a valuable resource as a collection of Branch history up to that time.

Branch Officers

Branch Presidents of the 1980s

There were seven presidents (the thirty-first through the thirty-seventh) who served the Branch during this decade:

> Joseph F. Pagano, Smith Kline and French Laboratories (1979-81*)
> John C. McKitrick, University of Pennsylvania (1981**)
> Henry R. Beilstein, Philadelphia Public Health Department (1981-83***)
> Toby Eisenstein, Temple University (1983-85*)
> Donald Stieritz, Hahnemann Medical College (1985-87*)
> James A. Poupard, Medical College of Pennsylvania (1987-89*)
> Paul Actor, Smith Kline & French, (1989-91*)
> (*July through June) (**July through October 1981)
> (***November 1981 through June 1983)

Other Branch Officers

President Elect (The bylaws were never amended to recognize the use of the term Vice President and it was dropped early in the decade.)
1979-81 John C. McKitrick, University of Pennsylvania
1981 Henry R. Beilstein, Manor Junior College
1982-83 Toby Eisenstein, Temple University
1983-85 Donald Stieritz, Hahnemann Medical College
1985-87 James A. Poupard, Bryn Mawr Hospital
1987-89 Paul Actor, Smith Kline & French Laboratories
1989-91 Alan Evangelista, Cooper Medical Center

Secretary
1979-81 Josephine Bartola, Pennsylvania Department of Health Laboratories
1981-83 Donald Stieritz, Hahnemann Medical College
1983-85 Alan Evangelista, Cooper Medical Center
1985-88 Eileen Hinks, Rolling Hill Hospital, United Hospitals Inc.
1988 March-1991 Linda Miller, Holy Redeemer Hospital

Treasurer
1979-81 Henry R. Beilstein, Philadelphia Public Health Department
1981-85 Bruce Kleger, Pa. Department of Health, Bureau of Laboratories

1985-87 Paul Actor, Smith Kline Beckman
1987-89 Alan Evangelista, Cooper Medical Center
1989-91 Donald Jungkind, Thomas Jefferson University

Councilor
1979-81 Donald Stieritz, Hahnemann Medical College
1981-83 James Poupard, Bryn Mawr Hospital
1983-85 Donald Jungkind, Thomas Jefferson University
1985-87 Walter Ceglowski, Temple University
1987-89 Irving Millman, Fox Chase Cancer Center
1989-91 Nick Burdash, Phila. College of Osteopathic Medicine

Councilor-Alternate
1979-81 James A. Poupard, Bryn Mawr Hospital
1981-83 Irving Millman, Institute for Cancer Research
1983-85 Walter Ceglowski, Temple University
1985-87 Page Morihan, Medical College of Pennsylvania
1987-89 Nick Burdash, Phila. College of Osteopathic Medicine
1989-91 Bruce Kleger, Pa. Department of Health, Bureau of Laboratories

Branch Events by Year

1980

In February 1980, a new set of bylaws was approved, and in April the dues were raised from $6 for two years to $10. In September, the newsletter editor, Linda Creeden, was replaced by co-editors Albert Giovenella of the Hospital of the University of Pennsylvania and Margaret Cook of Merck Sharp and Dohme. In June, Dr. Earle Spaulding was made an Emeritus Member of the Branch.

There were ten Meetings in 1980. In April, the fifth Stuart Mudd lecture was presented by Dr. Thomas F. Anderson, who summarized forty years of research using electron microscopy. The June meeting was a kick-off meeting for the "Clinical Helminthology" workshop.

There were three workshops: "Recent Advances in MIC Technology" in January, "Clinical Helminthology" in June, and "Blood Cultures: Update and Controversies" in December.

The twelfth November Symposium "New Horizons in Medical Mycology," was held 13-14 November at the Marriott Hotel on City Line Avenue. Helen Buckley and Richard Crowell were chairpersons.

1981

In May 1981, Dr. Harry Morton was made an Emeritus Branch member.

There were ten meetings in 1981. In May, the sixth Stuart Mudd lecture was to be presented by Dr. Hilary Koprowski on rabies, but was given by a substitute speaker on the same subject. The June meeting was a kick-off meeting for the "New Pathogens" workshop.

There was one workshop on "New Pathogens" held in June.

The fifth Basic Science Symposium, "Host Defenses to Intracellular Pathogens," was held 10-12 June at the Franklin Plaza Hotel. Toby K. Eisenstein and Herman Friedman co-chaired the symposium, and the book published by Plenum Press was edited by T. Eisenstein, P. Actor and H. Friedman.

1982

There were ten meetings in 1982. The January meeting was a kick-off lecture for the "Anaerobic Bacteriology" workshop. The seventh Stuart Mudd lecture was held in April. The June meeting was a kick-off lecture for the "Medical Mycology" workshop. On 30 October, the Branch held a special Saturday Clinical Microbiology one day seminar on "New Approaches to Workload Reporting" at Thomas Jefferson University.

There were two workshops: "Clinical Anaerobic Bacteriology" in January and "Medical Mycology" in June.

The thirteenth November Symposium "Hepatitis B: The Virus, the Disease and the Vaccine" was held 11-12 November at the Franklin Plaza Hotel. Baruch S. Blumberg and Irving Millman co-chaired the symposium, and the book published by Plenum Press was edited by I. Millman, T. Eisenstein and B. Blumberg.

1983

There were ten meetings in 1983. The January meeting was the kick-off lecture for the workshop on "Sexually Transmitted Diseases." The eighth Stuart Mudd lecture was held in April and was given by Moselio Schaechter. The June meeting was a kick-off lecture for the "Rapid Serologic Methods" workshop.

There were two workshops: "Sexually Transmitted Diseases" held in February and, in June, a workshop on "Rapid Serologic Methods."

The fourteenth November Symposium "Infections in the Compromised Host: Laboratory Diagnosis and Treatment" was held 17-18 November at the Franklin Plaza Hotel. Alan Evangelista, James Poupard and Paul Actor co-chaired the symposium, and the book published by Plenum Press was edited by P. Actor, A. Evangelista, J. Poupard, and E. Hinks.

1984

There were ten meetings in 1984. The January meeting was the kick-off lecture for the workshop on "Problems in Pediatric Microbiology." The ninth Stuart Mudd lecture was held in April. The June meeting was a kick-off lecture for the "Recent Advances in Laboratory Diagnosis of Diarrheal Disease" workshop.

There were two workshops: "Problems in Pediatric Microbiology" in January and "Recent Advances in Laboratory Diagnosis of Diarrheal Disease" in June.

The fifteenth November Symposium, "Urogenital Infections: New Developments in Laboratory Diagnosis and Treatment," was held 15-16 November at the Franklin Plaza Hotel. Joseph M. Campos, Donald Stieritz and Amedeo Bondi co-chaired the symposium, and the book published by Plenum Press was edited by A. Bondi, J. Campos and D. Stieritz.

1985

There were ten meetings in 1985. The January meeting was the kick-off lecture for the workshop on "Hospital and Community Acquired Respiratory Infections." February was the first Annual Poster Session night. The poster session was organized by an ad hoc committee headed by Rick Rest. The focus for this first meeting was on graduate student presentations, but it was open to all Branch members. The tenth Stuart Mudd lecture in April was given by Leonard Hayflick. For the May meeting, the keynote address was held for the workshop "Skin and Diseases with Skin Manifestations." For the November meeting, the keynote address was held for the workshop "Infections of the Central Nervous System."

There were three workshops: "Hospital and Community Acquired Respiratory Infections" in February, "Skin and Diseases with Skin Manifestations" in May, and "Infections of the Central Nervous System" in November.

The sixteenth November Symposium "Clinical Implications of Antimicrobial Resistance: Mechanisms, Testing Problems and Epidemiology" was held 21-22 November at the Franklin Plaza Hotel. Eileen T. Hinks, Gerald D. Shockman and Paul Actor co-chaired the symposium, and the book published by Plenum Press was edited by P. Actor, E. Hinks and G. Shockman.

1986

In June 1986, the Branch recognized the following Emeritus Members: Amedeo Bondi, Wesley G. Hutchinson, Morton Klein, Ralph Knight, Israel Live, Robert Mandle, Ruth Miller, Harry Morton, Helen R. Skeggs and Evan Stubbs. In the October Newsletter, it was announced that in addition to the M.S. Program in Clinical Microbiology at Thomas Jefferson University, a second M.S. Program in Clinical Microbiology was available at the Medical College of Pennsylvania. Since many of these graduates remained in the Philadelphia area, these programs increased the number of clinical microbiologists staffing laboratories in this geographical area.

There were ten meetings. The April meeting was the second Annual Poster Session. The eleventh Stuart Mudd lecture was in May. The June meeting was a kick-off lecture for the "Nosocomial Infections and Infection Control" workshop. The October meeting was the keynote address for the "Antimicrobial Susceptibility Testing Current Methods and Clinical Impact" workshop.

There were two workshops: "Nosocomial Infections and Infection Control" in June and "Antimicrobial Susceptibility Testing Current Methods and Clinical Impact" workshop in November.

The seventeenth November Symposium, "Host Defenses and Immunomodulation to Intracellular Pathogens," was held 19-21 November at the Adam's Mark Hotel. Toby K. Eisenstein, Ward E. Bullock and Nabil Hanna co-chaired the symposium, and the book published by Plenum Press was edited by T. Eisenstein, W. Bullock and N. Hanna.

1987

As summarized previously, a highlight of the year 1987 was the special celebration held in December to mark the 500th Branch meeting.

There were ten meetings in 1987. March was a Foundation for Microbiology Lecture. April was the third annual poster night. The twelfth Stuart Mudd lecture was in May. The June meeting was a kick-off lecture for the "Recent Advances in Parasitic Diseases" workshop.

There was one workshop on "Recent Advances in Parasitic Diseases" in June.

The eighteenth November Symposium, "Rapid Methods in Clinical Microbiology: Present Status and Future Trends," was held 12-13 November at the Adam's Mark Hotel. Bruce Kleger, Eileen Hinks, and Donald Jungkind co-chaired the symposium.

1988

At the start of the year 1988, the Branch authorized Dr. Donald Jungkind to revert to our early Branch method of printing our monthly meeting notices on postal cards. This freed the Branch from the need of publishing a newsletter each month. Alan Truant became the new newsletter editor, and Harry Smith took over responsibility for the actual printing of the postal cards and newsletter.

In September, Anna Rosen took over as newsletter editor with Lori Walsh taking over responsibility for formatting and publishing the newsletter. In the November newsletter, a list of the twenty-one current Emeritus Members of the Branch was published. In addition, Drs. Joseph Gots and Robert Norris, both emeritus professors at the University of Pennsylvania, as well as Henry Stempen, emeritus professors at Rutgers University, were made Branch Emeritus members. In the September newsletter, it was announced that James Poupard, the current Branch president, replaced Donald Shay as Chairman of the ASM Archive—Center for the History of Microbiology. This appointment facilitated the deposit of some select Branch archive material in the ASM Archives.

It was agreed to contribute $1,000 to the New ASM Headquarters Fund. However, on 22 November, Dr. Harry Morton died. He was not only a past-President of the Branch, but one of the most active and valuable Branch members. The decision was made to contribute the $1,000 to the ASM Center for the History of Microbiology in the name of Dr. Morton instead of to the New ASM Headquarters Fund. Dr. Morton dedicated his last years to documenting the Branch History as well as the history of the ASM. His presence was greatly missed since he was an active member since his arrival in Philadelphia in the 1930s and guided the Branch through many changes.

There were ten meetings in 1988. The thirteenth Stuart Mudd lecture was held in April at Smith Kline and French Laboratories in Upper Merion. In May, there was a special afternoon mini-symposium, "Antibiotic Susceptibility Testing—Pitfalls and Potential," consisting of four talks on antibiotic susceptibility testing available free to all members. This was followed in the evening by a reception, dinner and then the fourth Annual Poster Night. The June meeting was a kick-off lecture for the "Recent Advances in Gastrointestinal Tract Diseases" workshop. In June, there was a special one-day symposium sponsored by the Branch Education Committee and Neumann College.

The symposium, "Molecular Biology for Clinical Laboratory Scientists," was held at Neumann College.

There was one workshop on "Recent Advances in Gastrointestinal Tract Diseases" in June.

The nineteenth November Symposium, "Infection Control: Dilemmas and Practical Solutions," was held 3-4 November at the Adam's Mark Hotel. Kenneth R. Cundy, Eileen T. Hinks and Bruce Kleger chaired the symposium.

1989

There were ten meetings in 1989. The fourteenth Stuart Mudd lecture was in April. The fifth Annual Poster Night was held in May. The June meeting was a kick-off lecture for the "Nonculture Methods for the Diagnosis of Infectious Diseases" workshop.

There were two workshops: "Nonculture Methods for the Diagnosis of Infectious Diseases" in June and "Tips for Teaching Clinical Microbiology at the Bench" in October.

The twentieth November Symposium, "Emerging Developments in Infectious Diseases: Challenges for the '90s," was held 9-10 November at the Adam's Mark Hotel. Bruce Kleger, Olarae Giger and Donald Jungkind co-chaired the symposium.

Summary of the Branch in the 1980s

A review of the above material speaks for itself. It was a very busy decade with many accomplishments. Highlights included the 500th Meeting Celebration, the ten Stuart Mudd lectures, initiation of the Poster Session meetings, and a focus by the Education Committee to create innovative programs to introduce science teachers to microbiology. There was also a renewed effort to attract microbiology graduate students to become involved with the Branch.

CHAPTER ELEVEN

The Eastern Pennsylvania Branch: 1990-1999 New Lecture Series, Special Events, and a Changing Environment

Overview of the Branch in the 1990s

The Branch entered the new decade of the '90s as a highly organized, well-functioning organization, determined to maintain the momentum from the previous decade. Serving the needs of both the clinical microbiology and basic science members was still a challenge, especially when choosing the topics for the monthly meetings. The Branch took significant steps to satisfy both sides of the clinical-basic science issue by continuing the clinical microbiology programs initiated in the previous decade, and adding new programs to serve the needs of the basic scientists. In addition, as the decade progressed, there were changes in the medical community that produced significant alterations in the membership population served by the Branch, especially in the area of Clinical Microbiology. The main changes were not new to the 1990s, but they started to have visible effects during this period. These factors resulted in the clinical laboratory becoming more of a financial burden, rather than a revenue generator, due to changes in reimbursement policies by third-party-payers to hospitals. This resulted in staff reductions in clinical laboratories. One index of this was the sharp reduction in the Medical Technology Programs in the Philadelphia area. Another factor was the appointment of pathologists as directors of clinical microbiology departments, especially in community hospitals, to better take advantage of new reimbursement rules. The net result of all these changes for the Branch was a reduction in the number or availability of directors, supervisors and technologists to attend workshops and symposia that lasted several days. At the same time, competition from other educational sources, especially self-learning modules, had the effect of reduced attendance at several Branch programs.

Another problem that became more apparent as the decade progressed was the realization that as the older Branch members retired or became inactive, the relatively large pool of potential new members that previously existed was no longer available. There was a need for new, especially younger members, to serve on committees and to serve as officers. Just creating one or two well organized symposia each year required a very active committee to make this activity a success. The main challenge for the Branch in the 1990s was how to increase the number of active members to support the many activities that members came to expect from the Branch. There was limited success in meeting this goal.

In spite of all the difficulties outlined above, the Branch in the '90s, not only continued to carry out almost all the various activities initiated in the previous decade, but managed to start some new programs. These new programs were designed to address the changing environment and to serve the needs of the microbiology community, both the basic-science and clinically oriented members. The following are just some of the activities that characterized the Branch during this decade:

There were ninety-two regular monthly Branch meetings in the 1990s. In 1992, the Annual Infection and Immunity Forum was initiated and was held each year for the rest of the decade and continues to the present time. There was an annual Stuart Mudd Lecture each of the first five years of the decade. After twenty continuous years, it was decided to replace the Stuart Mudd Lecture series with **The Distinguished Branch Member Lecture Series** which started in 1996. There were also fifteen workshops and eleven symposia; all this, in addition to Branch poster sessions and special events celebrating the 600th Meeting and an event in honor of Smith Hall and Dr. Harry Morton.

For a summary of the Annual Infection and Immunity Forum, the Distinguished Branch Member Lecture Series, the 600th Meeting Celebrations and the event in honor of Smith Hall and Dr. Harry E. Morton, see Summary of Branch Meetings below.

Summary of Branch Meetings

There were nine or ten meetings each year, except for 1994, when there were eight meetings. Meetings were regularly held on the fourth Monday of every month, except July and August, at Alumni Hall of Thomas Jefferson University on Locust Street.

The following is an attempt to characterize the 92 meetings, (meeting numbers 528 to 619). Almost all meetings consisted of only one presentation. Attendance records were not routinely recorded. Non-local presenters accounted for 61 percent for this decade, which was almost a 10 percent increase from the previous decade of 52 percent. The local speakers at these meetings were from a variety of Philadelphia institutions.

The following institutions are represented as a percent of the ninety-two presentations: Eight percent of the presenters were from the University of Pennsylvania. Other well represented institutions were: Temple University, 4 percent; Thomas Jefferson University, 2 percent; Hahnemann Medical College, 2 percent; Saint Christopher's Hospital, 2 percent; Smith Kline Beecham, 2 percent; and Merck Sharp & Dohme, 2 percent. Other institutions included: Medical College of Pennsylvania, Saint Joseph's University, and Children's Hospital of Philadelphia.

The following is the percent for each category for the ninety-two meetings. The number in brackets indicates the percent increase or decrease when compared to the previous decade results. Items in bold print indicate a new category:

- bacteriology 22 [-3]
- clinical microbiology 16 [+11]
- virology 11 [-11]
- immunology/ vaccines 7 [-6]
- mycology 7 [+3]
- resistance antimicrobials and susceptibility testing 7 [+3]
- parasitology 6 [-1]
- history 4 [0]
- biochemistry 4 [+1]
- **education 3**

Two previous categories from the 1980s, venereal diseases, and mycoplasmas were eliminated due to small number of presentations.

Presentations in bacteriology included the following: streptococci, pneumococci staphylococci, other gram-positive cocci, venereal diseases, mycobacteria, mycoplasmas, chlamydia, intestinal infections, *E. coli, Haemophilus sp., Listeria sp., Salmonella sp.,* and *Vibrio sp.* Organisms that appeared for the first time in presentations during the 1990s included the following: *Campylobacter sp., Heliobacter pylori,* and *Ehrlichia sp.* Virology accounted for 11 percent of the talks and included a wide range of viruses or viral diseases including: AIDS, EB, RSV, CMV, herpes, hepatitis, influenza, smallpox, retroviruses, and tumor associated viruses. In 1991, the topic of biological warfare and bioterrorism was first introduced. Two other topics appeared in 1990, nosocomial and opportunistic infections. There was also an increase in educational topics during this decade. This is a result of increased activity of the Branch Education Committee.

Hot Topics of the 1990s. The symposia titles are one measure of the "hot topics" in microbiology as decided by the microbiology community in Philadelphia. These included the following subjects:

Symposia Held in the 1990s
1990 Second Conference on Candida and Candidiasis: Biology, Pathogenesis and Management
1990 Innovations in Antiviral Development and the Detection of Virus Infections
1991 Antimicrobial Susceptibility Testing: Critical Issues for the '90s
1992 The Migration of Infectious Diseases: Five Hundred Years after Columbus
1993 Antimicrobial Resistance—A Crisis in Health Care
1994 Vaccines: Preventive Strategies for the Twenty-first Century
1995 Resistance, Epidemiology, Laboratory Methods: The Gram-Positive Perspective
1996 Diagnosis of Infectious Diseases Using Molecular Methods: Impact on the Laboratory and Patient Care
1997 Chronic Infectious Diseases: Mechanisms, Diagnosis and Treatment
1998 Clinically Relevant Microbiology in the Era of Managed Care
1999 New Technologies Driving Microbiology into the Twenty-first Century: Applying Genomics, Microarrays, and Combinatorial Chemistries to Drug Discovery, Bacterial Pathogenesis and Vaccine Development

Special Meeting in Honor of Smith Hall and Dr. Harry Morton 1992
On a Tuesday afternoon, on 31 March 1992, there was an event in honor of Smith Hall and Dr. Harry E. Morton. The meeting started with Branch

members standing in front of Smith Hall for a group photograph. Smith Hall was originally called The Laboratory of Hygiene. This laboratory building was erected on a site on the south side of Smith Walk in 1892. The building was designed specifically for the study of hygiene and bacteriology, and was the first of its kind in the United States. It was intended not to mimic European models but to marry those models to the United States ideas of practical laboratory hands-on education. Its state-of the-art heating and ventilating systems were specifically designed to illustrate scientific principles and safeguard the health of those working with pathogenic bacteria. The first official course in bacteriology at the University of Pennsylvania began on the day the building opened. Bacteriologists, like Alexander C. Abbott, a co-founder of the Society of American Bacteriologists, and David H. Bergey, as well as many other early bacteriologists, got their "start" in the profession at the Laboratory. In 1920, Dr. Bergey brought together several Philadelphia bacteriologists in his office in the Laboratory of Hygiene, which resulted in the formation of the Eastern Pennsylvania chapter (Branch). He also completed work on the first edition of the famous *Bergey's Manual of Determinative Bacteriology* while working in this building. This was also the site, in 1909, where a school of hygiene and public health was organized, and resulted in awarding the degree of Doctor of Public Hygiene. The building also housed the Pennsylvania Health Department laboratories for many years. In its later years, it became known as Smith Hall and housed the History and Philosophy of Science Department. The building was demolished in 1995 to make room for new laboratory space to be named the Roy and Diana Vagelos Laboratories: Institute for Advanced Science and Technology. In 1992, a group of faculty and students at the University of Pennsylvania mounted an effort to save the building from demolition due to its historical significance, and the Branch decided to hold a meeting there to support the University group. It was also decided to use this meeting to honor long time Branch member and University of Pennsylvania Professor, Dr. Harry Morton.

The afternoon photo session was followed by a talk at the University of Pennsylvania Faculty Club, by Dr. Richard Thomson on "Contemporary Issues in Microbiology." He included some of the basic lessons learned while working with Dr. Morton. This was followed by a talk on the "History and Significance of Smith Hall" by Professor Robert Kohler. After Dr. Kohler's talk, there was time for hors d'oeuvres and for everyone to tour Smith Hall and absorb the essence and history that unfolded within these hallowed walls. A special dinner was then served back at the Faculty Club. The official program included the following:

Welcome, James A. Poupard, PhD, SmithKline Beecham Pharmaceuticals
Lecture in Honor of Dr. Harry E. Morton, "Contemporary Issues in Microbiology," Richard B. Thomson, Evanston Hospital, Evanston, IL.
History and Significance of Smith Hall, Professor Robert Kohler, History and Sociology of Science Dept., University of Pennsylvania.
Hors D'Oeurves and Tours of Smith Hall Conducted by Mark Hamel and Frederick Quivik, History and Sociology of Science Department, University of Pennsylvania
Dinner, including Honored Guest—Mrs. Harry Morton

Infection and Immunity Forum Initiated 1992

A new tradition was initiated on 10 April 1992. This year marked the start of a new annual series designed to get graduate students and faculty involved with the Branch. It was presented as an interdisciplinary forum for clinical and basic research scientists in the Philadelphia and Delaware Valley areas interested in microbial pathogenesis and host response. Richard Rest, of Hahnemann University, was in charge of the committee that initiated the first Philadelphia Infection and Immunity Forum that was held at Thomas Jefferson University. This first Forum included internationally known speakers, a poster session, two topical sessions and informal discussions. This first meeting established the format for future Forums, which became an annual event that continues to the present time. Once a Branch Student chapter was established, organizing this annual forum became included as part of their activities. They organized this event while working closely with an Executive Committee member. For more details see the **Infection and Immunity Forum** in the Appendix section.

The Distinguished Branch Member Lecture Series Initiated 1996

After twenty years (1976-1995) of the Stuart Mudd Lecture Series, it was decided that it was time for a change. There was a strong desire to maintain the general format that worked so well for the Stuart Mudd Lectures. This involved finding a well recognized speaker to give a lecture, followed by a dinner at a nearby restaurant for more informal discussions. It was decided that this new lecture series should be used to recognize Branch members who made special contributions to the success of the Branch over several years. The first Distinguished Branch Member Lecture was held on 22 April, 1996 at Thomas Jefferson University, with Dr. Harry E. Morton chosen as the first honoree. Like Dr. Morton, each honoree who followed was chosen for their various contributions to the Branch, and the lecturers associated with these events were chosen to reflect some aspect of the work or interests of the honorees. Like the Stuart Mudd Lecture Series, the Distinguished Branch Member Lectures became one of the most social, as well as scientific, highlights of the Branch academic year. It provided an opportunity for newer members to not only celebrate the contributions of the more senior members, but it gave all members an opportunity to become familiar with the accomplishments of the honoree. This was accomplished by providing a biographical sketch of the honoree, as well as providing a time for more informal remarks on the honoree at the dinner that followed. For a complete list of honorees and lectures see **Distinguished Branch Member Lecture Series** in the Appendix Section.

The 600th Branch Meeting Celebration 14 September 1998

On 14 September 1998, the Branch celebrated the 600th meeting with an afternoon session at Thomas Jefferson University entitled "A Reason to Celebrate Our 600th Meeting." This involved five Philadelphia Microbiologists "Looking Back" at the last 600 meetings and the microbiology represented at these meetings. The meeting was followed by a celebration dinner at Girasole Restaurant on Locust Street. The following topics were presented:

An Overview and Introduction—James A. Poupard, PhD, Director, Clinical Microbiology and Profiling, SmithKline Beecham Pharmaceuticals, Collegeville, PA.
A Look Back at Virology—Bruce Kleger, Dr. P. H., Director, Bureau of Laboratories, Pennsylvania Department of Health, Lionville, PA.
A Look Back at Mycobacterium—Adamadia Deforest, PhD, Director, Clinical Virology, St. Christopher's Hospital for Children and Allegheny University of the Health Sciences, Philadelphia, PA.
A Look Back at *Streptococcus pneumoniae*—Robert Austrian, MD, Professor and Chairman Emeritus, Department of Research Medicine, University of Pennsylvania School of Medicine, Philadelphia, PA.
Searching for Dr. Arrowsmith—James E. Prier, PhD, Professor of Microbiology, Phila., College of Osteopathic Medicine, Philadelphia, Pa.

Introduction to a Special Branch Membership Project, "100 Years of Philadelphia Microbiology"—James A. Poupard, SmithKline Beecham Pharmaceuticals.

Branch Officers

Branch Presidents of the 1990s

There were seven presidents (the thirty-seventh through the forty-second) who served the Branch during this decade (all terms were from July through June):

> Paul Actor, Smith Kline & French (1989-91)
> Alan Evangelista, Cooper Medical Center (1991-93)
> Linda A. Miller, Holy Reedemer Hospital (1993-95)
> Paul Cerwinka, MetPath Laboratories (1995-97)
> Richard Rest, MCP-Hahnemann Medical School (1997-99)
> Irving Nachamkin, University of Pennsylvania (1999-2001)

Other Branch Officers

President Elect
1989-91 Alan Evangelista, Cooper Medical Center
1991-93 Linda Miller, Holy Redeemer Hospital
1993-95 Paul Cerwinka, Corning Clinical Laboratories
1995-97 Rick Rest, MCP-Hahnemann University.
1997-99 Irv Nachamkin, University of Pennsylvania
1999-2001 Donald Jungkind, Thomas Jefferson University

Secretary
1988 March-1991 Linda Miller, Holy Reedemer Hospital
1991-93 Paul Cerwinka, MetPath Laboratories
1993-95 Olarae Giger, Episcopal Hospital
1995-99 Kathleen Beavis, Thomas Jefferson University
1999-2001 Anna Feldman-Rosen, Gwynedd-Mercey College

Treasurer
1989-93 Donald Jungkind, Thomas Jefferson University
1993-95 Irv Nachamkin, University of Pennsylvania
1995-2001 Olarae Giger, Episcopal Hospital

Councilor
1989-91 Nick Burdash, Phila. College of Osteopathic Medicine
1991-93 Toby Eisenstein, Temple University
1993-95 Carl Abramson, Pennsylvania College of Podiatric Medicine
1995-97 Alan Evangelista, MCP Hahnemann University
1997-99 Donald Jungkind, Thomas Jefferson University
1999-2001 Paul L. Cerwinka, Quest Diagnostics Inc.

Councilor-Alternate
1989-91 Bruce Kleger, Pa. Department of Health, Bureau of Laboratories
1991 (temporary) Ken Cundy, Temple University
1991-93 Carl Abramson, Pennsylvania College of Podiatric Medicine
1994-95 Post left vacant.
1995-99 Irv Nachamkin, University of Pennsylvania
1999-01 Richard Rest, MCP-Hahnemann Medical School

Branch Events by Year

1990

In 1990, the executive committee stressed the need for members to become more concerned with early science education. The activities that the education committee was performing to meet this objective was noted and encouraged. It was announced in the February newsletter that Dr. Richard Crowell was elected President of the National ASM. In April, the Branch was co-sponsor of a special afternoon meeting at the Albin O. Kuhn Library (where the ASM Archives are held). The topic was "The Biological Weapons Convention Under Siege: Disarmament Issues for the 1990s." It was one of the first open discussions within the ASM on biological weapons after several years of silence on this subject. It included a talk by James Poupard on "ASM and Biological Warfare: Activist or Ostrich?" In June, Marjorie Dole was made a Branch Emeritus Member. Marjorie worked with Dr. Austrian on the pneumococcal vaccine and, among other accomplishments, was head of the Clinical Microbiology Department at Pennsylvania Hospital for seventeen years.

There were nine meetings in 1990. The April meeting was the fifteenth Stuart Mudd lecture. The May meeting was the keynote address for the "Mycology 101: Identification Methods" workshop. In May, the sixth Annual Branch Poster Session was held.

There were two workshops: "Mycology 101: Identification Methods" held in June and "New Perspectives on AIDS" held in October.

There were two symposia held this year: A Basic Science symposium, "Second Conference on Candida and Candidiasis: Biology, Pathogenesis and Management," was held in April at the Adam's Mark Hotel. The chairperson was Helen R. Buckley. The twenty-first November Symposium on "Innovations in Antiviral Development and the Detection of Virus Infections" was held at the Adam's Mark Hotel. The symposium was chaired by Timothy M. Block, Richard Crowell and Donald J. Jungkind.

1991

In the February 1991 newsletter, it was announced that Dr. Eileen Randall had died in December. The death of Dr. Evan L. Stubbs, the eighth Branch President, 1934-35, was announced. He was one of our oldest Branch members. In September, the death of Thomas Anderson was noted. His pioneering work in electron microscopy contributed in making Philadelphia a center for these studies.

The National ASM recognized four Branch members who, as of 1990, were National ASM members for fifty or more years. These members were Ruth E. Miller, Earle H. Spaulding, Gladys L. Hobby, and Bernard Witlin. The following members became new Branch Emeritus Members: Samuel DeCourcy, Arthur Greene, Henry Kazal, Eugene Micklin and William Osborne. In May, several Branch members joined in a testimonial to Dr. Gerald Shockman to honor his seventeen years as Chairman of Microbiology at Temple University School of Medicine. In September, it was announced that Dr. Carlo Croce would be the new Chair of Microbiology at Thomas Jefferson University.

There were ten meetings in 1991. The April meeting was the sixteenth Stuart Mudd Lecture, and the May meeting was the seventh Branch Poster Night. June was the keynote address for the workshop on "Current Concepts in Clinical Parasitology." October was a special meeting in honor of Richard Crowell, ASM President 1991-1992. Dr. Crowell presented a lecture on "The Role of Recepters in Viral Infections." The November meeting was an ASM Foundation Lecture. The year ended with a wine and cheese party and a special lecture on "Death at an Early Age—Tuberculosis in Ancient Egypt."

There were two workshops: "Current Concepts in Clinical Parasitology" in June and "Medical Mycobacteriology in the 1990s" in October.

The twenty-second November symposium on "Antimicrobial Susceptibility Testing: Critical Issues for the '90s" was held at the Adam's

Mark Hotel. The symposium was chaired by James Poupard, Bruce Kleger and Lori Walsh. The book, published by Plenum Press, was edited by J. Poupard, L. Walsh, and B. Kleger.

1992

In 1992, the Branch elected five new Emeritus Members: Josephine Bartola, William Ball, Vern Pidcoe, Irving Millman and Connie Kruse.

This year marked the start of the **Infection and Immunity Forum** which was designed to get graduate students and faculty involved with the Branch. (See **Infection and Immunity Forum** in the **Summary of Branch Meetings** section above.) Richard Rest of Hahnemann University was in charge of the committee that produced the first Philadelphia Infection and Immunity Forum on 10 April 1992 at Thomas Jefferson University. This first Forum included internationally known speakers, (John Mekalanos from Harvard and Barry Bloom from Albert Einstein School of Medicine), a poster session, two topical sessions and opportunities for informal discussions. In addition to Rick Rest as organizer and chair, the following individuals were on the first organizing committee: Toby Eisenstein, Richard Johnston, Joel Mortensen, Irv Nachamkin, Daniel Portnoy and Norton Taichman.

There were nine meetings in 1992, with special events in March and December.

As described previously, the March 1992 meeting was a special event in honor of Smith Hall and Dr. Harry E. Morton. In December, there was a joint meeting sponsored by the Branch and Hahnemann University to honor Dr. Amedeo Bondi on his eightieth birthday and to recognize his forty-five-year career at Hahnemann University. The meeting was in the form of a colloquium on "Microbiology and Infectious Diseases in the Twenty-first Century Academic Medical Center."

April marked the seventeenth Stuart Mudd Lecture. The eighth Branch Poster Session was held in May. September was a Foundation for Microbiology lecture.

There were two workshops: "Recent Advances in Genital Tract Infections" was held in June and "Blood Cultures & Blood Borne Diseases" was held in October.

The twenty-third November symposium on "The Migration of Infectious Diseases: Five Hundred Years After Columbus" was held at the Adam's Mark Hotel. The symposium was co-chaired by Alan T. Evangelista, Carl Abramson and Bruce Kleger.

1993

In the February 1993 newsletter, the death of Morton Klein on 6 December 1992 was announced. Dr. Klein served as the twentieth Branch President in 1958-59. The Branch also lost another valuable member, Dr. Ralph Knight, who died on 28 September.

In the May newsletter, it was announced that Dr. Chris Platsoucas was named the new Chair of Microbiology at Temple University. In September, it was announced that Lori Walsh became the coeditor of the newsletter. In December, it was announced that the Education Committee had launched an exciting new program, The Laboratory Manual Project, an innovative program for biology teachers to stimulate interest in microbiology among their students.

There were three new Branch Emeritus Members: Robert Austrian, Herbert Heineman, and Barbara Lowery.

There were nine meetings in 1993. Two of the meetings were Foundation for Microbiology lectures (January and September). The March meeting was a panel discussion sponsored by the Education Committee on educational aspects and prediction of future needs in formulating microbiology educational programs. The eighteenth Stuart Mudd lecture, reception and dinner were held in April at the Korman Suites Hotel on Twentieth Street.

The second Philadelphia Infection and Immunity Forum was held 14 April at the Penn Tower Hotel at the University of Pennsylvania. Rick Rest was chairperson. The Forum included two distinguished speakers, a poster reception and topical discussion groups.

There were two workshops: "Laboratory Diagnosis of Fungal Infections" held in June and "Molecular Diagnosis in Clinical Microbiology" in October.

The twenty-fourth November symposium on "Antimicrobial Resistance—A Crisis in Health Care" was held at the Adam's Mark Hotel. The symposium was co-chaired by Donald Jungkind, Gary Calandra and Henry Fraimow. The book, published by Plenum Press, was edited by D. Jungkind, J. Mortensen, H. Fraimow and G. Calandra.

1994

The Branch lost three long time active members in 1994. Dr. Wesley G Hutchison died on 22 January. On 24 March, Theodore "Ted" Anderson died at the age of ninety-one, and in October, long time member Dr. Carl Clancy died.

Work was initiated to form an official student chapter. The new student chapter was to serve the needs of graduate students performing research at the Philadelphia medical schools. Among other tasks, the student chapter would help organize and run the annual infection and immunity forum.

In July, Hahnemann Medical College and the Medical College of Pennsylvania merged. Richard Crowell became Chair of the newly merged Microbiology Department.

In March, there was a mini-workshop sponsored by the Education Committee on "The Development of Critical Thinking Skills." The regular monthly meeting that followed was a keynote address on "Microbiology and the Undergraduate Science Curriculum: are we Trying to Produce a Marketable Product?" The evening program included a panel discussion moderated by Dr. Norm Willett.

There were nine meetings in 1994. The nineteenth Stuart Mudd lecture was held in May. The December meeting was a Foundation for Microbiology lecture.

The third Philadelphia Infection and Immunity Forum was held 15 April at the Penn Tower Hotel at the University of Pennsylvania. Rick Rest was chairperson.

There was a workshop on "Laboratory Diagnosis of Respiratory Infections" held in June.

The twenty-fifth November symposium was on "Vaccines: Preventive Strategies for the Twenty-first Century" and was held at the Adam's Mark Hotel. The symposium was co-chaired by Toby K. Eisenstein, Robert Austrian, and Jorg Eichberg

1995

The Branch lost two former past Branch presidents in 1995. Dr. Earle Spaulding, the thirteenth Branch President (1944-45) died on 2 February at age eighty-eight. He came to Temple University in 1936 and mentored many of our Branch members over the years. A memorial service was held on 11 March. Branch members were shocked to learn of the death of Dr. Robert Mandle on 3 April. He was the thirtieth Branch President (1977-79), directed the clinical microbiology master's program at Thomas Jefferson University, and was a significant influence in keeping the clinical microbiology section part of the Branch. A memorial service was held for him on 23 May.

The highly active and innovative education committee continued to formulate new programs. The Branch participated in the ASM-MicroCosmos training sessions. This involvement was initiated by Barbara McHale and Tom Rooney. As part of this program, funds were received from National ASM to support a high school teacher to pair with a scientist to attend special MicroCosmos training sessions at Boston University.

There were nine meetings in 1995. The January meeting was a Foundation for Microbiology lecture. The twentieth Stuart Mudd lecture was held in April. The June meeting served as the keynote address for the workshop on parasitology. The October meeting was held at McShain Hall at Saint Joseph's University.

There are no records in any area of the archives that document an Infection and Immunity Forum for this year; however, the organizers for the 1996 forum refer to it as the fifth forum. Although no records survived, it is assumed that there was a 1995 Forum; however it may have been more informal, since it was at a time when the Branch student chapter was being formed.

There were two workshops: "Clinical Parasitology-1995' was held in June and "New & Reclassified Organisms of Clinical Importance" was held in October.

The twenty-sixth November symposium on "Resistance, Epidemiology, Laboratory Methods: The Gram-Positive Perspective" was held at the Adam's Mark Hotel. The symposium was co-chaired by Linda A. Miller, Kathleen G. Beavis and Joel E. Mortensen.

1996

There were ten meetings in 1996. The first Distinguished Branch Member Lecture was held 22 April in honor of Dr. Harry E. Morton. The meeting was cosponsored by the World Affairs Council of Philadelphia. The speaker was Dr. Ray Zilinskas who spoke on "In Pursuit of Saddam Hussein's Biological Arsenal." See **The Distinguished Branch Member Lecture Series** in the **Summary of Branch Meetings** section.

May and November were Foundation for Microbiology lectures. June was the keynote address for the "What's New With Yeasts?" workshop.

The fifth Philadelphia Infection and Immunity Forum was held 12 April at the Penn Tower Hotel at the University of Pennsylvania. Rick Rest was chairperson.

There was a workshop on "What's New With Yeasts?" in June.

The twenty-seventh November symposium on "Diagnosis of Infectious Diseases Using Molecular Methods: Impact on the Laboratory and Patient Care" was held at the Adam's Mark Hotel. The symposium was co-chaired by Donald Jungkind, Alan Evangelista and David Persing.

1997

In December 1997, it was announced that the first Branch Web site was now "live." The Branch student chapter was officially recognized by the National ASM.

There were ten meetings in 1997. February and October were Foundation for Microbiology lectures. The second Distinguished Branch Member Lecture in honor of Dr. Earle H. Spaulding was held in April. Marie B. Coyle spoke on "Old Technologies Detect New Species in a Routine Laboratory" followed by a dinner at Bistro Bix on Twelfth Street. In December, there was the traditional wine and cheese reception following a lecture by Dr. Bennett Lorber on "Are All Diseases Infectious?"

The sixth Annual Philadelphia Infection and Immunity Forum was held 11 April at SmithKline Beecham Pharmaceuticals in Collegeville, Pa. Padmini Salgame was the forum coordinator.

There was a workshop on "Food & Water Borne Diseases of North America" in June.

The twenty-eighth November symposium was on "Chronic Infectious Diseases: Mechanisms, Diagnosis, and Treatment" held at the Adam's Mark Hotel. The symposium was co-chaired by Ian A. Critchley and Paul Actor.

1998

In August 1998, there was an executive committee retreat to evaluate the current state of the Branch and do some serious planning on the future of the Branch. The main theme was how to get new (young) members, not only to attend the monthly meetings, but to participate in the activities of the Branch.

There were nine meetings in 1998. January was a Foundation for Microbiology Lecture. The third Distinguished Branch Member Lecture in honor of Dr. Amedeo Bondi was held in April. Clyde Thornsberry spoke on "Changing Antimicrobial Resistance in the '90s: The Hot Spots" followed by a dinner at Lilies Restaurant on Twelfth Street. At the June meeting, Guy Webster gave the keynote address for the "Infectious Diseases with Skin Manifestations—Diagnosis & Pathogenesis" workshop. The September meeting was a special meeting in celebration of the 600th Branch Meeting. (For a full description of this event see the **600th Branch Meeting Celebration** in the **Summary of Branch Meetings** section.) The December meeting was the traditional wine and cheese reception following a lecture by Patrick Brennan on "TB: An Historical Perspective at the End of the Millennium."

The seventh Annual Philadelphia Infection and Immunity Forum was held 17 April at SmithKline Beecham Pharmaceuticals in Collegeville. Padmini Salgame was the forum coordinator.

There was a workshop on "Infectious Diseases with Skin Manifestions—Diagnosis & Pathogenesis" in June.

The twenty-ninth November symposium, "Clinically Relevant Microbiology in the Era of Managed Care" was held at the Adam's Mark Hotel. The symposium was co-chaired by Kathleen Gleason Beavis, Karen Carroll and Margaret M. Yungbluth.

1999

Dr. Gerald D. Shockman died on 27 October 1999 at the age of seventy-three. Dr. Shockman came to Temple University in 1960 and was chair of the microbiology Department there from 1974 to 1990.

Anna Feldman-Rosen and Lori Walsh turned over responsibility for the newsletter to the new editor, Julie Conaron.

There were nine meetings in 1999. The March and February meetings were Foundation for Microbiology lectures. The fourth Distinguished Branch Member Lecture in honor of Dr. Kenneth Cundy was held in April. Burton Wilcke spoke on "Microbiology and Its Continuing Impact on Public Health" followed by a dinner at Toto's Restaurant on Locust Street.

The eighth Philadelphia Infection and Immunity Forum was held 16 April at the SugarLoaf Conference Center on Germantown Pike. Padmini Salgame was the forum coordinator.

There was a workshop on "Practical Clinical Parasitology: Fundamentals, Review and Future Perspectives" in June.

The thirtieth November symposium on "New Technologies Driving Microbiology into the Twenty-first Century: Applying Genomics, Microarrays and Combinatorial Chemistries to Drug Discovery, Bacterial Pathogenesis and Vaccine Development" was held at the Adam's Mark Hotel. The symposium was co-chaired by David M. Mosser and Anna Feldman-Rosen.

Summary of the Branch in the 1990s

Two new lecture series were initiated in this decade, and there were several special meetings as well as the traditional symposia and workshops. By any standard, this was a decade full of activities and accomplishments.

However, as previously described, the pool of clinical microbiologists and medical technologists available to attend Branch functions, as well as younger members to replace retiring committee chairs and officers, was decreasing. This decrease would have a more significant effect in the following years.

CHAPTER TWELVE

The Eastern Pennsylvania Branch: 2000-2009 Adjustments to a New Century, ASM General Meeting, and the Laboratory of Hygiene

Overview of the Branch in the 2000s

In the opening statement of the new century, and essentially the new millennium, the Branch President, Irv Nachamkin, states in the first edition of the Branch Newsletter (January 2000): "As we begin the twenty-first century, communication and cooperation between our clinical and basic science members have never been more important. This was a message brought home in a big way at our thirtieth annual symposium: 'New Technologies Driving Microbiology into the Twenty-first Century,' held on November 4-5, 1999. The results of basic research in the areas of genomics, gene expression and computational biology will not improve human health unless there is a fundamental understanding by basic researchers about the role of the technologies in the clinical area and an understanding by clinical scientists of biological processes and pathogenesis. Why do I broach this subject? Where it [the Branch] has failed is in attracting individuals to attend programs that are not in their field of expertise. There has never become a more important time to think beyond our "box" and become more enlightened in other areas for us to maintain a competitive edge."

This was a valuable summary on the state of microbiology, not only as it relates to our Branch, but to microbiology in general, and was an essential message to the whole microbiology community. These remarks set the stage for the new century, and placed the focus on the most difficult problem the Branch has to face as the new century unfolds; namely, how to get new people involved in the Branch as active members. Without success in this area, the level of activity demonstrated in previous years will not be sustainable. It is a good summary for setting the biggest challenge for the Branch and for microbiology in Philadelphia. The Branch, with its monthly programs, in addition to the symposia, workshops, forums and many projects, still makes it the most active Branch of the ASM. The branch is struggling to maintain all these activities, but the incentive to succeed remains, and it is not the first time that the Branch had to adapt to a changing situation.

The first decade of the new century remained a very active time for the Branch. It not only maintained the extensive programs carried over from the previous century; in 2009 the Branch hosted the 109th ASM General Meeting, and the Laboratory of Hygiene of the University of Pennsylvania was designated as the third ASM Milestone in Microbiology site. By all standards, the Branch had a very active decade, and continued to represent the state of contemporary microbiology in the Philadelphia area.

The 109th ASM General Meeting. In May 2009 the ASM General Meeting was held in Philadelphia for the tenth time. This was significant, since the last time the Branch was host to a General Meeting was in 1972, and Branch members wanted to do everything possible to make this meeting a success. There were several unexpected problems that occurred in the spring of 2009. One problem was that the largest microbiology meeting in Europe, the annual ECCMID meeting, was held at the same time, which resulted in reduced General Meeting attendance. However, a more significant issue was an outbreak of "swine influenza" which imposed travel restrictions on many clinical, health department and government microbiologists. In spite of these difficulties, the meeting was successful, and there were several special events and projects associated with this meeting. The **Milestones in Microbiology Dedication Ceremony** was held prior to the start of the General Meeting (see below). Upon registration at the General Meeting, each attendee was provided with a tour guide brochure and walking map prepared by the Branch Education Committee. The brochure listed sites of historical interest relating to microbiology or science within walking distance of the Philadelphia Convention Center. In addition, James Poupard, working with the ASM Archives Committee, organized a special full-afternoon session that revised the tradition of focusing on some historical contributions of the host city to the field of microbiology. This session started with the History of Microbiology Lecture "Philadelphia Microbiology, Past, Present and Future: 1750 to 2050" by James Poupard. This lecture set the stage for a series of four Archive Session Lectures: "Philadelphia's Giants of Microbiology: Their Contributions, Past, Present and Into the Future." These included the following presentations: Ben Franklin: Microbiologist—by Douglas Eveleigh;

David Bergey's Legacy in Bacterial Classification—by William B. Whitman; Stuart Mudd and the Cold War Politics of Health—by William C. Summers; and Master of the Microscope: Joseph Leidy—by Joan Bennett. The ASM sponsored a reception following the presentations to enable attendees to interact with the speakers.

Milestones in Microbiology Dedication. "Milestones in Microbiology" is a joint project of the ASM Communications and Archives Committees. The program is designed to recognize sites of particular importance to the history of microbiology. Two sites have previously been recognized. The Laboratory of Hygiene of the University of Pennsylvania was the third site to receive this designation. On Friday 15 May 2009, prior to the start of the General Meeting, a dedication ceremony was held at the University of Pennsylvania to unveil a plague commemorating the site of the former Laboratory of Hygiene. James Poupard moderated the ceremony and discussed the significance of the Laboratory to the Branch, since the Branch was formed at a meeting in David Bergey's office, which was located on the second floor of the Laboratory. ASM President, Alison O'Brien, gave a brief talk welcoming the guests and explaining the Milestones Program. Arthur Rubenstein, Dean of the University of Pennsylvania School of Medicine, spoke on the history of the building and its significance to the University. William B. Whitman, representing the Beregy Trust, discussed David Bergey and his continuing influence on microbiology. The meeting was attended by Branch members, University of Pennsylvania and ASM staff, ASM Archive chair, Patricia Charache, and other ASM Archive Committee members. A reception, jointly sponsored by the Bergey Trust and the ASM, followed the official plaque presentation. Photographs from the Branch Archives were displayed at the reception. During the reception, various groups from several disciplines participated in lively discussions on the history of microbiology at the University and in Philadelphia.

Summary of Branch Meetings

There were seventy-five regular monthly Branch meetings in the 2000s, with a range of seven to ten meetings each year. Meetings were regularly held on the fourth Monday of every month, except July, August and November, at Alumni Hall of Thomas Jefferson University on Locust Street.

The following is an attempt to characterize the seventy-five meetings, (meeting numbers 620 to 702). Almost all meetings consisted of only one presentation. Attendance records were not routinely recorded. Nonlocal presenters accounted for 81 percent for this decade, which was a 20 percent increase from the previous decade. The local speakers at these meetings were from a variety of Philadelphia institutions.

The following institutions are represented as a percent of the seventy-five presentations: Eight percent of the presenters were from the University of Pennsylvania. Other well represented institutions were: Temple University, 4 percent; Drexel University, 4 percent; Pharmaceutical Companies, 4 percent (GSK, Wyeth, and Protez); Phila. or PA Health Department, 4 percent; and 1 percent from Thomas Jefferson University, Children's Hospital of Philadelphia, Swarthmore College, and the AI DuPont Research Institute.

The following is the percent for each category for the seventy-five meetings. The number in brackets indicates the percent increase or decrease when compared to the previous decade results. Items in bold print indicate a new category:

- bacteriology 36 [+14]
- virology 17 [+6]
- mycology 11 [+4]
- parasitology 11 [+5]
- clinical microbiology 9 [-7]
- immunology/vaccines 9 [+2]
- biochemistry 13 [+9]
- **bioterrorism 9**
- resistance—antimicrobials and susceptibility testing 8 [+1]
- **genomics 4**
- history 2 [-2]
- education 2 [-1]

Presentations in bacteriology included: streptococci, pneumococci, enterococci, staphylococci (MRSA), other gram positive cocci, mycobacteria, *E. coli, Salmonella sp. Vibrio sp. Klebsiella sp., Acinetobacter sp.* and intestinal infections. One organism, *Porphyromonas gingivali*, appeared for the first time in a presentation during this decade. Virology accounted for 17 percent of the presentations and included a wide range of viruses or viral diseases including: RSV, influenza, slow viruses, and bacteriophage. AIDS and retroviruses dominated the list; the topic of West Nile was first introduced in 2002 and SARS appeared for the first time in 2003. It is of interest that the topic of biological warfare, which was first introduced in 1991, now accounted for 9 percent of the presentations when combined with the subject of bioterrorism.

Hot Topics of the 2000s. The symposia titles are one measure of the "hot topics" in microbiology as decided by the microbiology community in Philadelphia. These included the following subjects.

Symposia Held in the 1990s
2000   Chronic Viral Hepatitis: Advances in Laboratory Testing, Therapy and Prevention
2001   Clinical Microbiology Update 2001
2002   Infections of the Central Nervous System
2003   Clinical Mycology Update: 2003
2004   Antimicrobial Resistance and Emerging Infections: A Public Health Perspective
2005   STDs and Other Genital Tract Infections: Current Status and Future Trends
2006   Emerging Community and Healthcare-Associated Infectious Diseases
2008   Challenging Issues in Antimicrobial Resistance
2009   Molecular Testing in Clinical Microbiology: Advancements, Challenges, Future Directions

Branch Officers

Branch Presidents of the 2000s

There were six presidents (the forty-second through the forty-seventh) who served the Branch during this decade (all terms were from July through June):

   Irving Nachamkin, University of Pennsylvania (1999-01)
   Donald Jungkind, Thomas Jefferson University (2001-03)
   Olarae Giger, Main Line Clinical Laboratories (2003-05)
   David Axler, Temple University (2005-06)
   Bettina Buttaro, Temple University (2006-09)
   Laura Chandler, Philadelphia VA Medical Center (2009-11)

Other Branch Officers

President Elect
1999-01 Donald Jungkind, Thomas Jefferson University
2001-03 Olarae Giger, Main Line Clinical Laboratories
2003-05 David Axler, Temple University
2005-06 Bettina Buttaro, Temple University
2006-09 Laura Chandler, Philadelphia VA Medical Center

Secretary
1999-09 Anna Feldman-Rosen, Gwynedd-Mercy College

Treasurer
1995-01 Olarae Giger, Episcopal Hospital
2001-03 David Axler, Temple University
2003-05 Bettina Buttaro, Temple University
2005-09 Barbara McHale, Gwynedd-Mercy College

Councilor
1999-01 Paul L. Cerwinka, Quest Diagnostics Inc.
2001-03 Richard Rest, MCP-Hahnemann University
2003-05 Irv Nachamkin, University of Pennsylvania
2005-09 Donald Jungkind, Thomas Jefferson University

Councilor-Alternate
1999-01 Richard Rest, MCP-Hahnemann Medical School
2001-03 Irv Nachamkin, University of Pennsylvania
2003-05 Donald Jungkind, Thomas Jefferson University
2005-09 Olarae Giger, Main Line Clinical Laboratories

Branch Events by Year

2000
Dr. William Ball, Branch long time member, died on 31 July. He served in World War II and had a particular interest in parasitology.

There were ten meetings in 2000. The fifth Distinguished Branch Member Lecture in honor of Dr. Bruce Kleger was held in April. Andre Weltman presented an "Introduction to Bioterrorism." The lecture was followed by a dinner at Toto's Restaurant on Locust Street. The May and

December meetings were Foundation for Microbiology lectures. June was the kick-off lecture for the workshop on "Antimicrobial Susceptibility Testing: Issues and Controversies."

The ninth Philadelphia Infection and Immunity Forum was held 4 May at the SugarLoaf Conference Center on Germantown Pike. Padmini Salgame was the forum coordinator.

There was a workshop on "A New Era of Susceptibility Testing: Challenges and Methods" in June.

The thirty-first November symposium on "Chronic Viral Hepatitis: Advances in Laboratory Testing, Therapy and Prevention" was held at the Adam's Mark Hotel. The symposium was co-chaired by Timothy M. Block, Mark A. Feitelson and Donald Jungkind.

2001

Long time active member, Dr. Helen Buckley, Professor of Mycology, at Temple University, died on 28 February. The Branch held a memorial lecture in her honor at the October meeting. On 5 April, the Branch also lost long time member Dr. Henry Stempen, Professor Emeritus at Rutgers University. He is especially remembered within the Branch as the designer of the Branch logo.

On 16 April, the new permanent exhibit on "Clinical Microbiology at the University of Pennsylvania" was dedicated in a ceremony at the exhibit location on the fourth floor of the Gates Building of the Hospital of the University of Pennsylvania. On 30 March, there was an executive committee retreat, the first of several during the decade, to discuss how to address low membership numbers and how to attract new active members.

There were nine meetings in 2001. February was a Foundation for Microbiology Lecture. The sixth Distinguished Branch Member Lecture in honor of Dr. Henry Beilstein was held in March. Stuart B. Levy spoke on "Antibiotic Resistance: Microbes on the Defense." The lecture was followed by a dinner at Toto's Restaurant on Locust Street. The June meeting was a keynote address by Lynne Garcia for the workshop on "Clinically Relevant Rapid Microbiology." The October meeting was a special remembrance in honor of Dr. Helen Buckley, Professor of Mycology, at Temple University.

The tenth Philadelphia Infection and Immunity Forum was held on 4 May at the SugarLoaf Conference Center on Germantown Pike. Bettina Buttaro was the forum coordinator.

There was a workshop on "Clinically Relevant, Rapid Microbiology" in June.

The thirty-second November symposium entitled "Clinical Microbiology Update 2001" was held at the Adam's Mark Hotel. The symposium was co-chaired by Irving Nachamkin and Donald Stieritz. This was the first joint symposium that relied on the National Laboratory Training Network for assistance.

2002

On 7 April, Dr. Ruth Emma Miller, the eighteenth Branch President (1954-55), died at the age of ninety-six. She received her PhD from the University of Pennsylvania in 1934 and was the first female president of the Branch. She was a long time faculty member, Professor and Chairperson of the Microbiology Department of the Medical College of Pennsylvania, and a long time active Branch member.

In September, the newsletter publication switched from an all print version to both a printed and electronic version. This was an effort to eventually go completely electronic to reduce expenses. Members could request a printed version if they did not want to make the switch to an electronic version.

There were eight meetings in 2002. February was a Foundation for Microbiology Lecture. The seventh Distinguished Branch Member Lecture in honor of Dr. Norman Willett was held in April. Caroline Genco spoke on "Bugs, Gums, and Heart Disease: Pathogenic Strategies of Periodontal Pathogen *Porhyromonas gingivalis* in Endothelial Cell Inflammation." The lecture was followed by a dinner at Toto's Restaurant on Locust Street. The June meeting was the keynote address for the workshop on "Strategies for the Identification of Gram-Positive Bacteria in the Clinical Laboratory."

The eleventh Philadelphia Infection and Immunity Forum was held 31 May at the MCP-Hahnemann, Queen Lane Campus. Bettina Buttaro was the forum coordinator.

There was a workshop on "Strategies for the Identification of Gram-Positive Bacteria in the Clinical Laboratory" in June.

The thirty-third November symposium on "Infections of the Central Nervous System" was held at the Adam's Mark Hotel. The symposium was co-chaired by Laura J. Chandler and Olarae Giger.

2003

Longtime Branch member, Eugene Micklin, died on 13 January at the age of seventy-seven. On 8 August, Dr. Joseph Pagano, the thirty-first Branch President (1979-81), died at the age of eighty-four. Dr. Bruno Bromke, an active member who served on the Symposium Committee, died on 25 July at the age of fifty-eight. Dr. Albert G. Moat, the twenty-fourth Branch President (1966-67), died at the age of seventy-seven.

There were eight meetings in 2003. The eighth Distinguished Branch Member Lecture in honor of Dr. Carl Abramson was held in April. Bennett Lorber spoke on "Snakes, Sex, Sushi and Saunas." The lecture was followed by a dinner at Toto's Restaurant on Locust Street. The June meeting was held at the Pennsylvania Department of Health Laboratories in Lionville, Pa., and it was the kick-off meeting for the workshop on "Presumptive Identification of Primary Agents of Bioterrorism and Proper Interaction with the Laboratory Response Network."

The twelfth Philadelphia Infection and Immunity Forum was held 30 May at the Drexel University, Queen Lane Campus. Bettina Buttaro and Richard Rest chaired the forum.

There was a workshop on "Bioterrorism and the Clinical Laboratory: Practical Information Needed in Every Microbiology Lab" in June.

The thirty-fourth November symposium entitled "Clinical Mycology Update: 2003" was held at the Biomedical Research Building at the University of Pennsylvania. The symposium was co-chaired by Donald Stieritz and Karin McGowan.

2004

On 10 July, Dr. Henry R. Beilstein died after a long illness. He was the thirty-third Branch President (1981-83), was a Distinguished Branch Honoree, served on the Executive Committee for many years and was a fifty-year member of the National ASM. He was influential in keeping all elements within the Branch together and would always have a calming effect in heated discussions. Dr. Lolita Daneo-Moore, of Temple University, died on 23 November.

There were eight meetings in 2004. The February and December meetings were ASM Waksman Foundation Lectures. The ninth Distinguished Branch Member Lecture in honor of Dr. James E. Prier was held in April. Mark Birenbaum spoke on "Are the Feds Waving Goodbye to CLIA Regulations of Clinical Laboratory Testing?" The lecture was followed by a dinner at Toto's Restaurant on Locust Street. The June meeting was the keynote lecture for the workshop on "Introduction to Molecular Diagnostic Techniques." The October meeting was preceded by three graduate student presentations in the afternoon.

The thirteenth Philadelphia Infection and Immunity Forum was held 30 May at the Drexel University, Queen Lane Campus. Bettina Buttaro and Rick Rest chaired the forum.

There was a workshop on "Introduction to Molecular Diagnostic Techniques" in June.

The thirty-fifth November symposium on "Antimicrobial Resistance & Emerging Infections: A Public Health Perspective" was held at the Biomedical Research Building at the University of Pennsylvania. The symposium was co-chaired by Irving Nachamkin and James Poupard.

2005

The year started with the announcement that the student chapter included sixty-five students and that they were incorporated into the activities of the Branch. The graduate students include members from the various Philadelphia Medical Schools. Their activities include organizing a special meeting to present their research projects and playing an active role in formulating the topics and organization for the annual Philadelphia Infection and Immunity Forum. The student chapter has become a significant element in making our Branch one of the most active Branches of the ASM. Toby Eisenstein provided a driving force in making the student chapter an intimate and crucial part of the Branch.

On 28 January, the Branch lost another significant microbiologist, Dr. Amedeo Bondi, sixteenth Branch President (1950-51). In 1947, he was appointed the first Chairman of the Department of Microbiology at Hahnemann Medical College and also served as Dean of the Graduate School. He was active with the Branch for his entire career and mentored many Branch members. On 25 February, Dr. Georganne K. Buescher, of Thomas Jefferson University, died. She was a graduate of the first class of the M.S. Clinical Microbiology Program at Thomas Jefferson University and was a driving factor in continuing that program, which supplied a continuing source of new graduates practicing Clinical Microbiology in the Philadelphia area.

There were eight meetings in 2005. The tenth Distinguished Branch Member Lecture in honor of Josephine Bartola was held in April. Caroline C. Johnson spoke on "A Perspective on Disease Control in Philadelphia." The lecture was followed by a dinner at McCormick & Schmick's on Broad Street. The October meeting was preceded in the afternoon by graduate student presentations on their research projects.

The fourteenth Philadelphia Infection and Immunity Forum was held 24 June at the Drexel University, Queen Lane Campus. Rick Rest and James Burns coordinated the forum.

There was a workshop on "Medical Mycology: A Refresher Course for Labratorians" in June.

The thirty-sixth November symposium entitled "STDs and Other Genital Tract Infections: Current Status and Future Trends" was held at the Biomedical Research Building at the University of Pennsylvania. The symposium was co-chaired by Richard Hodinka and Donald Jungkind.

2006

There were eight meetings in 2006. February was a Waksman Foundation Lecture. The eleventh Distinguished Branch Member Lecture in honor of Richard Crowell was held in May. Emilio Emini spoke on "The Importance and Challenge of Vaccine Development." The lecture was followed by a dinner at Varalli Restaurant on Locust Street. The June lecture was the keynote address for the workshop on "Food and Waterborne Intestinal Parasites." The December meeting was preceded by three local graduate student presentations on their thesis research.

The fifteenth Philadelphia Infection and Immunity Forum was held 23 June at the Drexel University, Queen Lane Campus. Simon Knight was forum coordinator.

There was a workshop on "Food and Waterborne Intestinal Parasites" in June.

The thirty-seventh November symposium on "Emerging Community and Healthcare-Associated Infectious Diseases" was held at the Biomedical Research Building at the University of Pennsylvania. The symposium was co-chaired by Linda A. Miller and Ebbing Lautenbach.

2007

On 25 March, one of the most prominent Philadelphia experts on infectious disease and microbiology, Dr. Robert Austrian, died. He was world famous for his work on pneumococcus. In 1976, he demonstrated that the pneumococcal vaccine he and his co-workers developed could significantly reduce mortality, especially in elder patients. He was a long time Branch member and made several presentations at the regular Branch meetings and functions. On 8 May, Dr. Russell Schaedler died. He was a Jefferson Medical College graduate in 1953 and long-time Chairman of the Department of Microbiology at Thomas Jefferson University. It was announced in

November that Dr. Herman Friedman, the twenty-sixth Branch president (1970-71), died. He was born, educated and remained in Philadelphia after getting his PhD. He was on the Microbiology Faculty at Temple University and Head of the Clinical Microbiology and Immunology Department at Albert Einstein Hospital. He was very active in the Branch. He initiated the first Branch Basic Science Symposium and chaired many of the symposia that followed. He was also active in forming the Clinical Microbiology Section of the Branch. In 1978, he left Philadelphia to become Chair of the Department of Medical Microbiology and Immunology at the University of South Florida, College of Medicine, but he returned to Philadelphia several times to participate in Branch functions.

There were nine meetings in 2007. The June meeting was held at the Public Health Laboratories in Lionville, Pa., and the lecture served as the keynote address for the workshop on "TB: Challenges for the Laboratory." The December meeting was preceded by three local graduate student presentations on their thesis research.

The sixteenth Philadelphia Infection and Immunity Forum was held 11 May at the Drexel University, Queen Lane Campus. Simon Knight was forum coordinator.

There was a workshop on "TB: Challenges for the Laboratory" in June.

2008

In January, Josephine Bartola died. Jo Bartola served the Branch as legal adviser for many years. She revised the Branch bylaws several times and served on almost all Branch symposia from the early years up to her retirement. She continued to be the Branch legal adviser after her retirement.

There were eight meetings in 2008. The Branch student chapter conducted a career roundtable session on 15 January. New bylaws were approved at the April meeting. The June meeting was presented by David Livermore, London, UK, on emerging resistance in gram negative organisms. The December meeting was preceded by three graduate student presentations on their thesis research.

The seventeenth Philadelphia Infection and Immunity Forum was held 9 May at the Drexel University, Queen Lane Campus. Simon Knight was forum coordinator.

The thirty-eighth November symposium on "Challenging Issues in Antimicrobial Resistance" was held at the Biomedical Research Building at the University of Pennsylvania. The symposium was co-chaired by Linda A. Miller and Anna Feldman-Rosen.

2009

At the February Executive Committee meeting, it was decided to eliminate the Branch Newsletter in the current form and replace it with some type of web based communication method. A sub-group to design a new website/blog to replace the newsletter was formed.

There were seven meetings in 2009. The December meeting was organized by the Branch student chapter. The regular meeting was preceded by four graduate student presentations on their thesis research.

The eighteenth Philadelphia Infection and Immunity Forum was held 8 May at the University of Pennsylvania School of Medicine Auditorium on Currie Blvd. Simon Knight was forum coordinator.

The thirty-ninth November symposium on "Molecular Testing in Clinical Microbiology: Advancements, Challenges, Future Directions" was held at the Dorrance H. Hamilton Building at Thomas Jefferson University. The symposium was co-chaired by Zi-Xuan Wang, Donald Jungkind and Laura Chandler.

Summary of the Branch in the 2000s

The Branch maintained the well established regular Monday evening meeting series throughout the decade. The Philadelphia Infection and Immunity Forum was held each year, and Branch student chapter members accepted almost full responsibility for organizing these meetings. They also organized special career development seminars, and by the end of the decade, also took responsibility for organizing one of the regular scheduled Monday evening meetings each year. In addition to an increase in the activities associated with the student chapter, the education committee was very active and initiated several innovative programs throughout the decade. There was a symposium held each year, except for the year 2007, which is the first year since 1969 that the Branch did not have at least one symposium. Also, starting in 2006, the symposium was reduced to a one day event. The change in the symposium series was due to competition from outside educational venues, a lack of support from government agencies that previously provided assistance for registration and continuing education credits, and less support from the pharmaceutical industry, which faced increased government restrictions on educational grants. There were seven workshops from 2000 through 2007. Workshops were not held in 2008 or 2009 due to organizational and scheduling issues. The Annual Distinguished Branch Member Lectureship was held each year through to 2006. This lecture series was initiated in 1996 as a continuation of the Annual Stuart Mudd Lecture Series, which was initiated in 1976. Therefore, 2006 marked the last of these annual meetings, which lasted for thirty years. In 2002, the decision was made to save mailing expenses by switching from a print version of the newsletter to an electronic version, with a printed version available to those members who did not want an electronic version. Then, in 2009, it was decided to eliminate the newsletter completely and replace it with some form of web based communication. A committee was formed to chart new ways to communicate with Branch members. Therefore, the last newsletter of 2008 marked the end of the most recent series of continuous newsletters that started in 1977.

It is apparent from this summary that many changes occurred, especially toward the end of the decade. There were various reasons for these changes; some were cost cutting efforts and others were related to a lack of volunteers on several committees. The need to redefine the level of activity that the Branch is willing to sustain, and the solution of expanding the committees with new active members, are the two biggest challenges which must be addressed in 2010.

CHAPTER THIRTEEN

Microbiology In Philadelphia: 2010 And Beyond

The Eastern Pennsylvania Branch

The first official Branch activity in 2010 was on 25 January with a presentation on the rotavirus vaccine at the first regular Monday evening meeting for this year. Therefore, 2010 started in the traditional way with the regular monthly meeting program. However, the start of this second decade brings with it a time for some critical decisions on the part of the Branch Executive Committee. The most important decision was related to replacing the newsletter. Until 2009, the newsletter was an important element in creating a sense of community among a diverse membership from several subspecialties of microbiology. In 2010, the question that had to be addressed was how to replace the newsletter with something that will improve communication and stimulate Branch members to become active in the organization. The variety of electronic communication systems currently available, combined with the resourcefulness of the committee, should result in a suitable replacement. It was anticipated that this will take some adjustment, especially with the older Branch members who prefer classic newsletters.

A second issue was how to attract new active members, especially from a variety of specialty areas within microbiology. This will be critical if the Branch is to continue with the extent of activities it has provided in the past. The symposia, workshops, forums, special seminars and social events, all require active members to make these events successful. In the past, a core of workers have filled this need; however, as these members have retired, or moved from the area, a sufficient number of new active members have not been recruited. The Branch must now decide on a way forward by either recruiting new members or curtailing some activities. This became apparent in the later years of the previous decade. One possible solution is to focus on the monthly meeting program, and replace programs, such as hands-on workshops, with virtual programs or web-based discussion groups. The possibilities are there, and it has become obvious the Branch will need to experiment to find the best solution for a changing environment.

Two current programs characterize the type of activities that are important to the future growth of the Branch. One is the student chapter and its role with the annual Infection and Immunity Forum. The student chapter has been very successful in bringing graduate and post-doctoral students into active Branch participation. The Infection and Immunity Forum has been very successful in serving the needs of the student chapter, providing them with a sense of what it takes to organize the Forum. It also provides them with a vehicle for presenting their research. A second program of note is related to the activities of the Branch Education Committee. This committee has been successful in serving the needs of teachers at all levels in the Philadelphia area, by providing material to stimulate interest in microbiology and by creating a sense of community among biology oriented teachers in the Philadelphia area. In addition, the Education Committee has initiated a community outreach program designed to stimulate appreciation of microbiology in the general community. The focus and success of this committee is based on getting young students interested in microbiology.

The combined efforts of these two committees are a model of what can be done to attract students to the field of microbiology, from early education through the post-doctoral level. There is confidence among the Branch Executive Committee that the Branch has had to adapt to a changing environment several times over its ninety-year history, and it will continue to do so in the future. Decisions made in 2010 will be important in determining the role of the Branch in shaping the history of microbiology in the twenty-first century.

The Pharmaceutical Industry

The pharmaceutical industry in the United States traces its roots to the city of Philadelphia. The Pharmaceutical industry evolved in complexity during the same period that microbiology was developing in Philadelphia. Men like Joseph McFarland helped move the industry from one that focused on classic medicines to an industry that included biologics, such as antitoxins and vaccines. During the twentieth century, Philadelphia served as a hub for such companies as Merck Sharp and Dohme, Smith Kline & French, Wyeth, and many others. These industries not only attracted a large workforce of microbiologists to the Philadelphia area; the industry was a source for many Branch presidents and active Branch Executive Committee members. Pharmaceutical companies also were a source of financial support for innumerable educational programs in the Philadelphia area. During the last decade of the twentieth century, and first decade of the twenty-first century, the pharmaceutical industry went through several significant changes. Mergers resulted in some companies being eliminated and others moving their headquarters, or R&D operations, to other geographical areas; thus reducing the number of industrial oriented microbiologists in Philadelphia. An additional factor is in the area of antimicrobials. Most Philadelphia pharmaceutical companies had a strong emphasis in developing new antimicrobial drugs. In the twenty-first century, most companies eliminated, outsourced or greatly reduced their antimicrobial research, which resulted in a reduction of the number of microbiologists in the Philadelphia area. In addition, regulations concerning pharmaceutical company support for educational programs have become more complex, and in some cases, such support has been eliminated. These changes have also resulted in the pharmaceutical industry supporting the formation of several biotechnology companies in the Philadelphia area. Although this has helped to some extent, these companies do not have large R&D departments. The net result is a significant reduction in a valuable resource that the Philadelphia community has relied on in the past. Only as the twenty-first century progresses will we discover if any of these trends will be reversed.

The following is a summary of the "Pharmaceutical Industry in Philadelphia" from the "Decade in Review" section of the Philadelphia Inquirer of 27 December 2009:

"A key and prestigious sector for the region, it saw a flourishing of small biotechnology firms, along with recent and rapid consolidation by some of the biggest players. The industry underwent a transformation in the explosive growth of generic medications as blockbuster patents expired and drug-development pipelines seemed to run dry. The trend was behind Merck & Co. Inc.'s $41 billion acquisition of Schering-Plough Corp. in November 2009. **What's Gone:** Wyeth, which employs several thousand in the region, was swallowed in October by Pfizer Inc . . . Also gone is the designation of GlaxoSmithKline's large Philadelphia operation as a corporate headquarters."

Medical and Clinical Microbiology

The presence of the Philadelphia medical schools, along with their related graduate and undergraduate programs, continues to represent a major source of a strong microbiology presence in Philadelphia. Of the medical schools and institutions described in previous chapters, the University of Pennsylvania School of Medicine, its School of Dental Medicine and School of Veterinary Medicine, which played such a critical role in the early days of Philadelphia bacteriology, continue to be an important resource for microbiology in Philadelphia. Two other institutions within the University played a significant role in the development of microbiology in Philadelphia: the Laboratory of Hygiene, which is gone, and the William Pepper Laboratory of Clinical Medicine which remains. Likewise, Thomas Jefferson University and the Jefferson Medical College also make major contributions. The Medical College of Pennsylvania and Hahnemann Medical College have been united to form the Drexel University College of Medicine. Needless to say, Temple University School of Medicine, which got its start during the first years of the last century, made up for its later start. However, taken as a group, the presence of these prestigious institutions, along with their graduate and undergraduate programs, account for the most significant single contribution to the microbiology community in Philadelphia.

Additionally, the hospitals associated with these institutions remain as a source of employment for a variety of microbiologists in Philadelphia. However, as noted previously, many of the community hospitals that had microbiology departments, have either consolidated these laboratories with other hospitals, or appointed a staff pathologist as director of the microbiology departments. Also, the number of medical technology training programs in the Philadelphia area in 2010 is a fraction of what it was previously.

When all these factors are considered, as we go beyond 2010, the area can rely on a significant base of academic, medical and clinical microbiologists, but not at the level or numbers that were present in the past.

Research Institutions and Health Departments

Of the independent research institutions presented in previous chapters, the Wistar Institute of Anatomy and Biology is the only institute that has adapted and expanded its research base in microbiology and infectious disease research. Other independent research institutes, such as the Henry Phipps Institute, have been incorporated into a University, or have become extinct.

The Philadelphia and the Pennsylvania Health Departments continue their important role in the Philadelphia area. Some of their emphasis has shifted from a classic public health role to include preparations for potential emergencies relating to bioterrorism. It is apparent that this role will continue beyond 2010.

Beyond 2010

It is obvious that many of the features of microbiology in Philadelphia that developed during the twentieth century will remain as dominant factors as the new century unfolds. It seems apparent that, for microbiology to remain a key player in the new century, and for the Philadelphia microbiology community to grow, microbiologists with relatively new subspecialties relating to environmental studies, new energy sources, public health and homeland security, will need to be incorporated into the Philadelphia microbiology community. The Eastern Pennsylvania Branch of the ASM, which essentially helped to define microbiology in the past, must take steps to continue to stimulate students to pursue careers in microbiology, and continue to establish a sense of a microbiology community in order for microbiology to maintain its strong presence in the Philadelphia area.

CHAPTER FOURTEEN

Summary And Conclusions

Philadelphia has a rich history in science and medicine dating back to colonial times and extending into the nineteenth century. The roots of bacteriology, separate from other disciplines of science and medicine, can be traced to the debates on the subject that occurred in the 1870s in Europe. However, it was not until the 1880s that several Philadelphia physicians could be identified as the first generation of Philadelphia bacteriologists. Two institutions were prominent in those early years: Blockley Hospital and, especially, the University of Pennsylvania. One would think that it would be very difficult to get agreement on who to identify as the first significant bacteriologist in Philadelphia; however, there is agreement that this honor belongs to Edward Orem Shakespeare, of Blockley Hospital. Shakespeare and the other first generation bacteriologists did not start their careers with the intent to become bacteriologists, but they all made contributions in establishing bacteriology as a definable subject of study in Philadelphia.

It was the second generation of Philadelphia bacteriologists who essentially laid the foundation for the teaching of bacteriology at various medical schools and for conducting research at several Philadelphia institutions. One second generation Philadelphia bacteriologist, Alexander Abbott, was cofounder of the Society of American Bacteriologists, and several Philadelphians were charter members of that organization. Another member of this second generation was Joseph MacFarland. He helped establish a biological division of one of the first major pharmaceutical companies, which later became a major American industry.

In 1912, a group of Philadelphians interested in bacteriology and pathology felt the need to form what became known as the Microbiological Club. This enabled a small group of Philadelphians to discuss and debate the latest developments in bacteriology. In 1920, David Bergey assembled members of the Microbiology Club to determine if Club members were interested in forming a chapter of the Society of American Bacteriologists.

There was agreement, and Bergey was to be first of a long line of chapter/branch presidents. Regular chapter meetings were held at a variety of Philadelphia institutions, starting in 1920.

In the 1930s, the membership in the Branch became open to all who were interested, and meetings continued to be held on a regular basis. It was decided to assign numbers to each regularly scheduled meeting and to publish the meeting abstracts in the Journal of Bacteriology. In the 1940s, with events leading up to the entrance of the U.S. into World War II, and with Branch members joining the Armed Services, there were thoughts of canceling the regular Branch meetings. However, with the influx of researchers who were assigned by the Armed Forces to research positions in Philadelphia, coupled with an interest to keep up with the latest technology, not only was the meeting schedule maintained, meeting attendance increased considerably. In 1947, the Branch hosted the Annual Meeting of the Society of American Bacteriologists. Branch members introduced several innovations at this meeting, and there was a special session on the history of bacteriology in Philadelphia. In the 1950s, activities centered on providing top quality presentations at the monthly meetings that reflected current developments taking place in microbiology. There was an increase in the use of outside speakers and a tendency to present several talks on a single topic for many of the meetings. There was also a tendency to have panel discussions at several meetings.

In 1960, the Branch again hosted the Annual Meeting of the Society of American Bacteriologists. In 1961, the name of the Eastern Pennsylvania chapter of the Society of American Bacteriologists was changed to the Eastern Pennsylvania Branch of the American Society for Microbiology, to be consistent with the name change of the national organization which became official late in 1960. A significant change occurred in the 1960s. Increasing numbers of members identified themselves as Clinical Microbiologists. In 1967, the Clinical Microbiology members decided that the Branch was not adequately serving their needs, and they decided to split from the Branch. It took great effort on the part of all parties involved to keep this new Clinical Microbiology Group as a functioning part of the Branch. This effort paid off since the compromise was to form the Clinical Microbiology Section, run by an independent steering committee, but still part of the Branch. The goal of the Clinical Microbiology Section was to serve the needs of all clinical microbiologists in the Philadelphia area. This decision resulted in major changes in the scope of activities offered to Philadelphia microbiologists. Starting in 1967, a series of Clinical Microbiology Section meetings were held on a different evening, independent of the regular Branch meetings, and with a separate numbering system. Starting in 1969, a series of workshops and symposia were initiated. These continued for the next forty years and became a reflection of the topics that Philadelphians perceived as current and important. Clearly, 1969 was a significant year for the microbiology community in Philadelphia since this new level of activity set the standard, and the challenge, to build on what was established that year.

In the early 1970s, it became apparent that attendance at the Clinical Microbiology Section meetings was increasing significantly, and there was a significant decrease in attendance at the regular Branch meetings. In 1973, there was agreement that the Branch would be restructured to accommodate the Clinical Microbiology members and that there would be one executive committee to direct all activities. This resulted in the elimination of the Clinical Microbiology Section and restoration of a unified Branch. The unified Branch produced two sets of highly successful symposia: the annual clinical microbiology oriented November Symposium Series and several Basic Science Symposia. Branch members worked with two national publishing companies to produce two successful series of books based on the symposia. During this period, the Branch produced an innovative and widely used audiotape program for Clinical Microbiologists. This program was distributed widely, both in the United States and elsewhere, generating significant royalties for the Branch. The needs of bench technologists were served by holding at least two workshops each year, targeted to this audience. A new annual lecture series to honor Dr. Stuart Mudd was initiated in 1976 and continued each year until 1995.

All these activities, initiated in the 1970s, were carried over to the 1980s. In 1987, the Branch took the occasion of the 500th regular Branch meeting to celebrate sixty-seven years of continuous activity. The celebration included presentations on: the history of the Branch, microbiology in general, and historical events that occurred during the past seven decades.

In 1992, the Annual Infection and Immunity Forum was initiated and has been held each year to the present time. Also, in 1992, there was a special event held in honor of Smith Hall, the former Laboratory of Hygiene of the University of Pennsylvania and to honor Dr. Harry E. Morton. The meeting was held to support efforts to keep the Laboratory of Hygiene from being demolished. In 1995, after twenty continuous years, it was decided to replace the Stuart Mudd Lecture series with the Distinguished Branch Member Lecture Series, which started in 1996 and continued each year through 2006. In 1998, the Branch celebrated the 600th regular meeting with an afternoon session that included five Philadelphia microbiologists "Looking Back" at the last six hundred meetings and the microbiology represented at those meetings.

There was considerable activity in the first decade of the new century. However, especially toward the end of the decade, there were signs of change. No workshops were held during the last two years of the decade, and a symposium was not held in 2007. Also, the number of days of the symposia was reduced. The Branch student chapter became increasingly active and took on most of the responsibility for conducting the Philadelphia Infection and Immunity Forum, which was held each year. A significant decision was made in 2009 to discontinue the newsletter and replace it with a web-based communication system in 2010. There are many factors that need to be considered in the design of an adequate replacement for the newsletter. There is also a significant need to attract new active Branch members to serve on committees if the high level of Branch activity is to continue. Many of these issues will be addressed early in the next decade.

In 2009, the ASM General Meeting was held in Philadelphia for the tenth time. There was a **Milestones in Microbiology Dedication Ceremony** held prior to the start of the General Meeting. This was a special ceremony to dedicate a plaque designating the Laboratory of Hygiene, of the University of Pennsylvania, as the third site given this distinctive honor. Upon registration for the General Meeting, each attendee was provided with a tour guide brochure, and walking tour map, prepared by the Branch Education Committee. There was a special full-afternoon session that presented several talks on the history of microbiology in Philadelphia, and there was a special session on the current H1N1 influenza pandemic that was receiving considerable attention in the contemporary scientific and lay literature.

Meeting Topics and Philadelphia Institutions

In an effort to summarize some of the information presented in this history of Philadelphia microbiology, I would like to attempt to identify what topics, and what institutions, made contributions to Philadelphia microbiology over the years. One way is to look at the institutions associated with the forty-seven Branch presidents. Another is to look at the titles of the 702 regularly scheduled Branch meetings listed in the Appendix. These meetings reflect the interests of Philadelphia Microbiologists and identify the institutions associated with these presenters.

Branch Presidential Institution Summary: 1920 through 2010

(The number of presidents from each institution are in parenthesis.)

University of Pennsylvania (10)
Temple University (8)
Jefferson Medical College (6)
Hahnemann Medical College (4)
Women's Medical College/MCP (2)
Henry Phipps Institute (2)
Sharp and Dohme Laboratories (2)
Smith Kline and French Laboratories (2)
And one each from the following:
Albert Einstein Medical Center
Bryn Mawr College
Cooper Medical Center
Holy Redeemer Hospital
Main Line Clinical Laboratories
MetPath Laboratories
Philadelphia General Hospital
Philadelphia and PA Public Health Departments
Philadelphia V.A. Medical Center
Wyeth Laboratories

Philadelphia Institutions that Provided Speakers for the Regular Branch Meetings: 1940 through 2009

Category One—well represented (in order of most to least):
University of Pennsylvania
Temple Medical School
Jefferson Medical College
Hahnemann Medical College
Albert Einstein Medical Center
Children's Hospital of Philadelphia
Merck Sharp & Dohme
Smith Kline & French Laboratories
Wistar Institute

Category Two—moderately represented
Woman's Medical College (Medical College of Pennsylvania)
Saint Christopher's Hospital
Fox Chase Cancer Institute
Bryn Mawr College and Hospital
Philadelphia General Hospital
Wyeth Laboratories
Graduate Hospital

Category three—occasionally represented:
Abington Hospital
Delaware Hospital
Drexel University / Institute
Haverford College
Jewish Hospital of Philadelphia
Lankenau Hospital
La Salle College
Mulford Laboratories
Pennsylvania Hospital
Philadelphia Naval Hospital
Philadelphia/Pennsylvania Health Departments
Protez Pharmaceuticals
Saint Joseph's University
Sterling Pharmaceuticals
Valley Forge General Hospital
Will's Eye Hospital

Regular Scheduled Branch Meeting Summary

Percent of the major topics, by decade, of the regular meetings—1940 through 2009 (meetings 143 through 702):

	1940s	1950s	1960s	1970s	1980s	1990s	2000s
Bacteriology	16	23	21	41	25	23	36
Virology	7	13	19	27	22	11	17
Immunology and Vaccines	5	14	21	25	13	7	9
Biochemistry	3	7	20	11	3	4	13
Mycology	2	5	<1	6	<1	<1	11
Parasitology	2	<1	<1	5	7	6	11

As would be expected, topics relating to bacteriology consistently dominated the presentations. It is interesting that subjects relating to viruses, even in the 1940s, received considerable attention. In addition to the above topics, it is interesting to note some of the percentage of topics that were greater than 1 percent, but showed shifting interests or trends over the years. The following are some additional topics of note:

> The topic of **antimicrobials** appears in the 1940s as 10 percent of the presentations, and also was 10 percent in the 1950s, with antimicrobial susceptibility testing also appearing in the 1950s and a measurable focus on antimicrobial resistance starting in the 1980s. **Electron microscopy** accounted for 4 percent of the presentations in the 1940s and 1950s, and then becomes less that 1 percent; likewise, **air-sampling** accounted for 3 percent of the presentations in the 1940s, then drops off as an identifiable subject. **Serology** was an identifiable subject in the 1940s, accounting for 4 percent of the presentations, with 2 percent in the 1950s, and then merges with immunology and other subjects. The subject, **PPLO,** accounts for 6 percent in the 1950s, and 4 percent in the early 1960s, but merges into **mycoplasmas** by the end of that decade. In the 1960s, two subjects, **tissue culture** and **interferon,** each account for 1 percent of the presentations for the first time period. In the 2000s, two new subjects are encountered: **bioterrorism** which accounted for 9 percent of the presentations, and **genomics** which accounted for 4 percent of the presentations.

In conclusion, although this work is an attempt to summarize some key aspects of bacteriology and microbiology as it evolved in Philadelphia from 1880, the most important goal of this work is to encourage microbiologists to become aware of the important role Philadelphia played, and will continue to play, as we enter the next phase in the development of microbiology. This current work focused on two aspects: the early years of 1880 to 1920, and starting in 1920, the role of the Eastern Pennsylvania Branch, in both contributing to and recording this rich history. The Appendices attached to this work provide additional details on specific subjects and provide an indication of what is now contained in the Branch Archives Collection for those who want to further explore this rich history. In closing, I would like to pay tribute to all those Branch members of the past who "stuck things in boxes" and hope that future Branch members will continue to do so!

A Chronology of the Eastern Pennsylvania Branch Related History

Unless otherwise noted, the source for items listed in this chronology can be found in the corresponding book chapter covered by the time period noted.

1840-1849
1846
Joseph Leidy identified Trichina larvae in pork and discussed its relevance to the proper cooking of meat.

1850-1859
1851
Joseph Leidy wrote *Flora and Fauna within Living Animals*, which was published in 1854. This massive and well illustrated work described, in remarkable detail, the plants, animals, protozoa and worms that live in and on various forms of life. In this work, as well as in other shorter reports, he identified many new species of both parasitic and free living life forms.

1853
Joseph Leidy, at the age of thirty, was appointed Professor of Anatomy at the University of Pennsylvania. This marked the trend of opening significant academic appointments at the University to more youthful professors, who replaced the old more indoctrinated professors. This also opened the door for the inclusion of new ideas, such as the importance of new developments in pathology, the biological sciences, hygiene and bacteriology.

1856
Joseph Leidy discovered and described canine heartworm.

1860-1869
1865
Following the Civil War, hygiene was added to the University of Pennsylvania Medical School curriculum, which eventually opened the way for bacteriology to gain in importance. Henry Hartshorne became the professor of hygiene.

1866

The natural sciences reentered the University of Pennsylvania Medical School curriculum. These courses were taught by an "auxiliary medical faculty" and a certificate was granted to the students who completed the course work. Students had to pass an examination in all five subjects designated as the natural sciences, including biology. Many students demanded more formal recognition for this extra work and in 1874, the trustees decided to employ the high European distinction of Doctor of Philosophy to those students who fulfilled all the requirements. This title was not previously employed in the United States. However, in 1879, when Johns Hopkins University began to confer the PhD according to German standards, the University abandoned this ill conceived PhD degree for this purpose. Therefore, starting in 1881, a Bachelor of Science was granted for this program in place of the PhD.

1870-1879
1876

In recognition of the importance of pathology at the University of Pennsylvania, James Tyson was awarded the first Chair of Pathology. This department became responsible for instruction of medical students in bacteriology.

1876

Joseph Lister made a visit to Philadelphia to speak on his ideas relating to antisepsis. As noted by Maulitz, Lister's visit to Philadelphia was a success and he "gained a limited degree of acceptance for the 'bacterian' thesis" among the medical community."

1878

The University of Pennsylvania School of Dental Medicine was founded. This later resulted in an expanded need for faculty to teach bacteriology to dental students. Formal courses in bacteriology for dental students were initiated in 1896.

1879

Joseph Leidy published *Fresh-Water Rhizopods of North America* which is still considered a reference for this aspect of protozology.

1880-1889

1881

Henry F. Formad had a laboratory in the the University of Pennsylvania Medical Building, where he conducted animal studies on tuberculosis. George M. Sternberg used Formad's laboratory to isolate and identify an organism, *Micrococcus pasteuri*, as an organism in human saliva that caused septicemia in rabbits.

1882

Edward O. Shakespeare, who became recognized as Philadelphia's first bacteriologist, was appointed Pathologist at Blockley Hospital.

1883

The previously established small laboratory by Henry Formad, in the University of Pennsylvania Medical Building, is described as supplied with a complete outfit of materials and apparatus for the investigation of bacteria in their relation to infectious diseases and the study of lower fungi in general. (From the 1883-1884 School Catalogue.)

1884

The University of Pennsylvania School of Veterinary Medicine was founded. Formal courses in bacteriology were initiated for veterinary students in 1886.

1884

Textbooks on various aspects of bacteriology started to reach Philadelphia. McFarland notes that Kline's book, *Microorganisms and Disease*, was the first book in the English language that reached Philadelphia.

1884-1889

William Osler joined the medical faculty of the University of Pennsylvania and set up a very small clinical laboratory in the University Hospital where he conducted clinical tests and microscopic examinations on patient specimens. He also established a clinical laboratory in a small brick building on the Blockley Hospital property where postmortem examinations were held. There he studied seventy cases of malaria. He was one of the first investigators in America to confirm the presence of the malaria parasite in the blood of patients with the disease. He also taught medical students how to look for the tubercle bacillus shortly after Koch published his findings on this bacillus.

1885

The University of Pennsylvania Department of Biology was created with Joseph Leidy as the first professor of biology. This new department accentuated the need for medical students to take preclinical studies in the biological sciences.

1886 Formal courses in bacteriology for veterinary students were initiated at the University of Pennsylvania School of Veterinary Medicine.

1886-1889

M. V. Ball, a medical student at Jefferson Medical College, states that there was no formal bacteriology taught there but, on one occasion, Jacob M. DaCosta introduced Julius Salinger, who had just returned from Berlin. He remembered Salinger drawing the "comma bacillus" on the blackboard during his presentation.

1888

Samuel G. Dixon, after a year studying with Robert Koch, was appointed University of Pennsylvania professor of hygiene and proceeded to set up a laboratory in the Medical Department to do research in bacteriology. This laboratory is most likely the first in America devoted to the teaching of hygiene.

1888-1891

Samuel G. Dixon, according to his assistant, Dr. Seneca Egbert, fitted and improved the University of Pennsylvania Medical Laboratory to make it better suited for bacteriology research.

1889

Edward O. Shakespeare was appointed bacteriologist at Blockley Hospital.

1889

Juan Guiteras replaced James Tyson as University of Pennsylvania Professor of Pathology. His reputation was built on his work on yellow fever and the pathology of infectious diseases.

1889

Simon Flexner became Chair of Pathology and strengthened the need for bacteriology at the University of Pennsylvania.

1889

William Pancoast, Professor of Surgery and founder of the Medico-Chirurgical College, while visiting Pasteur in Paris, asked Pasteur for advice on the future of bacteriology and for help in selecting a professor of bacteriology for his college. Pasteur recommended Ernest Laplace, who came to Philadelphia to fill the position, and E. B. Sangee became Assistant in Pathology and Bacteriology at the school.

1889

Willoughby Dayton Miller, one of the first graduates of the The University of Pennsylvania Dental School, published *Die Mikroorganismen der Mundhohle* (Micro-organisms of the Human Mouth) which initiated further studies in the role of bacteria in dental disease.

1890-1899

1890

Juan Guiteras, who had succeeded James Tyson as the University of Pennsylvania Professor of Pathology, secured leave to go abroad to study and select new apparatus for equipping a future laboratory of bacteriology for instruction of students.

1890

While Guiteras was away, the instruction in pathology and bacteriology at the University of Pennsylvania was carried out by Henry Formad, assisted by Allen J. Smith. Smith made several improvements in the culturing and staining of bacteria. Another Formad assistant, J. Leffingwell, improved techniques for isolating bacteria from blood.

1890

Edward O. Shakespeare presented his thousand-page publication, *Report on Cholera in Europe and India*, to Congress. The American Medical Association hailed it as a "great work" and Shakespeare was regarded as the leading American expert on cholera.

1890-1891

After finishing their medical studies at the University of Pennsylvania, Samuel Stryker Kneass took courses at the Pasteur Institute, and Joseph McFarland studied at Heidelberg with Paul Ernst. They then returned to the University to do laboratory studies in bacteriology.

1891

William Pepper Jr. established an endowment for the establishment of the Pepper Laboratory of Clinical Medicine, which was to be part of the University Hospital system.

1891

The Michael V. Ball book *Essentials of Bacteriology* was published by the W. B. Saunders Company of Philadelphia. This book is most likely the first book on the subject written by a Philadelphian.

1891-1892

Guiteras made the newly enhanced bacteriology laboratory at the University of Pennsylvania available to J. McFarland and S. Kneass, upon their return from Europe.

1891-1892

The school catalogue announced that "through the liberality of a number of citizens of Philadelphia, the University of Pennsylvania has been enabled to establish a Laboratory of Hygiene. A large building, especially planned and fitted for this purpose, is now nearing completion and will be completely equipped and ready for use on 1 February 1892."

1891

Joseph Leidy died on 30 April.

1892

The Alexander C. Abbott book, *The Principles of Bacteriology: a Practical Manual for Students and Physicians*, is published by Lea Brothers and Company of Philadelphia. The book sold for more than forty years.

1892

The new University of Pennsylvania Department of Hygiene was formed with John S. Billings, Director, A. C. Abbott, First Assistant, Albert

A. Ghriskey, Assistant in Bacteriology, and James Homer Wright, the Thomas A. Scott Fellow in Hygiene. Billings, through his assistants, focused on teaching practical hygiene, while Abbott and his staff focused on teaching bacteriology.

1892

The University of Pennsylvania Laboratory of Hygiene building opened on the second day of February. The course entitled, "An Elementary Course in Bacteriology," began on the day the building opened and continued for eight weeks, five days per week. This ushered in the real beginning of bacteriology as a separate course of study in Philadelphia, with a faculty dedicated to teaching bacteriology.

1892-1893

The first course in bacteriology at the University of Pennsylvania Laboratory of Hygiene had three students: Samuel Stryker Kneass, Adelaide Ward Peckham, and Mazyck Porcher Ravenel.

1892

W. M. L. Coplin, Demonstrator of Pathology, is listed in the Class Book of Jefferson Medical College as giving a lecture on "Hygiene and Bacteriology." The subject continued as part of the pathology department until 1909.

1892

Henry F. Formad, one of the first generation Philadelphia bacteriologists, died on 6 June.

1892

Lawrence Flick founded the Pennsylvania Society for the Prevention of Tuberculosis. This is the first tuberculosis society in the United States and was unique in that it combined lay and professional membership to concentrate activities against a single disease.

1892-1893

Bacteriology appears for the first time in the announcement of the University of Pennsylvania, Laboratory Medical Department, as a subject to be pursued during the third year. The catalogue states that there is one lecture a week for six weeks. Guiteras was most likely responsible for the course, and J. McFarland assisted through 1895.

1892-1897

The only link between the University of Pennsylvania Laboratory of Hygiene and the medical department was that J. S. Billings was Director of the Laboratory of Hygiene and professor of hygienee in the medical department. With the resignation of Billings, A. C. Abbott became professor of hygiene in the medical department.

1893

J. M. DaCosta and D. Brandon Kyle conducted a private clinical laboratory in which bacteriology examinations were made, and some private instructions given to medical students from Jefferson Medical College, but this laboratory was not part of the College course offered at the College.

1893

The Wistar Institute of Anatomy and Biology, the nation's oldest independent biomedical institute, opened at its current location. The building was enlarged in 1897.

1893-1894

The bacteriology course taught at the University of Pennsylvania, Laboratory of Hygiene, had six students, one being David Hendricks Bergey. A. W. Peckham was listed as a student in a new "Course in Advanced Bacteriology."

1893-1894

The staff of the University of Pennsylvania, Laboratory of Hygiene, consisted of John S. Billings, Director, A. C. Abbott, First Assistant, and M. P. Ravenel, the Thomas A. Scott Fellow in Hygiene.

1894

Joseph McFarland accepted the position as director of the Laboratory of Biology at the H. K. Mulford Company. The company soon became a major player in the new field of biological products.

1894-1895

Two bacteriology courses were offered at the University of Pennsylvania Laboratory of Hygiene, with a staff consisting of John S. Billings, Director, A. C. Abbott, First Assistant, M. P. Ravenel, Assistant in Bacteriology, and D. H. Bergey, the Thomas A. Scott Fellow in Hygiene. There were twenty-one students, including Robert L. Pitsfield.

1895

J. McFarland is listed as Lecturer on Bacteriology for the course offered at the University of Pennsylvania Medical Department. The course consisted of six lectures and also included six laboratory demonstrations.

1895

The University of Pennsylvania's new William Pepper Laboratory of Clinical Medicine building is erected with the most modern apparatus for chemical, bacteriological and clinical investigation. Samuel S. Kneass was placed in charge of the laboratory of bacteriology. The William Pepper Laboratory has the distinction to be the first clinical laboratory in the U.S. to be associated directly with a medical clinic, and it was under the control of the Department of Medicine.

1895

A Jefferson Medical College catalogue mentions one lecture a week and one demonstration a week in "Bacteriology and Clinical Microscopy" for second-year students. It is assumed that this course was given by W. M. L. Coplin and assisted by Alonzo H. Stewart, as Demonstrator of Clinical Microscopy.

1895

The Philadelphia Health Department initiated a bacteriology laboratory on the seventh floor of City Hall, with B. Meade Bolton as Director of Bacteriology, Herbert D. Pease as First Assistant in Bacteriology, and W. J. Gillespie, Second Assistant in Bacteriology.

1895

The new Philadelphia Health Department bacteriology laboratory became operational in May. The initial focus was on diphtheria diagnosis and production of antitoxin. The horses employed for antitoxin production were located at a Philadelphia Fire Department station at Frankford but, after about a year, they were moved to a larger fire station in West Philadelphia.

1895-1896

The University of Pennsylvania Laboratory of Hygiene staff consisted of John S. Billings, director; A. C. Abbott, first assistant; M. P. Ravenel, assistant in bacteriology, with D. H. Bergey listed as assistant in cemistry. There were thirteen students, including Lydia Rabinowitsch, PhD.

1896

This appears to be the year that regular instruction in bacteriology began at Jefferson Medical College. Coplin and Stewart conducted the instruction to the medical students. Coplin became professor of pathology and bacteriology later that year.

1896

Lydia Rabinowitsch, PhD, became director of the bacteriological laboratory at the Woman's Medical College of Pennsylvania. This was the first time any research or instruction was given in bacteriology at the college.

1896

The elective laboratory course in bacteriology for medical students at the University of Pennsylvania Laboratory of Hygiene, conducted by A. C. Abbott and his staff, became so popular in this year that the entire medical class was required to take this course.

1896

The following is stated in the University of Pennsylvania Dental School Faculty minutes, dated March 3: "Believing that a sufficient knowledge of Bacteriology is essential to the proper education of a dentist, it is recommended that the Trustees provide for a semi-popular course of about six lectures to be given to the senior class of the Department of Dentistry as soon as possible during the present session, which course should not be obligatory or include a final examination. But the ensuing year and thereafter, it is recommended and requested that instructions both in didactic and in laboratory work shall be given to third year dental students in Bacteriology so far as it is adapted to their needs, and that the study of it be made an obligatory part of the third year curriculum."

1896

At the University of Pennsylvania Dental School, arrangements were made with the Laboratory of Hygiene for A. C. Abbott to teach the bacteriology course to third year dental students. In later years, D. H. Bergey also taught in the dental school.

1896

At the Medico-Chirurgical College, Laplace became Professor of Surgery, and Joseph McFarland was appointed to the Chair of Pathology and Bacteriology. According to McFarland, his appointment marked a new era for the college "for it was accompanied by the establishment and equipment of modern laboratories for both pathology and bacteriology, with excellent microscopes for all of the students and a complete bacteriological outfit that enabled the students to make and sterilize their own media, plant their own cultures and perform all of the usual staining methods. With the erection of the new laboratory building a few years later, Robert L. Pitfield was appointed Demonstrator of Bacteriology.

1896

Joseph McFarland's book, *The Pathogenic Bacteria*, was published by W. B. Saunders Company of Philadelphia. It was a popular textbook that went through nine editions. The last edition was published in 1919.

1896

At the Philadelphia Polyclinic and College for Graduates in Medicine, H. D. Pease took over bacteriology duties after replacing Samuel Kneass.

1896-1897

The staff of the University of Pennsylvania Laboratory of Hygiene consisted of A. C. Abbott, director, and D. H. Bergey, first assistant. There were twenty students.

1896-1897

The first reference to bacteriology appears in the University of Pennsylvania Veterinary School Annual Catalogue as "Practical Bacteriology." The course for second year students was scheduled to meet after February first on Tuesday and Wednesday afternoons and, for third-year students, before February first on Monday mornings. Juan Guiteras was Professor in both the Medical and Veterinary Departments at that time, but the announcement did not state where or by whom the instruction was given.

1896-1899
David Reisman replaced J. McFarland as Lecturer for the bacteriology course taught in the medical department of the University of Pennsylvania.

1897
At the Philadelphia Health Department Bacteriology Laboratory, Director Bolton was forced to vacate his position, and A. C. Abbott became the new Director in February. He greatly expanded the work performed in the Department. Pease left his position and was succeeded by W. G. Gillespie. Alonzo H. Stewart was put in charge of the Widal testing laboratory and took the position vacated by W. G. Gillespie.

1897
At Jefferson Medical College, bacteriology was taught two days a week for six weeks by Coplin, assisted by David Biven and Randle Rosenberger.

1897-1898
At the University of Pennsylvania School of Veterinary Medicine, J. Guiteras taught the bacteriology course to third year students.

1898-1899
Third-year University of Pennsylvania School of Dental Medicine students take bacteriology for two, three hour periods per-week.

1898-1899
At the University of Pennsylvania School of Veterinary Medicine, M. P. Ravenel is listed as Demonstrator of Veterinary Bacteriology. At the end of 1899, Guiteras resigned his position to return to Cuba, and Simon Flexner succeeded him as Professor of Pathology in both the Medical and Veterinary Departments.

1898
At the Woman's Medical College of Pennsylvania, Lydia Rabinowitsch, PhD is appointed as the first Associate Professor of Bacteriology in April. In June, she is appointed Professor of Bacteriology. At the end of the teaching

session, she leaves for Germany to work with Robert Koch and does not return to Philadelphia.

1898

Adelaide Ward Peckham became Associate Professor of Pathology in charge of Bacteriology at the Woman's Medical College of Pennsylvania. Later that year, she became Professor of Bacteriology when Lydia Rabinowitch vacated that position. Peckham held that position until 1919.

1899

At the University of Pennsylvania, Simon Flexner succeeded Guiteras in charge of Pathology and insisted that pure bacteriology, that was still taught in the medical department, had to be taught by a bacteriologist. (The concept was that "practical" bacteriology was taught in the Laboratory of Hygiene.)

1899

At the University of Pennsylvania, A. C. Abbott became professor of hygiene and Bacteriology at the Laboratory of Hygiene and later that same year, he became Professor of Bacteriology in the medical department.

1899

L. Rabinowitsch officially resigned her position at the Woman's Medical College of Pennsylvania after returning to Berlin to continue work on tuberculosis at the Koch Institute.

1899

Alexander C. Abbott's book, *The Hygiene of Transmissible Diseases*, was published by W. B. Saunders of Philadelphia. There was a second edition published in 1901.

1899

The Society of American Bacteriologists was founded, with A. Abbott one of if the three founders of the new Society, and Philadelphians, like D. Bergey and J. McFarland, were charter members.

1899-1900

At the University of Pennsylvania School of Veterinary Medicine bacteriology was taught to the second year students at the medical department, most likely by A. C. Abbott, with the laboratory instructions given by Ravenel in the Veterinary Department.

1900-1909
1900
Joseph McFarland resigned his position as Laboratory Director of the Laboratory of Biology at the H. K. Mulford Co. and was replaced by J. J. Kinyoun, of the US Public Health Service, when the laboratories moved to Glenolden and Arthur P Hitchens joined the laboratory staff.

1900
Edward Orem Shakespeare, Philadelphia's first identifiable bacteriologist, died on 1 June at the age of fifty-four.

1903
Simon Flexner, as chair of pathology, University of Pennsylvania, separated bacteriology from pathology with A.C. Abbott in charge of bacteriology. The medical student classes were transferred to the Laboratory of Hygiene for their instructions, and D. H. Bergey was advanced to the position of Assistant Professor of Bacteriology to oversee their training in bacteriology.

1903
Randle C. Rosenberger was made Associate in Bacteriology at Jefferson Medical College and Assistant Professor of Bacteriology the next year.

1903
The work carried out by Lawrence Flick resulted in the founding of the Henry Phipps Institute for the Study, Prevention and Treatment of Tuberculosis.

1903
The Annual Meeting of the Society of American Bacteriologists was held in Philadelphia for the first time.

1903
Alexander Abbott was appointed Chief of Philadelphia's Bureau of Health, a position he held until 1909.

1904

The Annual Meeting of the Society of American Bacteriologists was held in Philadelphia for the second time.

1905

Arthur P. Hitchens replaced J. J. Kinyoun as Director of the Laboratory of Biology at the H. K. Mulford Co. and held that position until 1918.

1906

Samuel Dixon served as Commissioner of Health of Pennsylvania from 1906 until his death in 1918. Under his tenure as Health Commissioner, the Division of Laboratories was established at the Medical School of the University of Pennsylvania.

1907

The book by Robert Lucas Pitfield, *A Compendium on Bacteriology Including Animal Parasites*, was published.

1908

Robert Koch, Albert Calmett, and others gave brief talks in Philadelphia on 25 September as part of the International Congress on Tuberculosis, which was held in Washington, D. C. as well as in Philadelphia.

1909

Bacteriology was separated from pathology at Jefferson Medical College. W. Coplin remained in the Pathology Department, while Randle C. Rosenberger became Professor of Bacteriology, and he was first to hold the Chair of Bacteriology and Hygiene, a position he held until his death in 1944.

1910-1919

1910

The Henry Phipps Institute ownership is transferred to the University of Pennsylvania.

1910

The Department of Comparative Pathology and Parasitology is established at the University of Pennsylvania by Allen Smith.

1912

The Microbiological Club, a forerunner of the Eastern Pennsylvania Branch, is organized by A. Parker Hitchens, who assembled a group of twelve to fifteen individuals interested in either bacteriology or pathology.

1914

The Annual Meeting of the Society of American Bacteriologists was held in Philadelphia for the third time.

1915

Dr. David H. Bergey became president of the Society of American Bacteriologists.

1915

John A. Kolmer published his book, *Practical Text-book of Infection, Immunity and Specific Therapy*.

1916

A. P. Hitchins became managing editor of the Journal of Bacteriology during Dr. Bergey's term as President of the Society of American Bacteriologists.

1917

The Society of American Bacteriologists voted to have official Branches.

1918

The influenza pandemic, in which 13,000 Philadelphians died, demonstrates significant deficiencies in the Philadelphia Public Health and political systems in the city. (Reference 1)

1918

Samuel Gibson Dixon, one of the first generation Philadelphia bacteriologists died.

1920-1929

1920

On 24 February, Dr. Bergey called on several members of the Microbiological Club, and some new recruits, to his office in the Laboratory of Hygiene for the purpose of forming the Eastern Pennsylvania chapter/Branch.

1920

Dr. David H. Bergey, University of Pennsylvania, become the first president of the chapter and served in this office through 1922.

1920

At the second chapter/Branch meeting, held in March, Dr. C. P. Brown presented the first constitution and a set of bylaws.

1921

The Annual Meeting of the Society of American Bacteriologists was held in Philadelphia for the fourth time.

1921

The book by J. McFarland *Fighting Foes Too Small to See* was published.

1923

Bergey's Manual of Determinative Bacteriology was published.

1923

Claude P. Brown became secretary-treasurer of the Branch and held that position until 1936.

1923

Dr. A. Parker Hitchens is listed as the first Branch guest speaker; his topic was "Emergency Production of Potable Water."

1923

Randle C. Rosenberger, Jefferson Medical College, became the second chapter/Branch President through 1924.

1924

A. Parker Hitchins became President of SAB.

1924

Ernest Laplace, one of the first-generation Philadelphia bacteriologists, died on 15 May.

1925
A. C. Abbott, University of Pennsylvania, became the third chapter/Branch president through 1926.

1926
The annual meeting of the Society of American Bacteriologists was held in Philadelphia for the fifth time.

1927
Eugene L. Opie, Henry Phipps Institute, became the fourth chapter President through 1928.

1928
Abbott retired from the University of Pennsylvania, and Bergey became professor of hygiene and bacteriology, a position he held until his retirement in 1931.

1928
Joseph L. T. Appleton, Jr., Univ. of Pennsylvania Dental School, became the fifth chapter President through 1929.

1930-1939
1930
Joseph D. Aronson, Henry Phipps Institute, became the sixth chapter President, and served until October 1931.

1931
The Department of Bacteriology at the University of Pennsylvania School of Medicine was reorganized with Dr. Stuart Mudd as acting chairman.

1931
Jefferson H. Clark, Philadelphia General Hospital, became the seventh chapter President through 1934.

1932
Dr. Bergey retired from the University of Pennsylvania and became the Director of Research in Biology at the National Drug Company, where he developed production of tetanus toxoid.

1933

The Annual Meeting of the Society of American Bacteriologists was held in Philadelphia for the sixth time.

1934

A provision is approved to have branches represented on the Society of American Bacteriologists Council.

1934

Evan L. Stubbs, University of Pennsylvania Veterinary School, became the eighth chapter President through 1935.

1935

Dr. Randle Rosenberger, of Jefferson Medical College, became the first Branch Councilor.

1935

A. C. Abbott, third chapter President (1925-26), died at age seventy-five.

1936

The first publication of the Proceedings of the Local Branches in the Journal of Bacteriology was authorized in 1935. The first publication of the Proceedings of the Eastern Pennsylvania chapter was in a 1936 issue of the Journal of Bacteriology (31: 439) which reported on the 28 January 1936 chapter meeting at Temple University School of Medicine.

1936

Credit is given to David Bergey, Randle Rosenberger and Claude Brown for keeping the Branch going during "these precarious times" prior to 1936. In 1936, two people did much to rejuvenate the Branch, Drs. Carl Bucher and Harry Morton were determined to bring about significant changes. Dr. Bucher was elected President in 1936, and Dr. Morton became secretary-treasurer in 1937. This ushered in a period of regrowth.

1936

Branch membership became open to anyone who was interested in bacteriology, immunology or the allied sciences, and a permanent location for the Branch meetings was arranged at the Philadelphia County Medical Society Building at the Southeast corner of Twenty-first and Spruce Streets.

1936
Carl Bucher, Thomas Jefferson Hospital, became the ninth chapter president through 1937.

1936
The first significant material is deposited in the Branch Archives Collection.

1937
In September, Harry Morton started what he referred to as an Annual Newsletter which he designated volume I, number 1. Between 1937 and 1946, eight issues were published.

1937
The first recorded time that the designation *symposium* was used for a Branch meeting was in March at the 123rd meeting.

1937
Dr. Morton initiated the practice of assigning a number to each meeting. By searching the past records he determined that the 25 May meeting was the 125th Branch meeting and was the first meeting to be listed by title and number, a practice that continues to the present time.

1937
Dr. Bergey, who is credited with forming the Branch, died at the age of seventy-seven.

1937
The Joseph McFarland paper, entitled "The Beginning of Bacteriology in Philadelphia," is published.

1938
Mr. Christopher G. Roos, Sharp and Dohme Laboratories, became the tenth chapter president. He was the first nondoctoral president and the first president from a pharmaceutical company. He served through 1939.

1938
A. P. Hitchens resigned his position in the U.S. Army to return to Philadelphia as Professor of Military Science and Tactics at the University of Pennsylvania.

1938

The names and addresses of all individuals on the chapter mailing list were put on address-o-graph plates for more efficient mailing. The programs were printed on U.S. penny postal cards and then run through the address-o-graph.

1938

Mr. Roos made the announcement that the original minutes of the first meetings of the Society of American Bacteriologists had been located at Sharp and Dohme Laboratories. It was known that some records had been left at the laboratories in Glenolden in 1917 by the society's secretary, Dr. Hitchens, but searches for the documents had been fruitless. In the spring of 1938, laboratories in the main building were remodeled, and provisions were made for safe keeping of valuable records. This resulted in the discovery of the minutes of the early Society's first meeting. A record book contained the original transcript of the Society's constitution, minutes of the meeting of organization, names of the Society's organizers with a list of the names of the charter members, the first to the eleventh programs presented at the annual meetings and the minutes of transactions at those meetings. The records covered the period from 16 October 1899 to 30 December 1909.

1938

Lawrence Francis Flick, the last member of the first generation of Philadelphia bacteriologists, died.

1939

To mark the 200th anniversary of the University of Pennsylvania, a Department of Public Health and Preventive Medicine was established, with A. P. Hitchens as Chairman of the Department as well as the George S. Pepper Professor of that Department.

1940-1949

1940

Stuart Mudd, University of Pennsylvania, became the eleventh chapter president, and served through 1941.

1940

The officers of the chapter consisted of a president, secretary/treasurer, councilor and a councilor-alternate. The only standing committee was a program committee.

1940

The Branch was asked by the Commonwealth of Pennsylvania to demonstrate laboratory methods at the Pennsylvania Medical Society meetings held at the Bellevue-Stratford Hotel from 30 September to 3 October.

1941

In preparation for the events of WWII, Dr. Morton noted in the Branch newsletter: "It is also felt that in view of emergencies which will arise from the unsettled condition of the world that it would be well to consider procedures which would be very important in such emergencies."

1942

William A. Kreidler, Jefferson Medical College, became the twelfth chapter president and served through 1943.

1942

Many Branch members served in the armed forces of World War II. Many were in positions where they had to train personnel in microbiology, and they had very little material to work with. Dr. Morton urged Dr. Selman A. Waksman, then President of the Society of American Bacteriologists, to appoint a committee to fulfill these needs. Waksman appointed the Committee on Materials for Visual Instruction in Microbiology with Dr. Morton as Chairman, a position he held until 1971. The committee collected materials for lantern slides, photographic prints, and motion picture films. Several years later, when the Committee was disbanded, its activities were placed under the Board of Education and Training (BET). The name for this board was proposed by Dr. Earle H. Spaulding.

1944

Dr. Morton noted the following: "In spite of the severe restrictions being placed upon the release of certain research data, the Program Committee has already arranged excellent programs for the first two meetings in 1944. As results of confidential research projects become available throughout the year, the Committee plans to have them reported promptly to the Branch. Because so many of our active members are serving in the Armed Forces at the present time, it will be difficult to obtain the full quota of papers by our own local members. The fields of Bacteriology and Parasitology are more important today than they ever were before. Therefore, it is appropriate, if not essential, that the Branch's activities in no way be reduced, even though individually the accelerated pace of our routine duties may leave us little time."

1944

Earle H. Spaulding, Temple University School of Medicine, became the thirteenth chapter President and served through 1945.

1944

Upon the death of Randle C. Rosenberger, William A. Kreidler was appointed temporary Chairman of Bacteriology at Jefferson Medical College.

1945

Stuart Mudd became president of the Society of American Bacteriologists.

1945

J. McFarland died at age seventy-seven.

1946

Harry E. Morton, University of Pennsylvania, became the fourteenth chapter President and served through 1947.

1946

Kenneth Goodner became the second chairman of bacteriology at Jefferson Medical College, a position he held until 1967.

1946

Claude P. Brown became director of the Bureau of Laboratories of the Pennsylvania Department of Health.

1947

The annual meeting of the Society of American Bacteriologists was held in Philadelphia for the seventh time. Several innovations were introduced at this meeting by the Local Organizing Committee.

1947

There was a "History of Bacteriology in the Philadelphia Area" session at the Annual Meeting, organized by Dr. Bucher, with seven presentations. These talks were transcribed, and a copy placed in the Branch Archives.

1947

It is noted that The Eastern Pennsylvania chapter is one of the most careful in regard to the collection and publication of abstracts in the Journal of Bacteriology, and the chapter periodically reprinted these abstracts as bound collections.

1947

A membership list was prepared as of January. Of the three hundred names on the list, ninety are starred to indicate that they are members of the Society of American Bacteriologists.

1948

A. P. Hitchens succeeded Claude P. Brown as Director of the Bureau of Laboratories of the Pennsylvania Department of Health.

1948

William Verwey, Sharp and Dohme Laboratories, became the fifteenth chapter President and served through 1949.

1950-1959

1950

Amedeo Bondi, Temple University School of Medicine, became the sixteenth chapter president and served through 1951.

1952

James Harrison, Department of Biology, Temple University, became the seventeenth chapter president and served through 1953.

1954

Ruth Miller, Women's Medical College, became the eighteenth chapter president. She is the first female chapter president and served through 1955.

1955

Alonzo H. Stewart died at the age of eighty-eight. His death takes on special significance since he is the last of the second generation Philadelphia bacteriologists to die, and this date marks the end of the period influenced by this second generation group of Philadelphia bacteriologists.

1955

Starting in October, the regular monthly meetings were moved from the Philadelphia County Medical Society building to Medical Alumni Hall in the Maloney Building of the University of Pennsylvania, at Thirty-sixth and Spruce Streets.

1956

Kenneth Goodner, Jefferson Medical College, became the nineteenth chapter president and served through 1957.

1958

Morton Klein, Temple University School of Medicine, became the twentieth chapter president and served through 1959.

1960-1969

1960

Joseph S. Gots, University of Pennsylvania, became the twenty-first chapter president and served through 1961.

1960

In May, the Branch was host for the Annual Meeting of the Society of American Bacteriologists for the eighth time. The meeting was quite successful and, once again, the Local Organizing Committee, chaired by Dr. Harry Morton, was innovative in creating events for teachers and high school students. A successful session was held on the History of Bacteriology in the Philadelphia Area which covered the period of 1947 to 1960. This encouraged Dr. Morton to prepare a document that identified the institutions in Philadelphia associated with microbiology. This included detailed information on microbiology in Philadelphia at seventeen educational institutions and departments, twelve clinical and research institutions, and two pharmaceutical companies. A copy of this report is on deposit in the Branch Archive Collection.

1960

In December, the Society of American Bacteriologists name was changed to the American Society for Microbiology.

1961

In October, the name of the Eastern Pennsylvania chapter of the Society of American Bacteriologists was officially changed to the Eastern

Pennsylvania **Branch** of the American Society for Microbiology. Until this time, it was the only Branch that called itself a Chapter, and this was looked upon with disfavor by the National Society.

1962

L. Joe Berry, Bryn Mawr College, became the twenty-second Branch President and served through 1963.

1962

Funding was received from seven industrial sponsors, each contributing $50, to initiate a guest lecturer series, the first to be held in November, and was designated the Industry Sponsored Lecture Series. These lectures were held each year through 1971.

1962

Christopher G. Roos, tenth Branch president (1938-39), died.

1964

Harold S. Ginsberg, University of Pennsylvania, became the twenty-third Branch President and served through 1965.

1966

Albert G. Moat, Hahnemann Medical College, became the twenty-fourth Branch President and served through 1967.

1967

A decision was made by several Branch members to bring microbiologists who were interested in clinical microbiology together to discuss how the Branch could better serve the needs of clinical microbiologists. On 25 September 1967, approximately seventy people attended a special meeting. The meeting resulted in the formation of a splinter group within the Branch, which was designated as the Clinical Microbiology Section.

1967

In December, the Clinical Microbiology Section started to hold their series of meetings on Monday evenings. The regular Branch meetings continued to be scheduled on Tuesday evenings. The Clinical Microbiology Section was run by a steering committee, which initiated several new programs. There was also a joint committee formed to plan Branch symposia.

1968

George Warren, Wyeth Laboratories, became the twenty-fifth Branch president and served through 1969.

1968

The Branch initiated a program to attract sustaining members.

1969

The Clinical Microbiology Section started its program of workshops that continue to the present time. The initial goal was to hold two each year. The first workshop was entitled "Fluorescent Antibody Techniques in the Clinical Laboratory."

1969

The first Branch symposium was held in November. The Branch symposia series continues to the present time. Almost all the symposia for the next fifteen years resulted in a book published by various national publishers. The first symposium was entitled the "Northeast Regional Conference on Rubella."

1969

The office of vice president was initiated. The name of this position was changed to president-elect in 1979. Dr. James Prier was the first member to hold the position of vice president.

1970-1979

1970

Herman Friedman, Albert Einstein Medical Center, became the twenty-sixth Branch president and served through 1971.

1970

The secretary/treasurer position was split into two separate positions.

1972

James Prier, Pennsylvania Department of Health Laboratories, became the twenty-seventh Branch president and served through 1973.

1972

Topics In Clinical Microbiology was described as "A Major Audio-Visual Publishing Event" that occurred with the release of the program by Williams

& Wilkins Publishing Company. This was a joint project between the Eastern Pennsylvania Branch and Williams & Wilkins. The topics were described as a complete course in Clinical Microbiology and an official publication of the Eastern Pennsylvania Branch of the American Society for Microbiology. This was the culmination of a project that was started in late 1970 by members of the Clinical Microbiology Section of the Branch. Publicity on the topics describe it as a set of "twenty-four compact C-30 cassettes with eight cassettes in each of three color-coded volumes, a complete explanatory manual with references above and beyond the taped material, and a set of colored slides."

1972

In April, the annual ASM Meeting was held in Philadelphia for the ninth time. The *Topics in Clinical Microbiology* tape series was introduced to national members at this meeting. Also, the Branch set a precedent for future National ASM meetings by sponsoring a pre-convention workshop on computerization for clinical microbiology laboratories. Dr. Morton prepared a "History of Microbiology in the Philadelphia Area, 1960-1972" for the Annual Meeting.

1972

The first in a series of basic science symposia was organized on the subject of Virus Tumorigenesis and Immunogenesis. These symposia were held in the spring and were developed independently of the clinical symposia, which were held each November.

1973

The need to merge the regular Branch with the Clinical Microbiology Section became apparent due to a substantial decrease in attendance at the regular meetings and a significant increase in attendance at the Clinical Microbiology meetings. To accomplish this merger, there was need to restructure the governance of the Branch and to take on the various functions of the Clinical Microbiology Section, which had committees for workshops, educational tapes, program, education and a number of specialized committees to address such issues as new legislation and regulations. To serve this need, a new official Executive Committee was formed to assist in unifying the Branch. The merger became official in August of 1973.

1974

Richard L. Crowell, Hahnemann Medical College, became the twenty-eighth Branch President and served through June 1975.

1974

Thomas Jefferson University graduated its first class of eleven clinical microbiology master's degree students. Of this first class of eleven, six took either supervisory or director positions in microbiology at Philadelphia area hospitals.

1975

Norman P. Willett, Temple University School of Dentistry, became the twenty-ninth Branch president in July and served through June 1977.

1976

The First **Stuart Mudd Lecture** was initiated. This lecture series was held every year thereafter, through 1995. This became a focal point for the Branch each spring with a prominent speaker, the awarding of the Stuart Mudd plaque, and the presence of Stuart Mudd family members.

1977

Robert J. Mandle, Thomas Jefferson Medical College, became the thirtieth Branch President in July and served through June 1979.

1977

A new series of newsletters began in November (volume 18, number 1) that continued, with various change in editors, until 2009.

1978

In November, a "Branch Past Presidents Night" was celebrated. Seventeen past presidents were invited, and nine were able to attend. Each was given a Silver Reserve Bowl inscribed with their names and dates of service.

1978

A competition was held to design an appropriate logo for the Branch. From many submissions, a design by Dr. Harry Stempin was chosen.

1979

Joseph F. Pagano, Smith Kline and French Laboratories, became the thirty-first Branch president in July and served through June 1981.

1980-1989
1980
Branch standing committees included the following: academic affairs, archives, education, finance, industrial affairs, newsletter, placement, program, publications, publicity, membership, legislative, legal, scholarship/awards, public and scientific affairs, symposium, industrial affairs and workshop. There were also several ad hoc committees that handled annual events like the Stuart Mudd lectures and special Poster Session meetings.

1981
John C. McKitrick, University of Pennsylvania, became the thirty-second Branch president in July and served until October when he resigned to move out of state.

1981
Henry R. Beilstein, Philadelphia Public Health Department, became the thirty-third Branch President in November and served through June 1983.

1983
Toby Eisenstein, Temple University, became the thirty-fourth Branch President in July and served through June 1985.

1985
The first Annual Poster Session night was initiated in February. The poster session was organized by an ad hoc committee headed by Rick Rest. The focus for this first meeting was on graduate student presentations, but it was open to all Branch members. This replaced the tradition of having one regular meeting each year for graduate student oral presentations.

1985
Donald Stieritz, Hahnemann Medical College, became the thirty-fifth Branch President in July and served through June 1987.

1987
James Poupard, Medical College of Pennsylvania, became the thirty-sixth Branch President in July and served through June 1989.

1987

On 15 December, the Branch took the occasion of the **500th Branch Meeting** to celebrate sixty-seven years of continuous activity. The celebration, at the University of Pennsylvania, started with an afternoon of presentations on the history of the Branch, microbiology in general and historical events that occurred during the past seven decades. The presentations were followed by a reception and a spectacular dinner attended by 164 guests at the Faculty Club of the University of Pennsylvania.

1988

Dr. Harry Morton died 22 November. He was not only a past president (1946-47) of the Branch, but one of the most active and valuable Branch members. He dedicated his last years to documenting the history of the Branch, as well as the history of the ASM. His presence was greatly missed since he was an active member since his arrival in Philadelphia in the 1930s and guided the Branch through many critical changes.

1989

Paul Actor, Smith Kline and French Laboratories, became the thirty-seventh Branch President in July and served through June 1991.

1990-1999

1990

Dr. Richard Crowell was elected president of the National ASM.

1990

In April, the Branch was cosponsor of a special afternoon meeting at the Albin O. Kuhn Library (where the ASM Archives are held). The topic was "The Biological Weapons Convention Under Siege: Disarmament Issues for the 1990s." It was one of the first open discussions within the ASM on biological weapons after several years of silence on this subject.

1990

Dr. Eileen Randall died in December. She organized the first Branch Clinical workshop and served the Branch for many years before leaving Philadelphia.

1991

Two long-time Branch members died: Dr. Evan L. Stubbs, one of the oldest Branch members, and the eighth Branch President (1934-35); and

Thomas Anderson who produced pioneering work in electron microscopy that was often first reported at the Branch meetings and whose work made Philadelphia a leading center for these studies.

1991

Alan Evangelista, Cooper Medical Center, became the thirty-eighth Branch President in June and served until July 1993.

1991

Dr. Carlo Croce became the new Chair of Microbiology at Thomas Jefferson University.

1992

In March, there was an event in honor of Smith Hall (the former Laboratory of Hygiene) and Dr. Harry E. Morton. A group at the University of Pennsylvania mounted an effort to save the building from demolition due to its historical significance, and the Branch decided to hold a meeting there to support the University group. It was also decided to use this meeting to honor long time Branch member and University of Pennsylvania Professor, Dr. Harry Morton. There was an afternoon photo session, followed by presentations at the University of Pennsylvania Faculty Club. There was time for hors d'oeuvres and for everyone to tour Smith Hall and absorb the essence and history that unfolded within its hallowed walls. A special dinner was then served back at the Faculty Club.

1992

In April, a new tradition was initiated, the **Philadelphia Infection and Immunity Forum**. This marked the start of a new annual series designed to get graduate students, post-doctoral fellows and faculty involved with the Branch. It was presented as an interdisciplinary forum for clinical and basic research scientists in the Philadelphia and Delaware Valley areas interested in microbial pathogenesis and host response. The series continues to the present time.

1992

Dr. Morton Klein, the twentieth Branch President (1958-59), died on 6 December.

1993

Linda A. Miller, Holy Redeemer Hospital, became the thirty-ninth Branch President in June and served until July 1995.

1993

Dr. Chris Platsoucas was named the new Chair of Microbiology at Temple University.

1993

Long-time active member, Dr. Ralph Knight, died on 28 September.

1994

Hahnemann Medical College and the Medical College of Pennsylvania merged. Richard Crowell became Chair of the newly merged Microbiology Department.

1994

The Branch lost three long-time active members: Dr. Wesley G. Hutchison died on 22 January; Theodore "Ted" Anderson, a pioneer in the area of antimicrobial susceptibility testing, died at the age of 91 on 14 March; and in October, long time member Dr. Carl Clancy died.

1995

Paul Cerwinka, MetPath Laboratories, became the fortieth Branch President in June and served until July 1997.

1995

The Branch lost two former past presidents: Dr. Earle Spaulding, the thirteenth Branch President (1944-45), died on 2 February at age eighty-eight. He came to Temple University in 1936 and mentored many of our Branch members over the years. Dr. Robert Mandle, thirtieth Branch President (1977-79), died on 3 April. He directed the Clinical Microbiology Master's Program at Thomas Jefferson University and was a significant influence in keeping the Clinical Microbiology Section part of the Branch.

1996

The Distinguished Branch Member Lecture Series was initiated, to replace the twenty year series of Stuart Mudd Lectures. Dr. Morton was the first honoree, and each honoree who followed was chosen for their various contributions to the Branch. The lecturers associated with these events were chosen to reflect some aspect of the work or interests of the honorees. Like the Stuart Mudd Lecture Series, the Distinguished Branch Lectures became one of the social, as well as a scientific highlight, of each Branch academic year until it was discontinued in 2006.

1997
Richard Rest, MCP-Hahnemann Medical School, became the forty-first Branch President in June and served until July 1999.

1997
The Branch Web site was initiated.

1997
The Branch student chapter was officially recognized by the National ASM.

1998
The Branch celebrated the 600th meeting with an afternoon session at Thomas Jefferson University entitled: **A Reason to Celebrate Our 600th Meeting**. This involved five Philadelphia Microbiologists "looking back" at the last six hundred meetings and the microbiology represented at these meetings. The meeting was followed by a celebration dinner.

1999
Irving Nachamkin, University of Pennsylvania, became the forty-second Branch President in June and served until July 2001.

1999
Dr. Gerald D. Shockman died on 27 October at the age of seventy-three. Dr. Schokman came to Temple University in 1960 and was chair of the microbiology Department there from 1974 to 1990.

2000-2009

2000
Dr. William Ball, long-time member, died on 31 July. He served in World War II and employed his particular interest in parasitology to assist in several Branch educational programs relating to parasitology.

2001
Dr. Helen Buckley, professor of mycology at Temple University and long-time Branch member, died on 28 February. On 5 April, the Branch also

lost long-time member Dr. Henry Stempen, professor emeritus at Rutgers University. He is especially remembered within the Branch as the designer of the Branch logo.

2001

A permanent exhibit on "Clinical Microbiology at the University of Pennsylvania" was dedicated in a ceremony on 16 April. The exhibit is located on the fourth floor of the Gates Building of the Hospital of the University of Pennsylvania.

2001

Donald Jungkind, Thomas Jefferson University, became the forty-third Branch President in July and served until June 2003.

2002

Dr. Ruth Emma Miller, the eighteenth Branch President (1954-55), died on 7 April at the age of ninety-six. She received her PhD from the University of Pennsylvania in 1934 and was the first female president of the Branch. She was a long time faculty member, Professor and Chairperson of the Microbiology Department of the Medical College of Pennsylvania and a long time active Branch member.

2002

In September, the Branch newsletter publication was switched from an all print version to both a printed and electronic version. This was an effort to eventually go completely electronic to reduce expenses.

2003

Dr. Joseph Pagano, the thirty-first Branch president (1979-81), died on 8 August at the age of eighty-four, and Dr. Albert G. Moat, the twenty-fourth Branch president (1966-67), died at the age of seventy-seven.

2003

Olarae Giger, Main Line Clinical Laboratories, became the forty-fourth Branch President in July and served until June 2005.

2004

Dr. Henry R. Beilstein, the thirty-third Branch president (1981-83), died after a long illness on 10 July. He was a Distinguished Branch Honoree, served on the Executive Committee for many years and was a fifty-year member of the National ASM. He was influential in keeping all elements within the Branch together and would always have a calming effect in heated discussions.

2005

Dr. Amedeo Bondi, sixteenth Branch president (1950-51), died on 28 January. In 1947, he was appointed the first chairman of the Department of Microbiology at Hahnemann Medical College and also served as Dean of the Graduate School. He was active in the Branch for his entire career and mentored many Branch members. On 25 February, Dr. Georganne K. Buecscher, of Thomas Jefferson University, died. She was a graduate of the first class of the MS Clinical Microbiology Program at Thomas Jefferson University and was a driving factor in continuing that program, which supplied a source of new graduates practicing Clinical Microbiology in the Philadelphia area.

2005

David Axler, Temple University, became the forty-fifth Branch President in July and served until June 2006.

2006

Bettina Buttaro, Temple University, became the forty-sixth Branch President in July and served until June 2009.

2007

On 25 March, one of the most prominent Philadelphia experts on infectious disease and microbiology, Dr. Robert Austrian, died. He was world famous for his work on pneumococcus. In 1976, he demonstrated that the pneumococcal vaccine he and his co-workers developed could significantly reduce mortality, especially in elder patients. He was a long time Branch member and made several presentations at the regular Branch meetings and functions. On 8 May, Dr. Russell Schaedler died. He was a Jefferson Medical College graduate in 1953 and long-time Chairman of the Department of Microbiology at Thomas Jefferson University. It was announced in November that Dr. Herman Friedman, the twenty-sixth Branch president (1970-71), died. He was born, educated and remained in Philadelphia after getting his PhD. He was on the Microbiology Faculty at Temple University and Head of the Clinical Microbiology and Immunology Department at Albert Einstein Hospital. He was very active in the Branch.

He initiated the first Branch Basic Science Symposium and chaired many of the Branch symposia. He was also active in forming the Clinical Microbiology Section of the Branch. In 1978, he left Philadelphia to become Chair of the Department of Medical Microbiology and Immunology at the University of South Florida, College of Medicine, but he returned to Philadelphia several times to participate in Branch functions.

2008

In January, Josephine Bartola died. Jo Bartola served the Branch as legal adviser for many years. She revised the Branch bylaws several times and served on almost all Branch symposia committees from the early years up to her retirement. She continued to be the Branch legal adviser after her retirement.

2009

In May, the ASM General Meeting was held in Philadelphia for the tenth time. Each attendee was provided with a tour guide and walking map prepared by the Branch Education Committee. There was a special full-afternoon session dedicated to the history of microbiology in Philadelphia.

2009

The Laboratory of Hygiene of the University of Pennsylvania was selected by ASM as the third location to receive the designation of **Milestones in Microbiology**. A dedication ceremony was held in May prior to the start of the ASM General Meeting, at the University of Pennsylvania to unveil a plaque commemorating the site of the former Laboratory of Hygiene. ASM President, Alison O'Brien, welcomed the assembled guests and explained the Milestones Program. Arthur Rubenstein, Dean of the University of Pennsylvania School of Medicine, spoke on the history of the building and its significance to the University. William B. Whitman, representing the Beregy Trust, discussed David Bergey and his continuing influence on microbiology. The meeting was attended by several Branch members, University of Pennsylvania and ASM staff, ASM Archive chair, Patricia Charache, and other ASM Archive Committee members. A reception, jointly sponsored by the Bergey Trust, followed the plaque dedication ceremony.

2009

Laura Chandler, Philadelphia V.A. Medical Center, became the forty-seventh Branch President in July.

2009

The Executive Committee decided to eliminate the Branch newsletter in its current form and replace it with some type of web based communication method. A sub-committee was formed to design a new website/blog to replace the newsletter.

Ref.1= Thomas Wirth, Urban Neglect: The Environment, Public Health, and Influenza in Philadelphia, 1915-1919. Penna. History, Vol. 73, No. 3, 2006 p. 316-342

REFERENCES

CHAPTER ONE: BACKGROUND INFORMATION

References

1. Jonathan Liebenau, *Medical Science and Medical Industry*, 1987, The Johns Hopkins University Press, Baltimore, p. 11.
2. Russell C. Maulitz, *Robert Koch and American Medicine*, Annals of Internal Medicine, 1982, 97:761-766, p.761.
3. John F. Marion, *Philadelphia Medica*, SmithKline Corporation, 1975, Philadelphia, PA, page 32
4. Paul F. Clark, *Pioneer Microbiologists of America*, 1961, The University of Wisconsin Press, Madison, page 7.

General References

5. Simon Finger, *An Indissoluble Union: How the American War of Independence Transformed Philadelphia's Medical Community and Created a Public Health Establishment*, Pennsylvania History, 2010, 77: 37-72.
   Note: In addition to a review of the general medical history of this period, Finger provides the following statistic—Between 1747 and 1800, more than one hundred Americans returned from Edinburgh alone bearing medical degrees. In 1765, as many Americans as Scots took medical degrees there. (p. 39) However, as noted in Chapter two, travel to Europe to gain knowledge on bacteriology did not occur until the 1880s.
6. Christopher Gradmann, *Laboratory Diseases*, 2009, The Johns Hopkins University Press, Baltimore, MD.
   Note: This is an excellent text that reviews of the separation of bacteriology from pathology and histology, especially from a German perspective. It demonstrates how bacteriology started to separate from other biological and medical science starting in 1870s. However, it should be noted that this separation did not start until the 1880s-1890s in the United States.

7. Joseph McFarland, *The Beginning of Bacteriology in Philadelphia*, Bulletin of the Institute of the History of Medicine, 1937. 5:149-198.
8. Linda. A. Miller, H. E. Morton and J. A. Poupard (ed) *A History of the Eastern Pennsylvania Branch of the ASM 1920-1987.* 1987 Branch Publication.
9. Russel F. Weigley, ed, *Philadelphia A 300 Year History*, 1982, W.W. Norton & Company, NY.

CHAPTER TWO: EARLY PHILADELPHIA BACTERIOLOGY

References

1. John P. Swann, *Academic Scientists and the Pharmaceutical Industry*, 1988, The Johns Hopkins University Press, Baltimore, p.20.
2. Jonathan Liebenau, *Medical Science and Medical Industry*, 1987, The Johns Hopkins University Press, Baltimore, p.37.
3. D. A. Grimes and K. F. Schulz, Lancet, 2002, 359:145-149.
4. Russell C. Maulitz, *Robert Koch and American Medicine*, Annals of Internal Medicine, 1982, 97:761-766, p.761.
5. Joseph McFarland, *The Beginning of Bacteriology in Philadelphia*. Bulletin of the Institute of the History of Medicine, 1937. 5:149-198.

General References

6. George W. Corner, *Two Centuries of Medicine, A History of the School of Medicine*, University of Pennsylvania, 1965, J. B. Lippincott Company, Philadelphia.
7. Esmond R. Long, *A History of American Pathology*, 1962, Charles C. Thomas Publishers, Springfield.
8. Paul F. Clark, *Pioneer Microbiologists of America*, 1961, The University of Wisconsin Press, Madison.
9. George Rosen, *A History of Public Health*, 1993, The Johns Hopkins University Press, Baltimore.

References Specific to Each Biographical Sketch

Joseph Leidy, References: 6, 7, 8 and

10. Leonard Warren, *Joseph Leidy: The Last Man Who Knew Everything*, 1998, Yale University Press, New Haven.

11. Joseph Leidy Jr., *Researches in Helmenthology and Parasitology by Joseph Leidy*, 1904, Smithsonian Publication, Washington DC.

Edward Orem Shakespeare, References: 5, 7, 8 p.195, and
12. Website:[htpp://www.collphyphil.org/FIND_AID/S/ ShakespeareeoMSS2_0338_01.htm] MSS 2/0338-01 Acc. 2001-005.

Henry F. Formad, References: 5, 6 and
13. The Diphtheria Plant, *The Philadelphia Times*, October 12, 1881.
14. Martha L. Sternberg, *George Miller Sternberg: A Biography*, 1920, American Medical Association, Chicago, page 74.
15. George M. Sternberg, *A Fatal Form of Septicaemia in the Rabbit Produced by the Subcutaneous Injection of Human Saliva*, 1881, Bulletin of the National Board of Health 2:781-783.
16. Louis Pasteur, 1881, Su une maladie nouvelle provoquee par la salive d'un enfant mort de rage. Compt. rend Academ. d sc. de Paris 92:159.
17. *Philadelphia Ledger*, July 14, 1885.
18. Website: [*http://hss.sas.upenn.edu/microbio/panel2.html*].

Lawrence Flick, References: 4, 5, 6 p.219-220, 9 p.364-5 and
19. Website: [*http://libraries.cua.edu/achrcua/flick.html*]
20. Website: [*www.phila.gov/Health/units/tb/Flick_Biography.html*]
21. The New York Times, September 26, 1908.

Ernest Laplace, References 5.
Samuel Gibson Dixon, References : 5, 6 p.128, 181 and
22. Website [*www.dsf.health.state.pa.us/health/cwp/view.asp?A=191&Q=242483*]

Lydia Rabinowitsch, References : 5 and
23. Lori R. Walsh and James A. Poupard, *Lydia Rabinowitsch, PhD, and the Emergence of Clinical Pathology in Late 19th-Century America*. 1989. Archives of Pathology and Laboratory Medicine, 113:1303-1308.
24. JAMA Oct. 12, 1935

Institution References: 5, 6, 7 and
25. D. Bruce Cowgill, *A Brief History of the William Pepper Laboratory of Clinical Medicine*. Typed manuscript dated 3 March 1969, Branch Archive Collection.
26. Mazyck P. Ravenel, *The Old Pepper Laboratory*. *The General Magazine and Historical Chronicle*, University of Pennsylvania. July 1929.

27. Donald S. Young, Mary Cregar Berwick, and Leonard Jatett. *Evolution of the William Peppper Laboratory.* Clinical Chemistry 43:1 174-179, 1997.
28. Pamela Knight, *Wrecker's Ball Aimed at Birthplace of Bacteriology.* ASM News, 58(6):304-5, 1992.
29. Schools of the University of Pennsylvania Web site.
30. Russel Schadler, "Department of Microbiology" chapter 7, p 203-220, in Fredrick B. Wagner, Jr., Thomas Jefferson University: Traditions and Heritage, 1989, Lea & Febiger, Philadelphia.
31. Letter of H. K. Mulford to Joseph McFarland, 24 February 1937, ASM Archive McFarland Collection.
32. Barnett Cohen, *Chronicles of the Society of American Bacteriologists 1899-1950.* Society of American Bacteriologists, 1950.

CHAPTER THREE: A TRANSITIONAL OR SECOND GENERATION GROUP OF EARLY PHILADELPHIA BACTERIOLOGISTS DURING THE LATE NINETEENTH AND EARLY TWENTIETH CENTURY.

References:

1. Joseph McFarland, *The Beginning of Bacteriology in Philadelphia.* Bulletin of the Institute of the History of Medicine, 1937. 5:149-198.
2. George W. Corner, *Two Centuries of Medicine, A History of the School of Medicine*, University of Pennsylvania. 1965 J. B. Lippincott Company, Philadelphia.
3. Paul F. Clark, *Pioneer Microbiologists of America.* 1961, The University of Wisconsin Press, Madison.

Individual references:

Alexander Abbott: 1, 2 p.182, 3 p. 95 and 102, and
4. Alexander C. Abbott, *The Principles of Bacteriology; a Practical Manual for Students and Physicians,* 1892, Lea Brothers and Company, Philadelphia.
5. Alexander C. Abbott, *The Hygiene of Transmissible Diseases*, 1899, W. B. Saunders, Philadelphia.

Michael Ball: 1, and
6. Michael V. Ball, *Essentials of Bacteriology*, 1891, W. B. Saunders Company, Philadelphia.

David Bergey: 1, 2, and

7. David H. Bergey, *Bergey's Manual of Determinative Bacteriology: A Key for the Identification of Organisms of the Class Schizomycetes*, 1927, Williams and Wilkins Company, Baltimore.

Albert Ghriskey: 1, and

8. Letter, L. G. Leilith to Joseph McFarland, 22 March 1935(?), A.S.M. Archives,

Samuel Kneass: 1, 2.
Joseph McFarland: 1,2

9. Joseph McFarland, *The Pathogenic Bacteria*, 1896, W. B. Saunders Company, Philadelphia.

Adelaide Peckham: 1.

10. Letter, Adelaide Peckham to Joseph McFarland, 21 March 1936, ASMArchives.

11. *The Influence of Environment upon the Biological Processes of the Various Members of the Colon Group of Bacilli: An Experimental Study*. The Journal of Experimental Medicine, 1987, Vol 2, 549-591.

Robert Pitfield: 1.

12. Robert Lucas Pitfield, *A Compendium on Bacteriology Including Animal Parasites*, P. Blakiston's Sons and company, Philadelphia.

Mazyck Ravenel: 1, 2.
Randle Rosenberger:

13. Russel Schadler, "Department of Microbiology" chapter 7, p 203-220, in Fredrick B. Wagner, Jr., Thomas Jefferson University: Traditions and Heritage, 1989, Lea & Febiger, Philadelphia.

Ernest Sangree: 1.
Allen Smith: 1.
Alonzo Stewart: 1, 2, 3

14. Letter to Joseph McFarland from unknown author received by McFarland 1 January 1937, ASM Archives

CHAPTER FOUR: FOUR PERSPECTIVES ON ENTERING THE NEW CENTURY: PHILADELPHIA, THE SOCIETY OF AMERICAN BACTERIOLOGISTS, EARLY BACTERIOLOGISTS and PHILADELPHIA INSTITUTION UPDATES.

References:

1. Nathaniel Burt and Wallace Davis, *The Iron Age*, in Davis, p.471-523, (pages sited 471,474, 481, 494 and 496) in Russel F. Weigley, ed, *Philadelphia A 300 Year History*, W.W. Norton & Company, NY, 1982.
2. Barnett Cohen, *Chronicles of the Society of American Bacteriologists 1899-1950*. Society of American Bacteriologists, 1950.

The Continuation of the First Generation Philadelphia Bacteriologists: 1900-1938, and The Continuation of the Transitional Generation of Bacteriologists: 1900-1955.

3. Joseph. McFarland, *The Beginning of Bacteriology in Philadelphia*. Bulletin of the Institute of the History of Medicine, 1937. 5:149-198.
4. Russel Schadler, "Department of Microbiology" chapter 7, p 203-220, in Fredrick B. Wagner, Jr., Thomas Jefferson University: Traditions and Heritage, 1989, Lea & Febiger, Philadelphia.
5. George W. Corner, *Two Centuries of Medicine, A History of the School of Medicine*, University of Pennsylvania. 1965 J. B. Lippincott Company, Philadelphia.

Henry Phipps Institute for the Study, Treatment, and Prevention of Tuberculosis.

5. Same as above.
6. Florence B. Sierbert, *The History of Bacteriology of Tuberculosis in Philadelphia*, in History of Bacteriology in the Philadelphia Area, presented at the Forty-seventh General Meeting of the Society of American Bacteriologists, 13 May 1947, Eastern Pennsylvania Branch Archives Manuscript Collection.

Jefferson Medical College

7. Russel Schadler, "Department of Microbiology" chapter 7, p 203-220, in Fredrick B. Wagner, Jr., Thomas Jefferson University: Traditions and Heritage, 1989, Lea & Febiger, Philadelphia.

Temple College/ University School of Medicine

8. Temple Review, Winter 2009, p. 16

CHAPTER FIVE: THE MICROBIOLOGICAL CLUB AND THE FOUNDING OF THE EASTERN PENNSYLVANIA BRANCH OF THE AMERICAN SOCIETY FOR MICROBIOLOGY: 1912 TO 1935.

1. H. E. Morton, *A History of the Eastern Pennsylvania Branch of the ASM, A Personal Review*, in L. A. Miller, H. E. Morton and J. A. Poupard (ed) A History of the Eastern Pennsylvania Branch of the ASM 1920-1987. 1987 Branch Publication, Branch Archives Manuscript collection.
2. C. J. Bucher, *Introductory Remarks, Roundtable on Bacteriological History*, Society of American Bacteriologists, General Meeting 13 May 1947. Branch Archives Manuscript collection.
3. Barnett Cohen, *Chronicles of the Society of American Bacteriologists 1899-1950*. Society of American Bacteriologists, 1950.

Chapters Six and Seven
Unless otherwise noted, information for these chapters are based on meeting notices, minutes, newsletters, and correspondences deposited in the Branch Archives for the specific year cited.

Chapters Eight through Thirteen
Unless otherwise noted, information for these chapters are based on meeting notices, meeting minutes, newsletters, executive committee meeting minutes and correspondences deposited in the Branch Archives for the specific year cited.

Chapter Nine
1. James A. Poupard, *A Comparison and Evaluation of Relevance of Clinical Microbiology Master's Degree Programs in the United States*. 1982 University Microfilms International Ann Arbor MI p. 81-86.

LIST OF APPENDICES

APPENDIX:

I: Branch Presidents: 1920 to 2011

II: Branch Nonpresident Officers: 1920 to 2009

III: A Listing of Branch Meetings: 1936 through 2009

IV: The Clinical Microbiology Section of the Eastern Pennsylvania Branch: 1967 to 1973

V: Eastern Pennsylvania Branch Sponsored Symposia: 1969 to 2009

VI: Eastern Pennsylvania Branch Sponsored Workshops: 1969 to 2009

VII: Philadelphia Infection and Immunity Forum: 1992 to 2009

VIII: The Stuart Mudd Memorial Lecture Series: 1976 to 1995

IX: The Distinguished Branch Member Lectureship: 1996 to 2006

X: The Annual Industry Sponsored Guest Lectureship Series: 1962 to 1971

XI: Celebration of the 500th and 600th Branch Meetings: 1987 and 1998

XII: History of the Branch Newsletter: 1937 to 2009

Appendix I

Branch Presidents: 1920 To 2011

Presidents of the Eastern Pennsylvania Chapter of the Society of American Bacteriologists which became the Eastern Pennsylvania Branch of the American Society for Microbiology.

1	Dr. David Bergey	1920-1922	Began 24 February
2	Dr. Randle C. Rosenberger	1923-1924	
3	Dr. A. C. Abbott	1925-1926	
4	Dr. Eugene L. Opie	1927-1928	Served Until May 1928
5	Dr. J. L.T. Appleton, Jr.	1928-1929	Began in October
6	Dr. Joseph D. Aronson	1930-1931	Served Until October 1931
7	Dr. Jefferson Clark	1931-1934	November 1931 Until May 1934
8	Dr. Evan L. Stubbs	1934-1935	Began in October
9	Dr. Carl J. Bucher	1936-1937	
10	Mr. Christopher G. Roos	1938-1939	
11	Dr. Stuart Mudd	1940-1941	
12	Dr. William A. Kreidler	1942-1943	
13	Dr. Earle H. Spaulding	1944-1945	
14	Dr. Harry E. Morton	1946-1947	
15	Dr. William Verwey	1948-1949	
16	Dr. Amedeo Bondi	1950-1951	
17	Dr. James Harrison	1952-1953	
18	Dr. Ruth E. Miller	1954-1955	
19	Dr. Kenneth Goodner	1956-1957	
20	Dr. Morton Klein	1958-1959	
21	Dr. Joseph S. Gots	1960-1961	
22	Dr. L. Joe Berry	1962-1963	
23	Dr. Harold S. Ginsberg	1964-1965	
24	Dr. Albert G. Moat	1966-1967	
25	Dr. George Warren	1968-1969	
26	Dr. Herman Friedman	1970-1971	
27	Dr. James Prier	1972-1973	
28	Dr. Richard L. Crowell	1974-1975	Jan 1974 Until June 1975
29	Dr. Norman P. Willett	1975-1977	July 1975 Until June 1977
30	Dr. Robert J. Mandle	1977-1979	July 1977 Until June 1979

31	Dr. Joseph F. Pagano	1979-1981	July 1979 Until June 1981
32	Dr. John C. McKitrick	1981	July 1981 Until October 1981
33	Dr. Henry R. Beilstein	1981-1983	November 1981 Until June 1983
34	Dr. Toby K. Eisenstein	1983-1985	July 1983 Until June 1985
35	Dr. Donald D. Stieritz	1985-1987	July 1985 Until June 1987
36	Dr. James A. Poupard	1987-1989	July 1987 Until June 1989
37	Dr. Paul Actor	1989-1991	July 1989 Until June 1991
38	Dr. Alan Evangelista	1991-1993	July 1991 Until June 1993
39	Dr. Linda A. Miller	1993-1995	July 1993 Until June 1995
40	Mr. Paul Cerwinka	1995-1997	July 1995 Until June 1997
41	Dr. Richard Rest	1997-1999	July 1997 Until June 1999
42	Dr. Irving Nachamkin	1999-2001	July 1999 Until June 2001
43	Dr. Donald Jungkind	2001-2003	July 2001 Until June 2003
44	Dr. Olarae Giger	2002-2005	July 2003 Until June 2005
45	Dr. David A. Axler	2005-2006	July 2005 Until June 2006
46	Dr. Bettina Buttaro	2006-2009	July 2006 Until June 2009
47	Dr. Laura Chandler	2009-2011	July 2009 Until June 1011

Branch Presidents by Decade and with Institutions

1920s

1- David H. Bergey, University of Pennsylvania (1920-1922)
2- Randle C. Rosenberger, Jefferson Medical College (1923-1924)
3- A. C. Abbott, University of Pennsylvania (1925-1926)
4- Eugene L. Opie, Henry Phipps Institute (1927-1928)
5- Joseph L. T. Appleton, Jr., Univ. of PA Dental School (1928-1929)

1930s

6- Joseph D. Aronson, Henry Phipps Institute (1930-Oct. 1931)
7- Jefferson H. Clark, Philadelphia General Hospital (1931-1934)
8- Evan L. Stubbs, Univ. of Pennsylvania Veterinary School (1934-1935)
9- Carl Bucher, Jefferson Hospital (1936-1937)
10- Christopher G. Roos, Sharp and Dohme Laboratories (1938-1939)

1940s

11- Stuart Mudd, University of Pennsylvania (1940-1941)
12- William A. Kreidler, Jefferson Medical College (1942-1943)
13- Earle H. Spaulding, Temple University School of Medicine (1944-1945)
14- Harry E. Morton, University of Pennsylvania (1946-1947)
15- William Verwey, Sharp and Dohme Laboratories, Glenolden Pa. (1948-1949)

1950s
16- Amedeo Bondi, Temple University School of Medicine (1950-51)
17- James Harrison, Department of Biology, Temple University (1952-53)
18- Ruth Miller, Women's Medical College (1954-55)
19- Kenneth Goodner, Jefferson Medical College (1956-57)
20- Morton Klein, Temple University School of Medicine (1958-59)

1960s
21- Joseph S. Gots, University of Pennsylvania (1960-61)
22- L. Joe Berry, Bryn Mawr College (1962-63)
23- Harold S. Ginsberg, University of Pennsylvania (1964-65)
24- Albert G. Moat, Hahnemann Medical College (1966-67)
25- George Warren, Wyeth Labotatories, Radnor (1968-69)

1970s
26- Herman Friedman, Albert Einstein Medical Center (1970-71)
27- James Prier, Pennsylvania Department of Health Laboratories (1972-3)
28- Richard L. Crowell, Hahnemann Medical College (1974-75**)
29- Norman P. Willett, Temple University School of Dentistry (1975-77*)
30- Robert J. Mandle, Thomas Jefferson Medical College (1977-79*)
31- Joseph F. Pagano, Smith Kline and French Laboratories (1979-81*)
(**Jan 74 through June 1975) (* July through June)

1980s
31- Joseph F. Pagano, Smith Kline and French Laboratories (1979-81*)
32- John C. McKitrick, University of Pennsylvania (**1981)
33- Henry R. Beilstein, Phila. Public Health Department (1981-83***)
34- Toby Eisenstein, Temple University (1983-85*)
35- Donald Stieritz, Hahnemann Medical College (1985-87*)
36- James A. Poupard, Medical College of Pennsylvania (1987-89*)
37- Paul Actor, Smith Kline & French, (1989-91*)
(*July through June) (**July 1981 through October 1981)
(***November 1981 through June 1983)

1990s (Term: July through June)
37- Paul Actor, Smith Kline & French (1989-91)
38- Alan Evangelista, Cooper Medical Center (1991-93)
39- Linda A. Miller, Holy Reedemer Hospital (1993-95)
40- Paul Cerwinka, MetPath Laboratories (1995-97)

41- Richard Rest, MCP-Hahnemann Medical School (1997-1999)
42- Irving Nachamkin, University of Pennsylvania (1999-01)

2000s (Term: July through June)
42-Irving Nachamkin, University of Pennsylvania (1999-01)
43-Donald Jungkind, Thomas Jefferson University (2001-03)
44-Olarae Giger, Main Line Clinical Laboratories (2003-05)
45-David Axler, Temple University (2005-06)
46-Bettina Buttaro, Temple University (2006-09)
47-Laura Chandler, Philadelphia V.A. Medical Center (2009-11)

Presendential Institution Summary
(Number of presidents from each institution in parenthesis.)
University of Pennsylvania (10)
Temple University (8)
Jefferson Medical College and University (6)
Hahnemann Medical College (4)
Women's Medical College /Medical College of Penna. (2)
Henry Phipps Institute (2)
Sharp and Dohme Laboratories (2)
Smith Kline and French Laboratories (2)

And one each from the following:
Albert Einstein Medical Center
Bryn Mawr College
Cooper Medical Center
Holy Reedemer Hospital
Main Line Clinical Laboratories
MetPath Laboratories
Philadelphia General Hospital
Philadelphia Public Health Department
Philadelphia V.A. Medical Center
Pennsylvania Health Department
Wyeth Laboratories

Appendix II

Branch Nonpresident Officers: 1920 TO 2009

Secretary/Treasurers
1920-22 C.Y. White, University of Pennsylvania
1923-36 Claude P. Brown, Temple University School of Medicine
1937-45 Harry Morton, Univ. of Penn. School of Medicine
1946-47 Amedeo Bondi, Temple University
1948-50 W. G. Hutchinson, University of Penn
1951-53 Ruth E. Miller, Woman's Medical College
1954-57 Theodore G. Anderson, Temple University School of Medicine
1958-61 Elizabeth H. Fowler, Temple University School of Medicine
1962-65 Leonard Zubrzycki, Temple University School of Medicine
1966-69 Herman Friedman, Einstein Medical Center

Secretaries
1970-73 Eileen Randall, Jefferson Medical College (Until April 1973)
1973-75 Elizabeth Free, Hahnemann Medical College (Starting May 1973)
1975-79 Toby Eisenstein, Temple University
1979-81 Josephine Bartola, Pennsylvania Department of Health Laboratories
1981-83 Donald Stieritz, Hahnemann Medical College
1983-85 Alan Evangelista, Cooper Medical Center
1985-88 Eileen Hinks, Rolling Hill Hospital, United Hospitals Inc.
1988 March-1991 Linda Miller, Holy Redeemer Hospital
1991-93 Paul Cerwinka, MetPath Laboratories
1993-95 Olarae Giger, Episcopal Hospital
1995-99 Kathleen Beavis, Thomas Jefferson University
1999-09 Anna Feldman-Rosen, Penn State, Abington

Treasurers
1970-73 Norman P. Willett, Temple University School Dentistry
1974-77 Joseph F. Pagano, Smith Kline Diagnostics

1977-79 Carl Abramson, Pennsylvania College of Podiatric Medicine
1979-81 Henry R. Beilstein, Philadelphia Public Health Department
1981-85 Bruce Kleger, Pa. Department of Health, Bureau of Laboratories
1985-87 Paul Actor, Smith Kline Beckman
1987-89 Alan Evangelista, Cooper Medical Center
1989-91 Donald Jungkind, Thomas Jefferson University
1993-95 Irv Nachamkin, University of Pennsylvania
1995-01 Olarae Giger, Episcopal Hospital
2001-03 David Axler, Temple University
2003-05 Bettina Buttaro, Temple University
2005-09 Barbara McHale, Gwynedd-Mercy College

Councilors
1935-37 Randle C. Rosenberger, Jefferson Medical College.
1938-39 Carl J. Bucher, Jefferson Hospital
1940-41 A. Parker Hitchens, University of Pennsylvania
1942-45 Harry E. Morton, University of Pennsylvania
1946-47 W. Verwey, Sharp & Dohme Laboratories
1948-49 James A. Harrison, Temple University 1950-55 Unknown
1956-57 Amedeo Bondi, Hahnemann Medical College
1958-59 Ruth Miller, Womans Medical College
1960-61 Carl Clancy, Pennsylvania Hospital
1962-63 George Warren, Wyeth Laboratories
1964-65 Albert Moat, Hahnemann Medical Center
1966-67 Joseph Gots, University of Pennsylvania
1968-69 Leonard Zubrzycki, Temple University
1970-71 Richard L. Crowell, Hahnemann Medical College
1972-73 Joseph F. Pagano, Smith Kline and French Laboratories
1974-75 Robert J. Mandle, Thomas Jefferson University
1975-77 Ralph A. Knight, Medical College of Pennsylvania
1977-79 John McKitrick, University of Pennsylvania
1979-81 Donald Stieritz, Hahnemann Medical College
1981-83 James Poupard, Bryn Mawr Hospital
1983-85 Donald Jungkind, Thomas Jefferson University
1985-87 Walter Ceglowski, Temple University
1987-89 Irving Millman, Fox Chase Cancer Center
1989-91 Nick Burdash, Phila. College of Osteopathic Medicine
1991-93 Toby Eisenstein, Temple University
1993-95 Carl Abramson, Pennsylvania College of Podiatric Medicine

1995-97   Alan Evangelista, MCP Hahnemann University
1997-99   Donald Jungkind, Thomas Jefferson University
1999-01   Paul L. Cerwinka, Quest Diagnostics Inc.
2001-03   Richard Rest, MCP-Hahnamann University
2003-05   Irv Nachamkin, University of Pennsylvania
2005-09   Donald Jungkind, Thomas Jefferson University

Councilor Alternates
1937   David Bergey, National Drug Company
1940-41   Fred Boerner, Graduate Hospital
1942-43   Earle H. Spaulding, Temple University
1944-45   Ruth E. Miller, Woman's Medical College
1946-47   James A. Harrison, Temple University
1948-49   Amedeo Bondi, Hahnemann Medical College
1958-59   Bernard Briody, Hahnemann Medical College
1960-61   George Warren, Wyeth Laboratories
1962-63   Albert Moat, Hahnemann Medical College
1964-65   Herman Friedman, Einstein Medical Center
1966-67   Samuel Ajl, Einstein Medical Center
1968-69   Richard Crowell, Hahnemann Medical Center
1970-71   Joseph F. Pagano, Smith Kline and French Laboratories
1972-74   Ralph A. Knight, Medical College of Pennsylvania
1975-77   John McKittrick, University of Pennsylvania
1978-79   Burt Landau, Hahnemann Medical College
1979-81   James A. Poupard, Bryn Mawr Hospital
1981-83   Irving Millman, Institute for Cancer Research
1983-85   Walter Ceglowski, Temple University
1985-87   Page Morahan, Medical College of Pennsylvania
1987-89   Nick Burdash, Phila. College of Osteopathic Medicine
1989-91   Bruce Kleger, Pa. Department of Health, Bureau of Laboratories
1991   (temporary) Ken Cundy, Temple University
1991-93   Carl Abramson, Pennsylvania College of Podiatric Medicine
1994-95   Post left vacant.
1995-99   Irv Nachamkin, University of Pennsylvania
1999-01   Richard Rest, MCP-Hahnemann Medical School
2001-03   Irv Nachamkin, University of Pennsylvania
2003-05   Donald Jungkind, Thomas Jefferson University
2005-09   Olarae Giger, Main Line Clinical Laboratories

President Elect (pre 1980 Vice-Presidents)
1970-71 James Prier, Pennsylvania Department of Health Laboratories
1972-73 Richard Crowell, Hahnemann Medical College
1974-75 Norman P. Willett, Temple University School Dentistry
1975-77 Robert J. Mandle, Thomas Jefferson Medical College
1977-79 Joseph F. Pagano, Smith Kline &French Laboratories
1979-81 John C. McKitrick, University of Pennsylvania
1981   Henry R. Beilstein, Manor Junior College
1982-83 Toby Eisenstein, Temple University
1983-85 Donald Stieritz, Hahnemann Medical College
1985-87 James Poupard, Bryn Mawr Hospital
1987-89 Paul Actor, Smith Kline & French Laboratories
1989-91 Alan Evangelista, Cooper Medical Center
1991-93 Linda Miller, Holy Redeemer Hospital
1993-95 Paul Cerwinka, Corning Clinical Laboratories
1995-97 Rick Rest, MCP-Hahnemann University.
1997-99 Irv Nachamkin, University of Pennsylvania
1999-01 Donald Jungkind, Thomas Jefferson University
2001-03 Olarae Giger, Main Line Clinical Laboratories
2003-05 David Axler, Temple University
2005-06 Bettina Buttaro, Temple University
2006-09 Laura Chandler, Philadelphia VA Medical Center

Appendix III

A Listing of Branch Meetings: 1936 Through 2009

The Branch has held meetings on a regular basis since its founding on 24 February 1920. Although records were kept of the meetings that were held during the period from 1920 to 1935, most of these records have been lost. Complete records exist in the Branch Archives for all meetings (with the exception of four meetings) that occurred starting in 1936 and continuing to the present time. The Branch Archives contain the mailed penny-post-card meeting notices that were initiated in 1938 and employed for several decades. For meetings in 1936 through 1947 abstracts of each meeting were published as the Proceedings of the Eastern Pennsylvania chapter in the *Journal of Bacteriology*.

The meeting that occurred on 1 May 1937, as published in the Journal of Bacteriology, was designated as the "124th Meeting" of the Branch. This is the earliest meeting found that has a number designation and was determined by records that had been collected by the Branch Secretaries over the years and certified by Dr. Harry Morton. It should be noted that it was later discovered that this was actually the 126th meeting. Over the years many meeting notices have incorrect number designations. All meeting number designations in the list that follows are consecutive, starting with the first meeting held in 1920.

It should be noted that starting in September of 1967 and continuing through June of 1973 there were a series of meetings conducted by the Clinical Microbiology Section of the Branch. While the regular meetings were held on Tuesday evenings, these meetings were held on Monday evenings. These Clinical Microbiology meetings receivd a separate meeting number independent of the regular meetings. They are included in this list with a CM number designation, and are independent of the regular meeting numbering system.

Meetings 1936 to 1939—No. 118 through 142
Meetings 1940 to 1949—No. 143 through 211
Meetings 1950 to 1959—No. 212 through 281
Meetings 1960 to 1969—No. 282 through 348

Meetings 1970 to 1979—No. 349 through 427
Meetings 1980 to 1989—No. 428 through 527
Meetings 1990 to 1999—No. 528 through 619
Meetings 2000 to 2009—No. 620 throu®gh 702

LIST OF BRANCH MEETINGS:
1936 [No. 118] TO 2009 [No.702]

No.	Date	Title and Speaker
118	1-28-36	The Advantages of Vacuum Dried Complement for use in the Routine Wassermann Reaction.—Fred Boerner, Marguerite Lukens
119	3-24-36	Phagocytic Activity of Circulating Cells in Leukemias.—Max M. Strumia, Frederick Boerner The Effect of Sonic Vibrations on (A) Pure Proteins, and (B) Bacterial Constituents.—Earl W. Flosdorf, Leslie A. Chambers The Phagocytosis of Streptococci by Neutrophiles in Whole Human Blood.—David Lackman, Fred Boerner, Stuart Mudd Antigenic Compostion and Immunological Behavior of Hemolytic Streptococci—A Progress Report.—Stuart Mudd, E. J. Czarnetzky, Horace Pettit, David Lackman Some Observations and Correlations in the Dissociation of the Diphtheria Bacillus.—Harry E. Morton
120	10-27-36	The Incidence of Tuberculous Infection in American Colleges.—Esmond R. Long Chemical and Immunological Relationships between the Proteins from Different Acid-Fast Bacilli.—Florence B. Seibert A Serological Comparison of the Proteins of Various Strains of Leprosy Bacilli and other Acid-Fast Bacteria.—Howard J. Henderson Further Studies on the Mechanism of Immunity to Tuberculosis.—Max B. Lurie

| 121 | 11-24-36 | Immunity Response in Guinea-pigs to Alum Toxoid.—W. P. Knerr, G. A. Hottle, F. L. Rights
Hemolysis Test for Gas Gangrene Toxin and Antitoxin.—G. A. Hottle, C. Okono
Observations on the Protection Test for Anti-Meningococcic Serum.—J. Fertig
Meningococcic Meningitis in an Eight-day Old Infant.—J. L. Ingham, F. O. Zillessen
Cholecystitis Caused by Haemophilus influenzae.—S. E. Weintraub, F. O. Zillessen
Coexisting Typhoid and Paratyphoid "B" Infection.—J. Kincov, F. O. Zillessen
Clostridium welchii Infection in Dog Treated Successfully with Gas Gangrene Antitoxin.—A. F. Millar, W. B. Rawlings, W. G. Love
Unconcentrated Versus Concentrated Antitoxin in the Treatment of Gas Gangrene.—F. O. Zillessen
The Teaching of Bacteriology to Pre-Medical Students.—W. R. Hunt |
|---|---|---|
| 122 | 1-26-37 | Essential Immunizing Antigen of Pheumococci.—Lloyd D. Felton |
| 123 | 2-23-37 | Relation of Viruses to Neoplastic Diseases .—Baldwin Lucke
Studies in Active Immunization Against Human Influenza.—Joseph Stokes
Studies on Concentration and Preservation of the Viruses of Poliomyelitis and Influenza.—H. W Scherp, I. L. Wolman, E. W. Flosdorf, D. R. Shaw
Unsuccessful Attempts to Cultivate the Virus of Acute Anterior Poliomyelitis in Various Living Culture Media.—John A. Kolmer, Clara Kast, Anna M. Rule
Attempts to Transmit Acute Anterior Poliomyelitis to Rabbits, Guinea-pigs, Rats, Mice, Chickens and Ferrets with and without Depression by X-rays.—John A. Kolmer, Clara Kast, Anna M. Rule |
| 124 | 3-23-37 | The Colony Form of Hemolytic Streptococci from Clinical Sources.—David Lackman
Surface Composition of Mucoid, Glossy and Rough Variants of Hemolytic Streptococci.—Stuart Mudd, Horace Pettit |

		Isolation and Specificity of the Labile Antigen of Streptococcus hemolyticus (Group A).—E. J. Czarnetzky, Stuart Mudd, H. Pettit, D. Lackman A Stable Hemolysin-Leucocidin from B-hemolytic Streptococci.—E. J. Czarnetzky, Isabel Morgan Convalescent Scarlatinal Serum in the Prevention and Treatment of Streptococcal Diseases.—Aims C. McGuinness, Joseph Stokes Chemotherapy in Streptococcal Infections.—D. Sergeant Pepper, E. J. Czarnetzky, I. S. Ravdin
125	5-1-37	Bacteriophage.—John H. Northrop The Coccobacilliform Bodies of Fowl Coryza and Mouse Catarrh.—John B. Nelson Isolation of Mild Strains of Aster Yellows from Heat-treated Leaf Hoppers. —L. O. Kunkel
126	5-25-37	Bacterial Type Transformation.—Hobart A. Reimann The Effect of Dietary Minerals Upon Host Resistance.—Charles F. Church Staphylococcus Studies. I. Toxin Production.—E. P. Casman Concentration of Staphylococcus Toxoid by the Kidneys.—E. P. Casman Degenerative Changes of the Neutrophiles in Clinical and Experimental Observations.—Max M. Strumia
127	10-26-37	Dr. Bergey As I Knew Him.—Randle C. Rosenberger The Role of Inherited Natural Resistance to Tuberculosis.—Max B. Lurie The Nature of Inherited Natural Resistance to Tuberculosis.—Max B. Lurie A Study of the Agar Cup Plate Method. —Ruth E. Miller, S. Brandt Rose A Study of Mercury Antiseptics by the Agar Cup Plate Method.—S. Brandt Rose, Ruth E. Miller The Application of Sintered Glass Filters to Bacteriological Work.—Harry E. Morton, E. J. Czarnetzky
128	11-23-37	Virus Activities in Relation to Cancer.— Peyton Rous

129	1-25-38	Notes on Steam Pressure Sterilization.—Carl J. Bucher
130	2-22-38	Bacteriophagic Service in Septic Conditions.—Ward J. MacNeal
131	3-22-38	Bacteriological Studies on Rheumatic Fever Cases (680 Cases).—Rose Ichelson Demonstrations of the Methods for Measuring Sanitary Ventilation.—William F. Wells, Ruth Blumfeld
132	4-26-38	A Comparative Study of Media Employed in the Isolation of Typhoid Bacilli from Feces and Urine.—Cora B. Gunther, Louis Tuft The Chemistry of Nucleoprotein Antigen—'Labile Antigen' from Hemolytic Streptococci or Lancefield Group A.—M. G. Sevag, D. Lackman, J. Smolens Some Reactions of Sulfanilamide with Nuceleoproteins and Suggested Mechanism of the Action of Sulfanilamide.—E. J. Czarnetzky, H. E. Calkins
133	5-24-38	Molecular Weight, Electrochemical and Biological Properties of Tuberculin Protein and Polysaccharide Molecules.—Florence B. Seibert, Kai O. Pedersen, Arne Tiselius Studies on Tetanus Toxoid.—Herman Gold
134	10-18-38	The Original Minutes of the Society of American Bacteriologists.—C. G. Roos The Pathogenesis of Local Tetanus.—William Chalian Recent Advances in our Knowledge of Tetanus.—Warfield M. Firor
135	11-22-38	A Simplified Complement Fixation Technic for the Serological Diagnosis of Syphilis.—Fred Boerner, Marguerite Lukens A Simplified Method for the Preparation of an Antigen for use in the Complement Fixation Test for Syphilis.—Fred Boerner, Charles A. Jones, Maguerite Lukens Report of the Conference on the Laboratory Diagnosis of Syphilis.—A. Parker Hitchens

136	1-24-39	A New Filterable Agent Associated With Respiratory Infections.—Drs. Joseph Stokes, Hobart A. Reimann, Mrs. Dorothy R. Shaw Further Studies on the Value of Routine Anaerobic Cultures.—Dr. E. H. Spaulding, William Goode Demonstration of an Improved Anaerobic Apparatus.—Dr. E. H. Spaulding.
137	2-28-39	Nucleoproteins from Streptococcus pyogenes: Some Chemical and Serological Properties and Changes in Both Caused by Certain Enzymes.—Dr. Charles A. Zittle The Action of Sulfanilamide on Hemolytic Streptococci in Human Blood and Serum.—Dr. John S. Lockwood, Helen M. Lynch Some Observations on the Etiology of Vaginal Infections.—Dr. A. E. Rakoff
138	3-28-39	Hunting Tubercle Bacilli Fifty Years Ago.—Dr. Joseph McFarland The Introduction of Agar Agar into Bacteriology.—Dr. A. P. Hitchens, Morris Leikind Notes on the History of Pure Culture Methods.—J. R. Shramm
139	4-25-39	A Symposium on Lymphopathia venereum: History, Bacteriology, Frei Test and its Evaluation.—Harry E. Bacon, Francis D. Wolfe. Gross Lesions.—Collier F. Martin Pathology.—Eugene Case, M. S. Hwang Lymphatics.—Oscar V. Batson Blood Chemistry Findings.—Charles A. Jones Roentgenographic Interpretations of Rectal Stricture and Lymphopathia venereum.—Arthur Finkelstein
140	5-23-39	A Method for Making Bacterial Counts in a Test Tube.—Edward Redowitz The Pathogenesis of Rheumatic Fever.—Dr. Mark P. Schultz Bacterial Allergy.—Dr. Paul H. Langner A Typhoid-like Infection Caused by a Slow Lactose-Fermenting Organism.—Dr. John Eiman, Russell H. Fowler

141	10-24-39	A Summary of the Papers and Demonstrations on Rickettsial Diseases and Rickettsia Presented at the Symposium on Virus and Rickettsial Diseases by the Harvard School of Public Health, June 12-17, 1939.—Dr. Harry E. Morton A Summary of the Papers and Demonstrations on Virus and Viral Diseases Presented at the Symposium on Virus and Rickettsial Diseases by the Harvard School of Public Health, June 12-17, 1939.—Dr. Betty Lee Hampil A Summary of the Papers on Influenza Presented at the Third International Congress for Microbiology, New York, Sept. 2-9, 1939.—Dr. Joseph Stokes, Jr. A Summary of the Papers on Chemotherapy Presented at the Third International Congress for Microbiology, New York, Sept. 2-9, 1939.—Dr. John S. Lockwood
142	11-28-39	Tuberculosis in Sheep.—Dr. E. L. Stubbs The Fate of Tubercle Bacilli Phagocyted by Cells Derived From Normal and Immunized Rabbits.—Dr. Max B. Lurie The Preparation and Use of the BCG Vaccine.—Dr. J. D. Aronson, Erma I. Parr, Robert Saylor Natural Quantitative Respiratory Contagion of Tuberculosis.—Wm. F. Wells, Dr. Max B. Lurie
143-211	1940s	Meetings 1940 to 1949—No. 143 through 211
143	1-23-40	Immunologic and Chemotherapeutic Studies in Syphilis.—Dr. Harry Eagle Simplified Microscopic and Macroscopic Flocculation Tests for the Diagnosis of Syphilis.—Dr. Fred Boerner, Dr. Charles A. Jones, Marguerite Lukens
144	2-27-40	The Eastern Pennsylvania chapter of the Society of American Bacteriologists—The Completion of Two Decades.—Dr. Harry E. Morton Swine Influenza.—Dr. Joseph P. Scott Enzootic Tularemia.—J. F. Bell Serological Studies with Black-leg Aggressin.—Dr. I. Live Periodic Ophthalmia.—Dr. E. L. Stubbs, Dr. W. G. Love

145	3-26-40	A Rapid Slide Agglutination Test for the Diagnosis of Tularemia.—Cora Gunther, Dr. Agnes Beebe Vogt The Nucleic Acid—Pneumococcic Horse Serum Reaction.—Drs. David Lackman, Stuart Mudd, M. G. Sevag The Inhibition of Proteolytic Enzymes of Pathogenic and Non-pathogenic Clostridiae.—Louis DeSpain Smith Quantitating Gordon's Bacterial Test for Estimating Pollution of Air.—W. F. Wells, Dorothy Wells
146	4-23-40	An Agent Isolated from a Soil Bacillus which Inhibits Capsule Formation of Friedlander's Bacterium and is Highly Bactericidal for Gram Positive Microorganisms.—Dr. J. C. Hoogerheide Hemophilus pertussis Antigenicity.—C. Roos, Dr. H. P. Bellew Antigenic Studies on Various Phases of Hemophilus pertussis.—Dr. Earl W. Flosdorf, A. C. Kimball, T. F. Dozois The Physiology of Luminescence and Respiration of Luminous Bacteria.—Dr. Frank H. Johnson
147	5-28-40	Swine Influenza.—Dr. Joseph P. Scott The Rabies Situation in Pennsylvania.—Dr. M. F. Barnes Brief Summary of Results of Calfhood Vaccination in One Herd Against Bang Disease During a Five Year Period.—Dr. M. F. Barnes The Agglutination-lysis Test for Leptospira canacola. A Preliminary Survey of a Series of Healthy Dogs.—Dr. Clara Raven, Mrs. Kathryn Barnes Leptospirosis, A Public Health Hazard.—Drs. W. Paul Havens, Carl J. Bucher, Hobart A. Reimann
148	10-22-40	The Electron Optics of the Electron Microscope.—James Hillier Structural Differentiation Within the Bacterial Cell.—Dr. Stuart Mudd, David B. Lackman Pictures of Bacterial Forms Taken With the Electron Microscope.—Katherine Polevitzky Electron Micrographs of Further Biologic Specimens.—Thomas F. Anderson

149	11-26-40	The Relation of Sex to the Course of Experimental Tuberculosis in Mice.—Drs. Esmond R. Long, Agnes Beebe Vogt Relationship of Hereditary Constitution and Immunological Factors in Resistance to Tuberculosis.—Dr. Max B. Lurie Improvements in the Preparation of Purified Protein Derivative Tuberculin.—Dr. Florence B. Seibert, John T. Glenn Incidence of Coccidioides immitis Infection in Some Parts of the United States.—Dr. Joseph D. Aronson, Robert M. Saylor, Erma I. Parr
150	1-28-41	The Etiology of Infectious Stomatitis in Children.—Dr. T. F. McNair Scott Electron-Microscope Studies of the Reduction of Potassium Tellurite by Corynebacterium diphtheriae.—Dr. Harry E. Morton, Dr. Thomas F. Anderson
151	2-25-41	Experience in an Epidemic of Poliomyelitis in the Middle West.—Dr. Milton J. Rose Bacterial Symbiosis: Inoculation Studies in Man.—Dr. S. S. Greenbaum Bacteriology of the Dental Pulp.—Drs. J. L. T. Appleton, J. H. Gunter Anaerobic Culture Methods Demonstration—8 demonstrations.
152	3-25-41	Antigenic Structure of the Group A Hemolytic Streptococci: Some Properties of the Type-Specific Protein and the Group-Specific Polysaccharide.—Dr. Charles A. Zittle Isolation and Properties of Pigmented Heavy Particles From Streptococcus pyogenes.—Dr. M. G. Sevag, Mr. J. Smolens, Dr. Kurt G. Stern Action of Crystalline Pepsin on Diphtheria Antitoxin and Pneumococcus Antibody from the Horse.—Dr. A. M Pappenheimer, Jr., Dr. M. L. Petermann A Study of the Relationship Between the Virus of Influenza A and Filterable Components of Normal Lungs.—Dr. L. A. Chambers, Dr. Werner Henle Analytical Diffusion of Some Animal Viruses.—Dr. Jaques Bourdillon

| 153 | 4-22-41 | Culture Studies of Gonorrheal Infection in Women.—Dr. J. F. Mahoney
Comparison of Media and Laboratory Results in Gonococcus Cultures.—Dr. C. J. Van Slyke
Gonorrheal Infection in the Male.—Dr. P. S. Pelouze |
|---|---|---|
| 154 | 5-27-41 | The Pigments from Variants of Micrococcus tetragenus.—Dr. Hobart A. Reimann
Titration of Tuberculin Sensitivity and its Relation to Tuberculous Infection.—Dr. Waldo E. Nelson, Michael L. Furcolow, Barbara Hewell, Carroll E. Palmer
Mold Allergy.—Dr. G. I. Blumstein
A Confirmed Test for Streptococci From Air.—Dorothy Wells |
| 155 | 10-28-41 | Toxins in the Several Phases of H. pertussis.—Dr. E. W. Flosdorf, Mr. A. Bondi, Dr. T. Dozois
The Effect of Sulfanilyguanidine on Brucella abortus.—T. H. Grainger, E. V. Gibson, Dr. E. P. Campbell
Mechanism of Sulfanilamide Action-Simultaneous Inhibition of the Respiration and Growth of Streptococcus pyogenes by Sulfonamide Drugs.—Dr. M. G. Sevag, Miss Myrtle Shelburne
Demonstration by the Electron Microscope of the Combination of Antibodies With Flagellar and Somatic Antigens.—Dr. Stuart Mudd, Dr. Thomas Anderson |
| 156 | 11-25-41 | Electrophoretic Analysis of the Blood Proteins in Tuberculosis.—Florence B. Seibert, W. E. Nelson
Mechanisms of Immunity in Tuberculosis. The Fate of Tubercle Bacilli Ingested by Mononuclear Phagocytes Derived From Normal and Immune Animals.—Dr. Max B. Lurie, Peter Zappasodi
Risk of Tuberculous Infection in General Hospitals.—Dr. Harold L. Israel
The Demonstration of Tubercle Bacilli by Fluorescence Microscopy.—Dr. Earle H. Spaulding, Janet Raefler |

157	1-27-42	Studies in Surgical Bacteriology. I. Distribution of Bacteria Around the Nails and on the Skin in Relation to Disinfection of the Hands.—Russell H. Fowler Studies on the Nature of the Virus of Influenza.—Dr. Leslie A. Chambers, Dr. Werner Henle The Moving Picture: "Unseen Worlds" Illustrating the RCA Electron Microscope.—Dr. Harry E. Morton
158	2-24-42	A Study of Microaerophilic Organisms with Special Reference to the use of a Microaerophilic Incubator.—Dr. S. Brandt Rose Studies in the Prevention of Air-Borne Infections.—Dr. Werner Henle, Miss Harriet E. Sommer, Dr. Joseph Stokes, Jr. Microbial Antagonism and Brucella abortus.—Dr. Walter Kocholaty
159	3-24-42	Some Practical Applications of Bacteriology to Medical Nursing.—Dr. Carl J. Bucher Mold Inhibition in Various Food Products Through the Use of Inhibitory Chemicals.—Dr. D. K. O'Leary The Germicidal Properties of Soaps.—Dr. Werner Leszynski
160	4-28-42	A Simple Tube Battery for Use in Bacterial Air Analysis.—W. F. Wells The Boerner-Lukens Wassermann Test With Spinal Fluid With Special Reference to the Use of Egg Albumin.—Dr. Fred Boerner, Marguerite Lukens, Alice Ellis The Boerner-Jones-Lukens Macroflocculation Test With Spinal Fluid.—Dr. Fred Boerner, Marguerite Lukens, Alice Ellis Spirochetal Antigens in the Serum Diagnosis of Syphilis.— Dr. John A. Kolmer A Moving Picture: "A Lecture on the Spirochetes."—Dr. Theodor Rosebury
161	5-26-42	The Production of Antibiotic Agents by Microorganisms and its Significance in Natural Processes.—Dr. Selman A. Waksman

162	10-27-42	Studies on the Allergic and Antigenic Activity of Sonic Filtrates of Brucella abortus.—Dr. I. Live The Antigenic Structure of H. pertussis.—Drs. E. W. Flosdorf, A. Bondi, Harriet M. Felton, A. C. McGuinness Clinical Results of the use of Agglutinogen from Phase 1 H. pertussis as a Skin Test for Susceptibility to Whooping Cough.—Drs. Harriet Felton, E. W. Flosdorf The Antigenic Structure of H. parapertussis and its Probable Clinical Significance.—Drs. A. Bondi, Jr., Harriet M. Felton, E. W. Flosdorf
163	11-24-42	The Use of Syrian Hamsters as a Laboratory Animal.—Dr. Harry E. Morton The Use of Promin on Experimental Tuberculosis in the Rabbit.—Dr. Max B. Lurie Characteristics and Therapeutic Uses of Succinylsulfathiazole.—Dr. W. F. Verwey Laboratory Studies in Spotted Fever Immune Serum.—Florence Fitzpatrick
164	1-26-43	Electron-Microscope Study of Bacteriophage.—Dr. Thomas F. Anderson Swine Influenza Associated with Hog Cholera.—Dr. Joseph P. Scott Ultraviolet Blood Irradiation Therapy in Acute Pyogenic Infections.—Dr. G. P. Miley Observations of the Effects of Ultraviolet Irradiation (Knott technic) on Bacteria and Their Toxins.—Drs. G. P. Blundell, L. A. Erf, H. W. Jones, R. T. Hoban
165	2-23-43	Report on the Work of the Committee on Materials for Visual Instruction in Microbiology of S.A.B.—Dr. Harry E. Morton The Showing of a Motion Picture Prepared by Dr. AdrianusPijper, Pretoria, South Africa, on Dark-Field Studies on the Motility and Agglutination of Typhoid Bacilli.—Dr. Harry E. Morton Somatic and Flagellar Typhoid Antibodies and Pneumococcal Capsular Antibodies as Shown by the Electron Microscope.—Dr. Stuart Mudd Study of Flagella of a Fresh Water Bacterium by Motion Micro Photography and Electron Micrography.—Dr. W. G. Hutchinson, Mary R. McCracken

166	3-23-43	Some Recent Developments in the Study of the Staphylococci.—Dr. John E. Blair
167	4-27-43	Penatin, The Second Antibacterial Substance Produced by Penicillium notatum Westling "77".—Dr. Walter Kocholaty Effective Blood and Spinal Fluid Levels of Sulfonamide Drugs in the Treatment of Meningitis.—Dr. John A. Eiman, Mr. Haraold W. Fowler An Improved Technic for the Cultivation of Anaerobic Micro-organisms.—Dr. Harry E. Morton Factors Influencing the Propagation of Influenza A Virus in the Developing Chick Embryo.—Drs. Werner Henle, Gertrude Henle
168	5-15-43	Note on Antibiotic Substances Elaborated by an Aspergillus flavus Strain and by an Unclassified Mold.—Dr. Arthur E. O. Menzel, Dr. O.Wintersteiner, Dr. Geoffrey Rake. Synthesis of Pyridoxine by a "Pyridoxinless" X-ray Mutant of Neurospora sitophila.—Dr. J. L. Stokes, Dr. J. W. Foster, Mr. C. R. Woodward Jr. Some Evidence on the Etiology of Cancerous Properties as Exemplified in Plant Cells.—Dr. Phillip R. White, Dr. Armin C. Braun An Analysis of the Antagonistic and the Synergistic Action of Acetone, Ethyl Alcohol, Butyl Alcohol, Chloroform, Ether, and Urethane on Suflanilamide Inhibitions.—Dr. Frank H. Johnson, Dr. Henry B. Eyring, Mr. Wallace Kearns The Action of an Antibiotic Substance (Penatin) On Bacteriophage.—Dr. Thomas Anderson Increased Incidence of Virus Inclusion Bodies in Human Throats.—Dr. Jean Broadhurst, Miss Estelle MacLean, Inez Taylor Preparedness for Defense Against Influenza.—Dr. Ward J. MacNeal, Miss Ernestine R. Parker Test of Anti-Dysentery Agents in Embryonated Eggs.—Dr. Ward J. MacNeal, Miss Anne Blevins, Dr. Macello Pacis

169	5-25-43	Bacillary Dysentery.—Dr. Joseph Felsen
170	10-26-43	Characteristics of the Gonococcus and Recent Developments in the Laboratory Diagnosis of Gonorrhea.—Dr. Harry E. Morton Management of Venereal Diseases in the Navy.—Dr. Paul Leberman The Oxidase Test. (Motion picture film, colored).—Dr. Harry E. Morton A Microscopic Alkali-Solubility Test for Identifying the Gonococcus.—Dr. A. Cantor
171	11-23-43	Virus Reactions Inside of Bacterial Host Cells.—Dr. Thomas F. Anderson A Microbiological Assay for Riboflavin B2 using Lactobacillus casei E.—Dr. Walter L. Obold The Riboflavin Content of Luncheon Meats.—Miss Lilliam C. Berle, Dr. Walter L. Obold Stage of Suspension of Air-Borne Infection.—Mr. W. F. Wells, Miss Elizabeth Rigney
172	1-25-44	Protein Metabolism and Resistance to Infection.—Prof. Paul R. Cannon
173	2-22-44	The Prevention of Natural Air-Borne Contagion of Tuberculosis in Rabbits by Ultra-Violet Irradiation.—Dr. Max B. Lurie, Miss Helen Tomlinson, Mr. Samuel Abramson The Mechanism of Color Productin in Escherichia coli Cultures Containing Sulfonamides.—Mr. R. J. Strawinski, Dr. W. F. Verwey, Mr. J. L. Ciminera Sulfadiazine, Sulfamerazine, and Sulfamethazine in the Therapy of Experimental Infections in Mice. —Dr. W. F. Verwey, Miss Dorothy N. Sage
174	3-28-44	Randle C. Rosenberger MD (1873-1944). —C. J. Bucher Interference of Inactive Influenza Viruses with the Propagation of the Active Agent.—Drs. Werner and Gertrude Henle Epidemic Nausea, Vomiting, and Diarrhea Possibly Caused by an Enterotropic Virus.—Drs. Hobart A. Reimann, John Hodges, Alison H. Price

		Bacteriological Findings in Gastro-Intestinal Disturbances Occurring During the Winter Months.—Drs. George P. Blundell, William D. Beamer, Mrs. Ruth W. Crouch
175	5-23-44	Complement—No. 1 Jig-Saw Puzzle. —Dr. Michael Heidelberger
176	10-24-44	The Effect of Tuberculosis and Sensitized Sera and Serum Fractions on the Development of Tubercles in the Chorio-Allantoic Membrane of the Chick Embryo.—Drs. Emily E. Emmart, Florence B. Seibert The Role of the Lymphocyte in Antibody Formation.—Drs. T. N. Harris, W. E. Ehrich
177	11-28-44	Some Studies on Infectious Hepatitis. —Capt. John R. Neefe, MC, A.U. Development of Penicillin Production. —Mr. G. Raymond Rettew
178	1-23-45	The Action of Certain Poorly Absorbed Sulfonamides on the Intestinal Flora of Rats.—R. J. Strawinski, W. F. Verwey, R. Munder Antigenic Structure and Specificity of Luminous Bacteria.—George H. Warren Antigenic Variation in Pure Cultures of Paramecia.—James A. Harrison, Elizabeth H. Fowler
179	2-27-45	Fulminating Obstructive Laryngitis and Septicemia due to Hemophilus influenzae, Type B.—Dr. John Eiman, Mr. Russell Fowler An Improved Liquid Culture Medium for the Growth of Hemophilus pertussis.—Dr. W. F. Verwey, Miss Dorothy Sage The use of Hemophilus pertussis Agglutinogen in the Preparation of Hyper Immune Serum for the Treatment of Whooping Cough.—Dr. Harriet Felton, Mr. J. Smolens, Dr. Stuart Mudd, M. Carr, I. Lincoln, L. Walker
180	3-27-45	The Chemical Alteration of a Bacterial Surface with Special Reference to the Agglutination of Proteus OX-19. —Dr. Seymour S. Cohen

		In vivo and in vitro Studies on the Bacteriostatic and Bactericidal Actions of Mercurial Disinfectants on Hemolytic Streptococci.—Dr. Harry E. Morton, Mr. Leon L. North, Mr. Frank B. Engley Jr. Remarks on the Typing of Typhoid Bacilli by the use of V1-Bacteriophage.—Mrs. Cora Gunther A Study of the Agglutination of Living Typhoid Bacilli by V1-Serum by Means of Dark Ground Illumination. A Motion Picture.—Dr. Adrianus Pijper
181	4-26-45	Immunity to the Larger Animal Parasites. —Dr. William Hay Taliaferro
182	5-22-45	The Disease Agents in Relation to Chemotherapy.—Dr. Stuart Mudd
183	10-23-45	Bacterial Penicillinase:Production, Nature, and Significance.—Dr. Amedeo Bondi, Jr., Miss Catherine Dietz The Combined Action of Penicillin and Sulfonamide in vitro: The Nature of the Reaction.—Dr. Morton Klein, Mr. Seymour S. Kalter Studies on Blood Agar. I. An Improved Blood Agar Base.—Lt. Cdr.Ezra P. Casman
184	11-27-45	Chick Embryo Technic for Intravenous and Chemotherapeutic Studies.—Dr. Henry F. Lee, Dr. Abram Stavitsky, Margaret P. Lee The Antigens of Epidemic Typhus Vaccine. I. The Distribution and Immunological Properties.—Dr. Leslie A. Chambers, Dr. Seymour S. Cohen, Jean R. Clawson The Antigens of Epidemic Typhus Vaccine. II. The Chemical Composition of Rickettsiae.—Dr. Seymour Cohen, Dr. Leslie A. Chambers, Jean R. Clawson
185	1-22-46	Cholera in Chungking in 1945.—Dr. Hobart A. Reimann The Shadow-Casting Technique in Electron Microscopy.—Dr. Thomas F. Anderson Phase Microscopy.—Dr. Oscar W. Richards
186	2-26-46	Clinical Bacteriology and Immunology of Ocular Infections.—Charles Weiss, M. D. Interference With the Antibacterial Action of Streptomycin by Reducing Agents. —Amedeo Bondi, Jr., PhD, Catherine C. Dietz, BS, Earle H. Spaulding, PhD

		Papers Presented at the Conf. on Antibiotics, NY Academy of Science 17-19 Jan 1946: Microbiological Aspects.—Albert N. Kelner, PhD Chemical Aspects.—J. M. Sprague, PhD Pharmacological and Clinical Aspects.—Stuart Mudd, M. D., Harry E. Morton, Sc. D.
187	3-26-46	The Tubercle Bacillus as an Indicator Organism in Quantitative Studies of Airborne Infection. I. Quantitative Aerosol Suspension of Tubercle Bacilli.—W. F. Wells II. Quantitative Enumeration of Tubercle Bacilli in vitro.—Cretyl Crumb III. Quantitative Enumeration of Tubercle Bacilli in vivo.—H.L.Ratcliff, Sc.D Development of Streptomycin Resistance of Shigellae.—Morton Klein, PhD, Leonard Kimmelman
188	4-23-46	Serology of Rheumatoid Arthritis.—A. D. Wallis, MD Allergy Against Insulin.—Mary Hewett Loveless, MD
189	5-14-46	Studies on Inhibition of Growth by Structural Analogues of Metabolites.—Dr. D. W. Woolley
190	10-22-46	Intravenous Infection of the Chick Embryo with Human Turbercle Bacilli: Inhibitory Effects of Streptomycin.—Dr. Henry F. Lee Reducing the Pyrogenicity of Concentrated Solutions of Protein.—Dr. Robert Pennell, Mr. William Elliott Smith Some Observations on Microbiology in the U.S.S.R.—Dr. Stuart Mudd
191	11-26-46	The Effect of Ultraviolet Irradiation on Various Properties of Influenza Viruses.—Drs. Werner, Gertrude Henle Familial Non-Specific Serological Reactions for Syphilis.—Dr. Arthur G. Singer, Dr. Fred Boerner Inhibition of Division in the Protozoan, Tetrahymena, by Antisera and Bacteria. Demonstration of Bacteria and Protozoa Under the Phase Microscope.—Miss Elizabeth H. Fowler, Dr. James A. Harrison

192	1-28-47	The Effect of Caronamide Upon Penicillin Therapy of Experimental Pneumococcus and Typhoid Infections in Mice.—Dr. W. F. Verwey, Dr. A. K. Miller A Mechanism for the Development of Resistance to Penicillin and Streptomycin.—Dr. Morton Klein One Incident of Failure to Remove Sensitizing Material From Berkefeld Filter in Spite of Careful Cleansing.—Ida Teller Histoplasmin Sensitivity and Non-Tuberculous Pulmonary Calcifications.—Dr. Robert H. High
193	2-25-47	Relation of Diet to Poliomyelitis in Mice.—Dr. J. H. Jones, C. Foster Studies on the Susceptibility to Typhus of Rats on Deficient Diets.—Florence K. Fitzpatrick Constitutional Physiology as Seen in Insanity.—Dr. S. de W. Ludlum
194	3-26-47	Constitutional Factors in Resistance to Tuberculosis: The Effect of Estrogen.—Dr. Max B. Lurie, Dr. Samuel Abramson, Mr. Marvin J. Allison The Life History and Cell Structure of Bacillus mycoides Studied with the Electron Microscope.—Dr. Georges Knaysi
195	4-22-47	Air Sampling Performances.—Prof. W. F. Wells, Cretyl Crumb Epidemic of Influenza A Among a Recently-Vaccinated Population: Isolation of a New Strain of Influenza A Virus.—Dr. M. Michael Sigel, Dr. F. W. Shaffer, Dr. Werner Henle A Serological Type of Paracolon Bacillus as Probable Cause of an Epidemic of Gastroenteritis.—Dr. T. F. McNair Scott, Dr. Lewis L. Coriell, Dr. Harriett Davis, Dr. B. H. Boltjes Studies on the Virus of Epidemic Parotitis.—Dr. Gertrude Henle, Susanna Harris, Werner Henle The Relationship Between the Height of Naturally Acquired Antibody and the Agglutinative Response to Immunization with H. pertussis Vaccine.—Dr. Aims C. McGuinness, Dr. Harriet Felton, Cecelia Willard

196	10-28-47	The Synergistic or Additive Activity of Chemotherapeutic Compounds. —Dr. John A. Kolmer Studies on the Inhibition of a Bacteriophage Multiplication by Salicylate.—Dr. John Spizizen A Strain of Gliocladium Which Destroys Cellulose.—Scott V. Covert Tryptophane Metabolism by Sulfonamide-Resistant and Sulfonamide-Susceptible Strains of Staph. aureus.—Edward Steers, Dr. M. G. Sevag
197	11-25-47	Application of Microbiology in Physiological Research.—Lemuel D. Wright, Ph. D. Virus Adaptation to Different Host Species and Different Tissues.—R. A. Kelser, D. V. M., PhD An Improved Medium for the Isolation of the Gonococcus in Twenty-four Hours. —Charles M. Carpenter Evaluation of Culture Media for the Diagnosis of Gonococcal Infection. —T. C. Buck Jr.
198	1-27-48	The Survival of Salmonella in Reconstituted Egg Powder Subjected to Holding and Scrambling.—Mathilde Solowey, Eleanor J. Calesnick Studies on Pneumoccal Enzymes Involved in Resistance to Drugs.—J. S. Gots, M. G. Sevag Plasma Concentration Following Intramuscular Injections of Various Doses of Penicillin.—A. Katherine Miller, William P. Boger The Determination of the Molecular Weight of Shiga Neurotoxin by Diffusion Constants.—F. B. Engley Jr.
199	2-24-48	Present Status of B.C.G. Vaccination. —Dr. Joseph D. Aronson Studies on the Organisms of the Pleuropneumonia Group (L organisms) Isolated From Human Sources. —Dr. William E. Smith
200	3-23-48	The in vivo Reactivation by BAL (2, 3, dimercapto-propanol) of Influenza Virus and Pneumococci Inactivated by Bichloride of Mercury.—Dr. Morton Klein, J. E. Perez

		Methods Employed in the Hirst Reaction. —Dr. Betty Lee Hampil The Application of Mumps Complement-Fixation Tests to Clinical and Epidemiological Problems.—Dr. Gertrude Henle, Susanna Harris, Dr. Werner Henle
201		Missing
202	5-25-48	Histoplasmosis in Pennsylvania.—Hobart A. Reimann, MD, Alison H. Price, MD Studies on the Antigenicity of Toxic Extracts of Hemophilus pertussis.—W. F. Verwey, Sc.D., Elizaberth H. Thiele Influence of Amino Acids on Antibody Response in Man. (Preliminary report).—S. Brandt Rose, MD, Michael G. Wohl, MD, John G. Reinhold, Ph.D
203	10-26-48	The Complement-Fixation Test as a Tool in the Diagnosis of Viral and Rickettsial Diseases.—M. Michael Sigel, PhD An Experimental Study of Some Repository Dosage Forms of Penicillin.— A. Katherine Miller, PhD, W. F. Verwey, Sc.D., Dorothy Wilmer Report on Study Session on Pleuropneumonia-like Organisms.—Louis L. Dienes, M. D.
204	11-23-48	Unusual Bacteriological Findings From an Infected Wound.—Dr. Carl F. Clancy A Possible Symbiotic Microorganism in the Cytoplasm of Paramecium.—Dr, John R. Preer Isolation of Enterococci from Human Saliva.—Dr. Ned B. Williams
205	1-25-49	Motility in Bacteria Produced in 1930. —Dr. R. P. Loveland Shape and Motility of Bacteria Produced in 1947.—Dr. Adrianus Pijper Electron Micrographs of Bacterial Flagella With a Critical Discussion of Bacterial Motility.—Dr. Stuart Mudd
206	2-22-49	A Possible Function of Hyaluronidase in Hemolytic Streptococcal Respiration. —Dr. Bennett Sallman General Aspects of Hyaluronidase. —Dr. M. G. Sevag Interrelationships of Time and Concentration in the Bacteriocidal Action of Penicillin.—Dr. W. F. Verwey

207		Missing
208		Missing
209	5-24-49	Clinical and Laboratory Studies of the Newer Antibiotics.—Dr. Harold A. Zintis Factors Affecting the Disappearance of Bacteria Placed on Normal Human Skin.—Mr. G. Rebell, Dr. D. M. Pillsbury, Miss M. de Saint Phalle, Miss D. Ginsburg
210	10-25-49	Extension of the Pure Culture Concept Through the Use of Germfree Animals. —Prof. James A. Reyniers
211	11-22-49	Use of Perchloric Acid in Bacterial Cytology.—William A. Cassel Inhibition of Yeast Hexokinase By Homologous Rabbit Immune Serum. —Ruth E. Miller, V. Z. Pasternak, M. G. Sevag Difference in the in vitro Susceptibilities of the Pleuropneumonia-like Organisms to Streptomycin and Dihydrostreptomycin. —Paul R. Leberman, Paul F. Smith, Harry E. Morton
212-281	1950's	Meetings 1950 to 1959—No. 212 through 281
212	1-24-50	Studies on Alcoholic Extracts of Enteric Organisms.—Robert Feinberg The Typhoid Research Program of Army Medical Department Research and Graduate School.—H. C. Batson, PhD
213	2-28-50	An Improved Staining Technique for the Nuclear Chromatin of Bacteria.—Andrew G. Smith, PhD Bacterial Nuclei as Vesicular Structures. —Stuart Mudd, MD, Andrew G. Smith, PhD Susceptibility to Typhus in the Progeny of Immune Animals.—Florence Fitzpatrick, PhD Sensitivity of Freshly Isolated Bacteria to Neomycin.—Carl F. Clancy, PhD
214	3-28-50	The Synthesis of Bacterial Viruses. —Seymour S. Cohen, MD Studies on the Propagation of Influenza Virus.—Werner Henle, MD
215	4-25-50	Biochemical Mechanisms of Cellulose Breakdown by Microorganisms.—R. G. H. Siu, PhD Relation of Vitamin B12 to Methionine Synthesis and Sulfonamide Inhibition in Escherichia coli.—Joseph S. Gots, PhD

		The Stimulation of Growth of Lactobacillus arabinosis 17-5 by Folic Acid Decomposition Products.—B. E. Koft, M.S., M. G. Sevag, PhD, E. Steers, Ph.D Experimental Studies on Aureomycin. —Morton Klein, PhD, Sonia Schorr, BS, Sylvia Tashman, BS, Andrew D. Hunt Jr., MD
216	5-27-50	Some Microbiological Aspects of Cancer Fungi Associated with Tumor Tissues. —Irene Corey Diller, Mycobacterial Forms Observed in Tumors.—V. Wuerthele-Caspe, E. Alexander-Jackson, L. W. Smith, J. Hiller and R. Allen. Antagonistic Action of Certain Neurotropic Viruses toward two Lymphoid tumors of Chickens.—G. Sharpless. Immunological Properties of Polysaccharides from Serratia marcescens.—H. J. Creech.
217	10-24-50	The Use of the Periodic Acid Schiff Reaction for the Demonstration of Fungi.—Albert M. Kligman, MD, PhD Studies on the Cultivation of Pleuropneumonia-like Organisms with a Description of Essential Growth Factor Present in Serum and Ascitic Fluid.—Paul Smith, M.S., Harry Morton, Sc.D., Paul R. Leberman, MD The Employment of Selective Bacteriostatic Agents in Isolating Pleuropneumonia-like Organisms from Clinical Material.—Harry E. Morton, Paul Smith, Paul Leberman The Susceptibility of Pleuropneumonia-like Organisms to Various Antibiotics and the Clinical use of Streptomycin and Terramycin.—Paul R. Leberman, Paul Smith, Harry Morton
218	11-28-50	Morphological Changes in Growing Cells of Proteus vulgaris OX-19.—Henry Stempen, W. G. Hutchinson, PhD Studies on the Fractionation of Hemophilus pertussis Extracts.—Elizabeth H. Thiele, M.S., Robert Pennell, PhD The Effect of Stretptomycin on the Growth and Glycolytic Potential of Escherichia coli.—Aaron E. Wasserman, Sc.D. Virus Diagnostic Laboratory, Analysis of Results in Sporadic and Epidemic Infections. —M. M. Sigel, PhD, Lillian P. Kravis, M. D.

219	1-23-51	Some Aspects of Laboratory Diagnosis: Recent Developments in the Laboratory Diagnosis of Tuberculosis.—Earle H. Spaulding, PhD Studies on the Hemagglutination Test for Tuberculosis.—W. L. Gaby, PhD Antibiotic Sensitivities as Determined by the Disc Test.—T. G. Anderson, Ph.D Present Status of Serological Tests with Streptococcal Antigens.—T. N. Harris, M. D.
220	2-27-51	Paper Chromatographic Studies on Products of the Enzymatic Degradation of Desoxyribonucleic Acid.—Helen R. Skeggs, B.A., W. Baumgarten, PhD, J. Spizizen, PhD, L.D. Wright, PhD Pyrimidine Synthesis in Lactobacillus bulgaricus 09 as Determined by Radioactive Isotope Technique. —L.D. Wright, PhD, C. S. Miller, Ph.D H. R. Skeggs, J. W. Huff, PhD, L. L. Weed, MD, D. W. Wilson, PhD The Critical Point Method for Drying Electron Microscope Specimens. —T. F. Anderson, PhD
221	3-27-51	Whither Immunization? A Panel discussion: Joseph Aronson, MD, Betty Lee Hampil, Sc.D., Raymond A. Kelser, D.V.M., Aims C. McGuinness, MD, Stuart Mudd, MD, Elizabeth Kirk Rose, MD, Kenneth Goodner
222	4-24-51	Evidence Suggesting That the Metachromatic Granules of Mycobacteria are Mitochondria.—Stuart Mudd, MD, Loren C. Winterscheid, Edward D. DeLamater, MD, Howard J. Henderson Studies on Infectious Hepatitis.—Miles E. Drake, MD, Gertrude Henle, MD, Joseph Stokes, Jr., MD, Werner Henle, MD Studies on the Psittacosis-L.G.V. Group. I. The Pattern of Multiplication in the Allantois of the Chick Embryo. —M. Michael Sigel, Anthony J. Girardi, Emma G. Allen Studies on the Psittacosis-L.G.V. Group. II. Survival of Meningopneumonitis Virus Under Various Conditions of Storage. —Emma G. Allen, Anthony J. Girardi, M. Michael Sigel, PhD

223	5-22-51	Mechanism of Action in Tuberculin Allergy.—Cutting B. Favour, MD
224	10-23-51	Germination of Spores of Genus Bacillus as seen with the Phase Microscope. —E. H. Fowler, J. A. Harrison, PhD Nicotinamide and Related Substances in Murine Tuberculosis. —Florence K. Fitzpatrick, PhD Tryptophane Metabolism and its Relation to Bacteriophage Resistance in Escherichic coli.—Joseph S. Gots, PhD, Won Young Koh, M. D. Present Status of Certification of Bacteriologists.—E. H. Spaulding.
225	11-27-51	Effect of Benemid and of Carinamide on Penicillin Therapy of Experimental Pneumococcus Infections in Mice. —W. F. Verwey, Sc. D, A. Katherine Miller, PhD, Dorothy W. Schlottman Studies on the Reduction of Tetrazolium Salts and Nitrofurans by Flavoprotein Systems.—Arnold F. Brodie, Joseph S. Gots, PhD Mechanism of Development of Resistance to Streptomycin.—M. G. Sevag, PhD, Eugene Rosanoff, M.S.
226	1-22-52	Some Aspects of Viral Diseases.—Frank M. Burnet, MD, Ph.D,., F.R.S.
227	2-26-52	Herpangina and Coxsackie Viruses.—Klaus Hummeler, MD, Lillian P. Kravis, MD, M. M. Sigel, PhD The Use of Chicken Antiserum in the Immuno-Chemical Studies of Edestin.—J. Munoz, PhD Studies on Spirochetes by Means of Phase Contrast Microscopy.—Edward D. DeLamater, M. D.
228	3-25-52	Current Emphasis in Research on the Etiology of Rheumatic Fever. —T. N. Harris, MD Studies on a Red Blood Cell Absorption-Agglutination Test with Streptococcal Culture Filtrate.—T. N. Harris, MD, Susanna Harris, Studies on Anti-Hapten Antibodies. —Fred Karush, PhD

		The Transfer of Cells From Rabbit Lymph Nodes Draining the Sites of Injection of Antigens.—Susanna Harris, PhD, T. N. Harris, MD, Miriam B. Farber, Henry D. Beale, MD
229	4-22-52	Symposium on Diagnostic Medical Microbiology—A Panel Discussion: Cora Gunther, PhD, Albert M. Kligman, MD, William G. Sawitz, MD, M. Michael Sigel, PhD, Earle H. Spaulding, PhD, Morton Klein, PhD
230	5-27-52	The Present Status of Immunization Against Rabies.—Hilary Koprowski, MD
231		Missing
232	11-25-52	Symposium on the Role of Microbiology in Biochemistry: The Use of Microbiological Methods for the Study of Biochemical Pathways.—Joseph S. Gots, PhD Contributions of Virus Research to Biochemistry.—Seymour S. Cohen, Ph.D Application of Microbial Nutrition in Biochemical Research.—Helen R. Skeggs
233	1-27-53	Some Observations on the Epidemiology of Rocky Mountain Spotted Fever.—Norman Topping, MD The Role of Bacterial Resistance in Antibiotic Synergism and Antagonism.— Morton Klein, PhD, Sonia E. Schorr, M.S. Mitochondria in Bacteria.—Philip E. Hartman, John C. Davis. BS, Stuart Mudd, MD
234	2-24-53	The Genesis of Primary Tuberculous Foci. Quantitative Relations as Controlled by the Resistance of the Host and the Virulence of the Microorganisms.—Max B. Lurie, MD, Peter Zappasodi, BS, Samuel Abramson, V. MD, Arthur M. Dannenberg, MD Studies on the Specificity of Complement Fixation Tests with Members of the Psittacosis-Lympho-granuloma Group: Treatment of Serum.—Ralph Pollikoff, PhD, M.M. Sigel, PhD Treatment of Antigen.—M. M. Sigel, PhD PhD, Ralph Pollikoff, PhD

235	3-24-53	Studies on the Sensitivity of Mice to Histamine Following Injection of Hemophilus pertussis.—J. Munoz, Ph.D, L.F. Schuchardt, W. F. Verwey, Sc.D. Localization of Enzymes in Microorganisms by Immunological Procedure.—Melva J. Derrick, PhD, Ruth E. Miller, PhD, M. G. Sevag, PhD Competitive Antagonism Between Antiphosphatase Antibody and Substrate for Phosphatase.—M. G. Savag, Ph.D, Melva J. Derrick, Ph. D, Ruth E. Miller, PhD
236	4-28-53	Inspection of Certain Phenomena in the Reaction of Bacteria to Electrical Impulses.—Sidney Tolchin, W. R. Hunt, PhD Studies on the Reversible Injury of Escherichia coli by Streptomycin.—Aaron E. Wasserman, James. M. Lessner, Margaret K. West Enzymologic Studies on the Mechanism of Softening of Tubercles.—Charles Weiss, PhD, MD, Joseph Tabachnick, PhD, Harold Cohen, Ph.D
237	5-26-53	The Binding of Penicillin in Relation to its Cytotoxic Action.—Harry Eagle, MD
238	10-24-53	The Inactivation of Some Bacteriophages by Specific Protein Reagents.—J. R. Stockton, Ph. D., John Spizizen, PhD, Betty Lee Hampil Passive Anaphylaxis in the Hemophilus pertussis Treated Mouse.—J. Munoz, PhD, Lee F. Schuchardt, Williard F. Verwey, Sc.D. Chemotherapy and the Bacterial Flora of the Intestinal Tract.—Earle H. Spaulding, PhD, N. U. Rao, PhD
239	11-24-53	Amino Acid Requirements of Yeasts in Relationship to Biotin Deficiency.—Albert G. Moat, PhD, Ellen Emmons, B. S. False Positive Serologic Reactions for Syphillis With Special Reference to the Treponema pallidum Immobilization Test.—John A. Kolmer, MD, Elsa R. Lynch

240	1-26-54	Adaptation of Influenza Virus to Mice.—W. A. Cassel, PhD, B. A. Briody, PhD, Jean Lytle, B. A., Mary Fearing, B. A. Effect of stilbamidine Therapy on Experimental Blastomyces dermatitidis Infections in Mice.—Margaret King West, W. F. Verwey, Further Studies on the Transfer of Lymph Node Cells.—T. N. Harris, M. D., Susanna Harris, PhD
241	2-23-54	Symbiotic Biosynthesis of Compound Replacing PABA and Folic Acid in Bacteria.—Bernard Koft, PhD AN Factor: Description, Isolation and Characterization.—Lemuel D. Wright Interrelationships Among the Microbic Populations of the Oral, Pharyngeal and Nasal Regions of Humans.—N. B. Williams, D.D.S., PhD, J. L. T. Appleton, D.D.S., Katherine Polevitzky, MD
242	3-23-54	The Cytology of C. diphtheriae.—John C. Davis, Stuart Mudd, MD Manganese and the Proteolytic Activity of Bacillus megaterium Spore Extracts in Relation to Germination.—H. S. Levinson, M. G. Sevag, Chelation in Microbiology.—Albert Schatz, PhD, Nicholas D. Cheronis, PhD, Vivian Schatz, BS
243	4-27-54	The Turbidimetric Determination of Bacterial Gelatinase Activity.—Harry Smith Studies on the Antigenic Components of Strains of Heterofermentative Lactobacilli.—Ned B. Williams, D.D.S., PhD, Eva Blau, M. S. The Biochemical Treatment of Dairy Wastes by Aeration.—Nandor Porges, PhD
244	5-25-54	Recent Advances in Immunology: Bacterial Diseases.—Kenneth Goodner, PhD Viral Diseases.—Morton Klein, PhD Use of Live Vaccines in Veterinary Medicine.—Raymond Fagan.
245	10-36-54	Pleuropneumonia-Like Organisms: Electron Microscopy of Pleuropneumonia-Like Organisms.—Harry E. Morton, James G. Lecce, PhD, John J. Oskay, Nettie H. Coy.

		Pleuropneumonia-like Organisms in Veterinary Medicine.—James G. Lecce, PhD The Association of a Corynebacterium and the Campo Strain of Pleuropneumonia-like Organisms.—Don M. Peoples The Role of Serum in the Nutrition of Pleuropneumonia-like Organisms. —Raymond J. Lynn Nutritional Differences Among Strains of Pleuropleumonia.-like Organisms. —Paul Smith
246	11-23-54	An Improved Leptospira Bacterin for the Control of Bovine Leptospirosis.—Albert Brown, PhD, Alan A. Creamer, Samuel F. Scheidy Dehydrogenase and Respiratory Activity in Streptomyces fradiae and Streptomyces nitrificans.—R. Ram Mohan, PhD, Albert Schatz, PhD, Gilbert Trelawny, Demonstration—Automatic Machine for Filling and Plugging Media Tubes —J. A. Harrison.
247	1-25-55	Symposium On Tissue Culture: Introduction and Methods in Tissue Culture.—Erling Jensen, PhD Virus Growth Studies in a Suspended Tissue Culture System.—Anthony Girardi, PhD A Study of Animal Sera for the Presence of Antibodies and the Viruses of Poliomyelitis.—Morton Klein, PhD, Pasquale Bartell Some Aspects of in vitro Tuberculin Sensitivity.—Warren Stinebring, Some Aspects of the Adaptation of a Strain of Dengue Virus to Embryonic Mouse Brain Tissue Culture.—Jack Frankel, PhD
248	2-22-55	Microbial Resistance to Antibiotic Agents: Metabolic Aspects of Resistance.—Joseph S. Gots, PhD Genetics of Microbial Resistance to Antibiotics.—Waclaw Szybalski, Clinical Laboratory Aspects of Resistance. —Amedeo Bondi, PhD
249	3-22-55	Changes in Staphylococcal Resistance to Streptomycin in Mice.—Eleanor Bliss, Sc.D. Influence of Host Metabolism on the Susceptibility to Bacterial Infections in Mice.—L. Joe Berry, PhD

		The Role of Histamine in Anaphylaxis of Pertussis-Treated Mice.—J. Munoz, PhD, Lee F. Schuchardt1
250	4-26-55	Glucose Dissimilation by Serratia marcescens.—Aaron Wasserman, W. J. Hopkins, T. S. Seibles Replicate Plating on Semi-solid Medium for Serological Study of Salmonella.—Elizabeth Fowler, James A. Harrison, PhD Antibacterial Activity of Mixed and Parotid Human Saliva: Preliminary Report.—B. J. Zeldow Effect of Penicillin on Oral Flora.—Ned Williams, PhD, D.D.S.
251	5-24-55	Symposium on Influenza Virus: Enzymatic Activity of the Virus.—J. F. McCrea, PhD The Susceptible Host.—R. R. Wagner, MD The Formation of 'Incomplete' Virus.—Kurt Paucker, PhD
252	10-25-55	The Effect of Mixed Bacterial Toxins on Transplanted Mouse Tumors.—H. Francis Havas, PhD, Marjorie E. Grosebeck Observations on Small Colonies (G-Variant) of M. pyogenes vs. aureus.—Robert Wise, PhD, MD Storage, Synthesis and Oxidation by Organic Substances from Skim Milk by Mixed Bacterial Cultures in Aeration Systems.—Nandor Porges, PhD, L. Jasewicz
253	11-22-55	Some Studies on Drug Allergy.—Merrill Chase, PhD
254	1-24-56	Occurrence of Blood Group Active Substances in Higher Plants and Bacilli.—Georg Springer, MD Sensitivity Testing to Antibiotics by the Disk Method—A Panel Discussion: E. H. Spaulding, Ph.D, T. G. Anderson, Ph. D., George Eisenberg, E. G. Scott, M. T, (ASCP)
255	2-28-56	Antibody Formation: Hepatic Injury and the Production of Antibody.—W. P. Havens, Jr. In vivo Studies on Precipitin Production by the Rabbit Spleen.—Kingsley Stevens, MD Transfer of Lymph Node Cells After in vitro Incubation with Dysentery Antigens.—T. N. Harris, MD, Susanna Harris, PhD Genetic Theory of Antibody Formation.—William E. Ehrich, M. D.

256	3-27-56	Microbial Genetics: Influence of Ploidy on Thermal Inactivation of Yeast.—Thomas Wood, Ph.D Genetic and Serologic Analysis of the Immobilizing Antigens of Paramecium.—Irving Finger, PhD Suppressor Mutations in Purine-requiring Salmonella.—Joseph Gots, PhD
257	4-24-56	Current Problems In Medical Microbiology: The Production of Poliomyelitis Vaccine.—John H. Brown, D.V.M. Studies of Certain Immunologically Significant Components of H. pertussis Cells.—Willard Verwey, Sc.D. An Approach to Fungal Chemotherapy.—Raymond C. Bard, PhD
258	5-15-56	Tuberculosis: Electron Scattering Granules and Reducing Sites in Mycobacteria.—Stuart Mudd, MD, Kinji Takeya, MD, H. J. Henderson, B.A. A Review of Chromogenic Acid-fast Bacilli.—Carl Clancy, PhD Constitutional Factors in Resistance to Tuberculosis. On the Role of the Thyroid Hormone.—Max Lurie, M. D., Peter Zappasodi, BS, Richard Levy, M.Sc.
259	10-23-56	A Rapid Method for Determining Susceptibility of Microorganisms to Antibiotics. (Film)—Dept. of Micro., U. S. Air Force Purine Biosynthesis in Yeast.—Herman Friedman, A. G. Moat, PhD Some Aspects of Amino Acid Metabolism by Pleuropneumonia-like Organisms.—Paul Smith, PhD A Study of Naturally Acquired Immunity to Poliomyelitis in the Philadelphia Population.—Morton Klein, PhD, Harvey Rabin, B.A., Phyllis Natale, BS, Elizabeth Earley, B.A.
260	11-27-56	A Medical Mission to Moscow.—Colin MacLeod, MD
261	1-22-57	Epidemiology: Epidemiological Investigation of Staphylococcal Infections.—Robert Wise, MD, Frank Sweeney, MD

		Epidemiological Investigation of Anthrax.—Phillip Brachman, MD Changing Nature of the Rabies Problem.—Raymond Fagan, D.V.M. Role of Adenoviruses in Respiratory Infection.—George Werner, PhD
262	2-26-57	Handling of Twelve Common Laboratory Animals—AMed. Res. Soc. Motion Picture. Management and Care of Guinea Pigs Used in Experimental Investigations.—Charles Weiss, MD, PhD Sulfaethylthiadiazole: in vitro and in vivo Evaluation.—Russel Rhodes,M.Sc., R. C. Bard, PhD, J. A. McCaughan, B,S, Reducing Sites in Clostridial Cells and Their Differentiation From Other Cell Structures.—John Davis, PhD, Stuart Mudd, MD
263	3-26-57	Synchronization of Division in Cultures of E. coli.—Dwight McNair Scott, PhD Prodiglosin Biosynthesis in Serratia marcesens.—Ursula Santer, M.S. Production of Receptor-destroying Enzyme by Vibrio chloerae.—William Feldman, PhD
264	4-23-57	The Combined Use of Antibiotic Agents—A Panel Discussion: Gladys Hobby, PhD,Harrison Flippin, MD, Earle Spaulding, PhD, Robert Wise, MD, Moderator—A. Bondi
265	5-21-57	Immunization in Brucellosis.—Israel Live, PhD Dermatomycosis in Animals Transmissible to Humans.—Frank Kral, D.V.M. Viremia and Neutralizing Antibodies in Canine Distemper.—Gunnar Rockborn, D.V.M.
266	10-22-57	Myxomatosis—The Evolution of a Virus Disease.—Frank Fenner, M.D,
267	11-26-57	Specific Release of Heterogenic Mononucleosis Receptor by Influenza Virus and Plant Proteases: Serological and Immunochemical Implications.—Georg Springer, MD, Abram Kaplan, B.A. Studies on the Nucleic Acids of Poliomyelitis Infected HeLa Cells.—E. L. Rothstein, M.S., L. A. Manson, PhD Human Skin-fixing Blocking Type Antibody.—N.C. Balczuk, J. A. Flick, MD

268	1-28-58	Symposium—Microbiology and Cancer Research: Biosynthetic Control in Microorganisms. —Joseph Gots, PhD Microbiological Screening of Potential Anti-Cancer Agents.—M. L. Rogers Chemistry of Homo-Fermentative Growth. —Gerrit Toennies, PhD Problems of Lymphomatosis. —Vittorio Defendi, MD
269	2-25-58	Panel Discussion—Changing Horizons in Medical Microbiology 1900-1965: How We Have Come to Where We Are. —Stuart Mudd, M. D. Discussion on the Forseeable Future—Panel with Morton Klein, Moderator.
270	3-25-58	Variation in Vaccinia Virus During Passage Through Ehrlich Ascites Tumor. —William Cassel, PhD Selection of Yeast Mutants Using Antibiotics.—Albert Moat, PhD Virulence Factors in Plague.—Stanley Jackson, PhD
271	4-22-58	Oral Microbiology: Human Parotoid Saliva as Sole Source of Nutrient for Oral Microorganisms.—Ned Williams, PhD Manometric Studies of Microorganisms in Human Parotoid Saliva.—Sterling Jackson Moniliasis as Complication to Antibiotic Therapy.—Albert Kligman Some Philosophical Introspections on Dental Caries.—Albert Schatz
272	5-27-58	Antibody Responses Following Infection and Vaccination With Asian Influenza Virus as Measured by Strain Specific Complement Fixation Tests.—Werner Henle, MD
273	10-28-58	Studies On Germ Free-Animals: Development of Germ-free Animal Research.—Martin Forbes, PhD Role of Intestinal Flora in the Growth Response of Chicks to Dietary Penicillin.—Martin Forbes, PhD, J. T. Park, PhD, M. Lev, Ph.D, The Problem of Natural Antibodies Studied in Germ-free and Conventional Chicks. —Georg F. Springer MD, R. E. Horton, MD, M. Forbes, PhD

274	11-25-58	Mechanisms Of Pathogenesis: Lysogeny and Toxinogeny in Corynebacterium diphtheriae.—Lane Barksdale, PhD The Importance of Infecting Dose on the Pathogenesis of Virus-Induced Rous Sarcoma.—Frank Rauscher, Ph.D, Vincent Groupe, PhD Cellular Factors in the Pathogenesis of Brucellosis.—Warren Stinebring, PhD
275	1-27-59	Virus Studies: Studies With Herpes Simplex and Pseudorabies Viruses.—Albert Kaplan, Ph.D Studies on the Structure of Infectious and Incomplete Influenza Virus.—Kurt Paucker, PhD Isolation of a Bovine Virus Neutralized by Human Gamma Globulin and Sera.—Morton Klein, PhD, Elizabeth Earley, Jack Frankel, PhD, Joseph Zellat, PhD
276	2-24-59	Immunology: Immobilization Antigens of Paramecium.—John Preer, MD Host Reaction to Transferred Lymph Node Cells.—T. N. Harris, MD, Susanna Harris, PhD, The Use of Agar-precipitin Analysis in the Study of Human Streptococcal Infections.—Seymour Halbert, M. D.
277	3-24-59	Symposium—Problems of Hospital Infections: Problems of Identification of Pseudomonas and Pseudomonas-like Organisms in the Clinical Laboratory.—William Gaby, PhD, Elizabeth Free, M.T. The Staphylococcal Problem.—Kenneth Schreck, M. D., Germicides: Their Value in Control of Infections.—Earle Spaulding
278	4-28-59	Cellobiose Phosphorylase of Clostridium thermocellum.—James Alexander, PhD Galactose Metabolism of Streptococcus fecalis.—T. T. Fukuyama, Daniel O'Kane, PhD Novel Amino-sugars and Their Role in the Formation of Bacterial Cell-walls.—Friedrich Zilliken, PhD
279	5-26-59	Taking Off From the Cell Surface.—Sven Gard, M. D.

280	10-27-59	Pleuropneumonia-Like Organisms: Role of Sterols in the Physiology of PPLO. —Paul Smith, PhD Decontamination of PPLO-infected Tissue Cultures.—Leonard Hayflick Infection of Germ-free Animals with PPLO.—Robert Smibert, PhD Possible Relationships of PPLO and Bacteria.—George Rothblat
281	11-24-59	Morphological and Biochemical Comparisons of Sporogenic Bacillus cereus with Asporogenic Mutants. —George Beskid, PhD The Role of Phospholipids in the Uptake of Amino Acids by Lymph Node Cells. —Herman Friedman, PhD, William Gaby, PhD Coagulases and Virulence in Staphylococcus aureus.—Frank Kapral
282-348	1960s	Meetings 1960 to 1969—No. 282 through 348
282	1-26-60	Symposium on Microbiological Problems in the Food Industry: The Identification and Role of Indicator Organisms in Frozen Foods.—Karl Kereluk, PhD The Role of Psychrophilic Bacteria in Defrost Spoilage of Frozen Foods. —Arthur Peterson, PhD Microbiological Aspects of Hot Food Vending.—Stanley Segall, PhD Microbiology of Maple Sap.—Hilmer Frank, PhD, C. O. Willits, PhD
283	2-23-60	The Role of Phospholipids in Amino Acid Transport.—Ihor Zajac Ronald Silberman, B. A., William Gaby, PhD Antibiotic Action and the Bacterial Surface.—G. D. Shockman, PhD Demonstration of Extracellular Material Associated With a Strain of Staphylococcus aureus.—T. Sall, PhD, S. Mudd, MD, F. F. Nightingale, N. A, Vali, M. D.
284	3-22-60	Symposium on Bacterial Endotoxins: Effects of Bacterial Toxins on Tumors and Mice.—H. Francis Havas Adrenergic Inhibition and Protection Against Escherichia coli Lipopolysaccharide. —Richard McLean Metabolic Effects of Bacterial Endotoxins in Mice.—Dorothy Smythe, L. Joe Berry, PhD

285	4-28-60	Physiological Control Mechanisms in the Bacterial Cell.—Herman Lichstein
286	5-24-60	Serial Propagation of Some Influenza Viruses in a Human Cell Line.—Samuel Wong, PhD The Immunological and Epidemiological Aspects of B Virus.—James Prier, D.V.M., PhD A Serumless Medium for the Continuous Cultivation of Cells of Normal and Malignant Origin.—Robert Neuman, PhD, Alfred Tytell
287	10-25-60	A Mechanism for Resistance to Purine Analogs.—G. P. Kalle, PhD, Joseph Gots, PhD Environment-induced Changes in the Ribonucleic Acid of E. coli.—Melvin Santer, PhD, David Teller, BS, Willard Andrews, B. S. The Biochemistry and Physiology of a Microbial Toxin.—Sam Ajl
288	11-22-60	Specific Interference Between Coxsackie Viruses in a Carrier Culture.—Richard Crowell, PhD The Induction of Immunological Tolerance in Adult Animals.—B. A. Rubin, PhD A Basis For Cell Injury to Adenoviruses. —Harold Ginsberg, M. D.
289	1-24-61	Effects of Ozone on Escherichia coli.—wight McNair Scott, PhD, Eilene Lesher, M. S. Effects of Deuterium Oxide on Virus Multiplication.—Lionel Manson, PhD, Richard Carp, D.V.M., David Kritchevsky, PhD Nutritional and Environmental Effects on DNA Replication in Bacteria.—Daniel Billen, PhD
290	2-28-61	Labeling Procedures Employing Chromatographically Pure Fluorescein Isothionate.—Charles Griffin, PhD Research Applications of Fluorescent Antibody Techniques.—Klaus Hummeler, MD Diagnostic Applications of Fluorescent Antibody Techniques.—Theodore Carski, MD

291	3-28-61	Inhibitory Effect of Ammonium Ions on Influenza Virus in Tissue Culture.—E. M. Jensen, PhD, E. E. Force, B. S., J. B. Unger, BS Specific and Differential Inhibition of Some Enteroviruses.—Normand Goulet, PhD Recent Advances in Viral Prophylaxis and Therapy.—O. C. Liu, M. D.,
292	5-23-61	Relationship Between Biotin and Phosphate Exchange.—W. W. Umbreit, Ph.D
293	10-24-61	Nuclear Accumulation of RNA During Drug Treatment of B. megaterium.—David Ezekiel, PhD Metabolic Stability of Ribonucleic Acids in Mammalian Cells.—Adrian Rake, PhD, A. F. Graham, PhD Analysis of the Intracellular Development of a DNA-containing Mammalian Virus (Pseudorabies) By Means of UV Irradiation.—Albert Kaplan, PhD
294	11-28-61	Symposium on Biology and Chemistry of Antibody Formation: Suppression of Transferred Lymph Node Cells in Neonatal and Adult Rabbits.—T. N. Harris, M. D., S. Harris, PhD Antibody Synthesis in vitro.—John McKenna, PhD A Role for Nucleoproteins in Antibody Formation.—Herman Friedman, Antibody Activity in a Cell-free System.—Dieter H. Sussdorf, PhD
295	1-23-62	The Relation of Fluorescence to Toxicity of the Toxin of Clostridium botulinum.—Daniel Boroff, PhD Exocellular Products of B. cereus and their in vivo and in vitro Effect on Blood Components—Abramo Ottolenghi, PhD, S. Gollub, Ph.D Further Studies on a Cell-bound Hemolysin of Group A Streptococci.—Isaac Ginsburg, PhD, T. N. Harris, M. D.
296	2-27-62	Newer Concepts of Influenza Vaccination.—Florence Lief, PhD The Use of Genetic Markers in the Study of Polio Virus Vaccination.—Richard Carp Studies on Lyophilized Smallpox Vaccine.—Pasquale Bartell, PhD, H. Tint, PhD

297	3-27-62	Symposium on Staphylococcal Infections: The Interaction of Human Sera With Various Serotypes of Staphylococcus aureus. —S. Mudd, M. D., Nancy Lenhart, M. A., Samuel DeCourcy, M. A. In vivo Growth of Staphylococci.—Stanley Levine, B.A.,R. J.Mandle Use of Diffusion Chambers in the Study of the Pathogenesis of Staphylococcal Infections.—Enoch Houser, M. A., L. J. Berry, PhD Observations on the Epidemiology of Staphylococcal Infections.—Kenneth Schreck, M. D.
298	4-24-62	Symposium on Bacterial Cell Wall and Lysis: Current Concepts of Bacterial Cell Walls. —Gerald Shockman, PhD Chemistry of Streptococcal Cell Walls. —Hans Haymann, PhD Activation of Autolytic Mechanism for the Production of Cell Wall and Spheroplasts of Escherichia coli and Bacillus cereus. —Raam Mohan
299	5-22-62	The Mechanism of Cell Disintegration by Ultrasonics.—David Hughes, Ph.D
300	10-23-62	Genetic Alterations of Purine Pyrophosphorylases.—J. S. Gots, PhD, G. P. Kalle, PhD, J. Adye, PhD The Effect of 5-Fluoro-2-Deoxyuracil on Experimental Herpes Simplex Infection.—Claire Engle, M. S. The Effect of Biotin Deficiency on the Synthesis of Nucleic Acids, Protein and Enzymes in Yeast.—Ahmad Fazal, PhD
301	11-27-62	The First Annual Industry Sponsored Guest Lecture: Experiments on the Role of Heredity in Experimental Sensitization. —Merrill Chase, PhD
302	1-22-63	Symposium on Biochemistry and Physiology of Protozoa: Introduction.—Robert Conner, PhD Studies on Phosphate Balance of Tetrahymena.—P. S. Leboy, PhD Recent Studies on the Killer Character in Paramecia.—John Preer Lipids in Aging Cell Populations. —R. R. Ronkin, PhD

303	2-26-63	Some Effects of Nitrogen Mustard on Synchronized Cultures of E. coli.—D. B. McNair Scott, PhD Lipid-Amino Acid Complexes in the Course of Antibody Synthesis.—Judy Spitzer Studies on Mima and Herellea Strains Encountered in a Routine Bacteriology Laboratory.—Eileen Randall Cytotoxic Effect of Hemolytic Streptococci on Ascites Tumor Cells.—H. Frances Havas, PhD
304	3-26-63	Immunoelectrophoresis and Qualitative Tests for Multiple Myeloma and Macroglobulinemia.—Leonard Korngold, PhD
305	4-22-63	Hydrocarbons: Substrates for Microbiological Utilization.—Andre Brillaud, PhD Neutralizing Potentialities of Crotalid Antivenin.—Eleanor Buckley Microbiological Considerations in the Licensing of a Human Biological.—E. S. Barclay
306	5-28-63	Microbiology and the Public Health Laboratory.—R. B. Hogan, M. D. Performance of Fluorescent Antibody Technique (FAT) in the Diagnosis of Gonorrhea.—Henry Beilstein A Panel Discussion on Susceptibility Testing.
307	10-22-63	Virus—Tumor Relationships: Studies on Murine Leukemia.—Marvin Rich, PhD Comparative Study of Tumorogenesis in vivo and in vitro with Polyoma Virus.—Vittorio Defendi, M. D. Viral Chromosome Relationships.—Warren Nichols, MD
308	11-26-63	The Second Annual Industry Sponsored Guest Lecture: Enzyme Induction and Catabolite Repression.—Boris Magasanick
309	1-28-64	Tuberculosis: Some Biochemical Aspects of Tuberculosis Infections.—Adam Bekierkunst, PhD Studies on Host-Parasite Relations in Tuberculosis.—Arthur Dannenberg, M.D Changing Concepts in Tuberculosis—The Anonymous Mycobacteria.—John Albertson, Jr.

310	2-25-64	The Biology and Chemistry of Endotoxin: Investigation of Endotoxin Structure.—Alois Nowotny, PhD A Comparison of Two Endotoxins.—Henry Freedman, PhD Some Characteristics of the Immune Response to Endotoxic Polysaccharides.—Maurice Landy, PhD
311	3-24-64	Scientific Reports on Studies Performed Abroad: Experiments on the Solublization of a Cell Antigen Reacting with a Lymph Node Cell Suppresive Antibody.—T. N. Harris, MD The International Standard for Antibiotic Susceptibility.—T. G. Anderson Some Aspects of Gluconeogensis.—Judy A. Spitzer
312	4-28-64	Graduate Student Reports: Endotoxins in Urinary Tract Infections of Rats.—Mary Ann Fritz Immunologic and Physiochemical Studies of Staphylococcal Hyaluronate Lyase.—Carl Abramson, M. S. Analysis of Mutagen-Induced Revertants of ade-A Mutants of Salmonella typhimurium.—Marilyn Bailin, BS Linkage of Markers Transformed Interspecifically Between Hemophilus influenzaae and Hemophilus parainfluenzae.—Lois Nickel Interaction of Mycoplasma (PPLO) and Murine Lymphoma Cells.—Paul Kraemer, Dr. P. H. Enzymatic Differentiation Between Coxsackie B3 and Poliovirus T1 Receptors of Living HeLa Cells.—Ihor Zajac, MD
313	5-26-64	Bench Marks in Biological Nitrogen Fixation.—Perry Wilson, PhD
314	10-27-64	Metabolism and Control Mechanisms in Microbiology: A Comparison of the Regulation of Serine Biosynthesis in Bacterial and Cultured Human Cells.—L. I. Pizer, PhD The Regulation of Fatty Acid Metabolism by Glyoxylate.—W. Wegner, PhD, H. C. Reeves, PhD, S. J. Ajl, PhD An Alternative Route of Nicotinic Acid Biosynthesis in Microorganisms.—A. G. Moat, PhD, F. Ahmad, PhD, A. Isquith, J. N. Albertson, Jr.

315	11-24-64	The Third Annual Industry Sponsored Guest Lecture: Growth of an RNA Bacteriophage.—Norton D. Zinder, PhD
316	1-26-65	Enzyme Formation in Synchronous Cultures.—Arthur Pardee, PhD
317	2-23-65	Staphylococcal Infections: Clinical Aspects.—Frank Sweeney, M. D. Epidemiologic and Laboratory Aspects.—Kenneth Schreck, M. D. Immunologic Aspects.—Frank Kapral, PhD Presentations followed by a Panel Discussion.
318	3-23-65	Graduate Student Program: Transduction of Escherichia coli Serotype 04.—Judith Green, M. A. Endotoxin Intoxication in Neonatal Rats.—Mark Taylor, B. S. Biosynthesis of Nicotinic Acid and NAD by Clostridium butylicum.—Alan Isquith, M. S. A Rapid Agglutination Test for the Diagnosis of Adenovirus Infections.—Jay Satz, B. A. Molecular Varieties of Equine Antibodies.—Norman Klinman, M. D. Mechanism of Intracellular Uncoating of Adenovirus.—William Lawrence
319	5-25-65	Immunologic Aspects of Virus-induced Tumors: Characterization of a Tumor-like Antigen in Type 12 and Type 18 Adenovirus-Infected Cells.—Z. Gilead, PhD, H. S. Ginsberg, M. D. The Induced Non-virion CF Antigen of SV40.—R. Gilden, PhD, R. Carp, PhD Prevention of SV40 Virus Tumorigenesis by Immunization with SV40 Transformed Human Cells.—A. Girardi, PhD
320	10-26-65	Structure and Synthesis of Antibodies: Relationship of Structural Heterogeneity of Antibodies to Problems of Synthesis.—Norman Klinman, MD Multiple Molecular Forms of Rabbit IgG Antibodies.—Peter Stelos Studies on the Synthesis of Gamma Globulin.—Matthew Scharff
321	11-30-65	The Fourth Annual Industry Sponsored Guest Lecture: Cellular Interaction and Metabolic Controls in Cultured Human Cells.—Harry Eagle, M. D.

322	1-25-66	Sweet and Sour DNA.—Salvador Luria, M. D.
323	2-22-66	Recent Advances in the Biology and Chemistry of Mycoplasmas and L—forms: Recent Developments in the Biology of the Mycoplasmas.—L. Hayflick, Ph.D Genetic Relatedness Among Mycoplasmas as Determined by Nucleic Acid Homology.—N. L. Somerson, PhD, P. R. Reich, MD, S. M. Weissman, MD, R. M. Chanock, M.D, Biochemistry of Cell Wall Inhibition in a Stable Streptococcal L Form.—C. Panos, PhD
324	3-22-66	Graduate Student Program: The Effect of Tryptophane and Alpha Methyl Tryptophane on the Response to Endotoxin in Mice.—Robert Moon, A.B. Changes in Tertiary Structure Accompanying Detoxification of Serratia marcescenes Endotoxins.—Daniel Tripodi, M. S. Comparative Studies of the Regeneration of HeLa Cell Receptors for Poliovirus Type I and Coxsackie Virus Type B3.—Neil Levitt, B.A. Biosynthesis of NAD and Nicotinic Acid by Clostridium butylicum.—Alan Isquith, M.S. Inhibition of Host RNA Synthesis in Cells Infected with Type 5 Adenovirus.—Arnold Levine, B.A. The Restriction of Phage T-2 Infection by E. coli, Strain W.—Helene Smith, B.S
325	4-26-66	Immunogenetics and Transplantation: Delayed Hypersensitivity Reactions to Transplantation Antigens in Hamsters.—R. E. Billingham, PhD Immunological Aspects of Carcinogenesis.—B. A. Rubin, PhD In vitro Demonstration of the Heterogeneity of Rabbit Leucocyte Antigens.—C. A. Ogburn, PhD, S. Harris, PhD, T. N. Harris, MD
326	5-17-66	Microbiology of Leukemias: Studies on Burkitt's Lymphoma.—Gertrude & Werner Henle, MD Etiologic Considerations of Bovine Leukemia.—Ray Dutcher, PhD, Edward Larkin, PhD, Joseph Tumilowicz, PhD, Keyvan Nazerian, PhD, Neale Stock Studies on Murine Leukemia.—Marvin Rich, PhD

327	10-18-66	Pathogenesis And Prevention Of Rubella: Rubella Epidemiology.—T. H. Ingalls, MD The Rubella Virus.—S. Plotkin. MD Rubella Immunization.—H. M. Meyer, MD
328	11-22-66	The Fifth Annual Industry Sponsored Guest Lectureship: The Interferons.—Robert Wagner, MD
329	1-24-67	Mechanisms of Specific Immunologic Tolerance.—Byron Waksman
330	2-21-67	The Role of Nucleic Acids in Antibody Production: The Role of Macrophage RNAs in the Immune Response.—M. Fishman The Conversion of Non-Immune Cells to Antibody-Forming Cells by RNA—The Specifically Receptive Cell.—E. P. Cohen, MD Transfer of Homograft Immunity to Normal Lymphoid Cells with RNA from Tissues of Specifically Immunized Donors.—D. B. Wilson, PhD
331	3-28-67	Research: Fact or Fancy—Michael Shimkin Graduate Student Program: The Immunoglobulin Response in Recurrent Herpes simplex Infection in Man.—Adamadia Deforest, M.S. Pronase Hdydrolysis Products of Human Gamma-globulin.—Bruce Sloan, B.A. The Effect of Endotoxin and Actinomycin D on Tryptophane Pyrrolase in Liver Slices.—Joyce Greene, M.A. Immunological Evidence of a Cytoplasmic Site for Synthesis of Adenovirus Structural Proteins.—Leland Velicer, D.V.W.
332	4-25-67	New Look at Infectious Diseases: Pneumococcal Disease in the Antibiotic Era.—Robert Austrian, MD The Laboratory Aspects of the Treatment of Drug-Resistant Tuberculosis.—Ralph Knight, PhD Melioidosis.—Murray Spotnitz, MD
333	5-23-67	Rabies: Antirabies Vaccines, Present and Future.—Tadeusz Wiktor Positive and Negative Aspects of Rabies Prophylaxis.—E.A.Hildreth Methods and Problems in Rabies Diagnosis.—J. E. Prier, PhD

CM1	9-25-67	Training in Clinical Microbiology.—Earle Spaulding; Convener Richard Clark
334	10-14-67	Recent Developments in Immunology: Recent Immunological Studies with Synthesis Polymers of Amino Acids. —P. H. Maurer, PhD Electron Microscopic Observations on Antibody-Producing Cells. —T. N. Harris, MD, K. Hummeler, MD, S. Harris, PhD Recent Studies in vitro on Delayed-type Hyper-sensitivity.—B. R. Bloom, PhD
335	11-28-67	The Sixth Annual Industry-Sponsored Guest Lecture: Adenovirus SV-40 Hybrid Viruses.—Wallace Rowe, MD
CM2	12-4-67	Microbial Aspects in the Diagnosis and Management of Urinary Tract Infections.—Panel Discussion, Moderator, Theodore Anderson, Panel: A. Rubin, F. J. Sweeney, L. Karafin, E. Randall
336	1-23-68	Ontogeny of the Immune Response. —Dennis Watson, PhD
337 & CM3	2-5-68	Joint Regular and Clinical Microbiology Section Meeting: Acute Respiratory Infections and Its Laboratory Management.—Panel Discussion, Moderator, James E. Prier, Panel: R. Austrian, M. Manko, S. A. Plotkin
338	2-27-68	Immunity to Smallpox.—Abram Benenson, MD Microbiological Transformations of a B-noogramr Steroid.—Louis Fare Virus Function Necessary for Induction and Stability of Cell Transformations. —Vittorio Defendi, MD
339	3-26-68	Graduate Education in the Basic Sciences —Robert Baldridge Graduate Student Program: Effect of Endotoxin and Cortisone on Isotope Incorporation in Liver RNA in Adrenalectomized Mice.—Thelma Shtasel Characterization of a Serum Globulin Factor which Facilitates Hemolytic Plaque Formation.—Edward O'Donnell Biochemical Analysis of Lysine Auxotrophs of Staphylococcus aureus.—Isabel J. Barnes

CM4	4-8-68	Chronic Respiratory Disease and Its Laboratory Management.—Panel Discussion, Moderator, Ralph Knight, Panel: N. Huang, R. Johnston, F. Blank
340	5-28-68	The Physical and Immunological Structure of Adenoviruses.—B.A.Rubin, Ph.D The Interferons.—A. Kirk Field, PhD Manifestation of Ferret Viral Rhinitis and Their Utility in Drug Evaluation.—Carl Pinto, Richard Haff, PhD
CM5	9-16-68	Application of Fluorescent Microscopy in the Clinical Laboratory.—Panel Discussion, Moderator, Richard Clark, Panel: J. Copeland, J. Quinn, R.A. Jakubowitch
341	10-29-68	The Seventh Annual Industry Sponsored Guest Lecture: A Re-Examination of the Germ Theory of Disease.—Gordon Stewart
CM6	11-4-68	New Techniques and Advances in Blood Culturing.—Panel Discussion, Moderator, Eileen Randall, Panel: G. Evans, B. Free, W. Kozub
342	11-26-68	The Relation of E.B. Virus to Infectious Mononucleosis and Burkitt Lymphoma.—Werner Henle, MD
CM7	1-13-69	Mycoplasma.—Moderator, Harry E. Morton Epidemiology of Mycoplasm pneumoniae Infections in Marine Recruits.—Richard R. Gutekunst Mycoplasm of the Uterine Cervix.—James E. Gregory Isolation and Identification of Mycoplasma.—Harry E. Morton
343	1-28-69	Cytolytic Toxins of Bacterial Origin.—Alan W. Bernheimer, PhD
344	2-25-69	Medical Applications of Laminar Air Flow Systems: Control of Airborne Infection.—Lewis Coriell Results with High and Low Flow Rate Systems.—Gerard McGarrity Smoke Tests of Air Flow Patterns.—Lewis Coriell Protection of Experimental Animals.—Gerald McGarrity

CM8	3-17-69	Universal Standards for Antimicrobial Susceptibility Testing.—Theodore Anderson
345	3-25-69	Graduate Student Program: Purification and Comparison of Alpha Toxins Produced by Three Strains of Staphylococcus aureus.—Robert Stockmal, M.S. Studies on a Bacteriophage-Associated Staphylolytic Enzyme.—Stephen Sonstein, M.S. Protein Initiation Following Virulent Bacteriophage Infection.—Margaret Miovic, A.B. Ribosomal Antigens in Immunity to Salmonella and Staphylococcal Infections.—Toby Eisenstein, A.B. The Growth of E. coli on Fatty Acids. —Philip Furmanski, BS
CM9	5-12-69	Role of the Clinical Laboratory in the Diagnosis of Viral and Rickettsial Disease.—Moderator, James E. Prier Technics for Isolation and Identification of Viruses.—Randolph Riley Technics for Serologic Identification of Virus Infection.—Walter Ceglowski Virus Diagnostic Services in the Hospital Laboratory.—Herbert Heineman Public Health Laboratories as Reference Agencies in Virus Diagnosis.—Vern Pidcoe
346	5-27-69	Cellular Formation of Antibody During Immunity and Tolerance: Cellular Changes During Immunity and Tolerance to Serum Proteins as Assessed by Immuno-fluorescence.—Jan Cerny, MD, PhD Antibody Forming Cells to a Hapten in Immune and Tolerant Rabbits.—Tomas Hraba, MD, PhD Cytokinetics of Bacteriolytic Antibody Plaque Formation in Mice Tolerant to the Lipopolysaccharide Antigen of E. coli. —Jerry Allen, Ph.D Enumeration of Vibriolytic Antibody Forming Cells to Cholera Antigens. —Robert McAlack, PhD

CM10	9-29-69	Advances in the Isolation and Identification of Mycobacteria.—Panel Discussion, Moderator, Ralph Knight, Panel: E. Micklin, W. Jackson, W. Fraimow
347	10-21-69	Ultrastructural Studies of Lymphoid Tissues From Leukemia Virus-Infected Mice. —Gloria Koo, Walter Ceglowski, PhD Immunofluorescent Studies on the in vitro Interaction Between Lymphoid Cells and Target Cells.—Junius Clark, PhD Tolerance Following Immunity in Adult Rabbits Injected with HeLa Cells.—Gail Miller, PhD Structure of Rabies Virus.—Frank Sokol, Ph.D
CM11	11-3-69	Practical Methods of Quality Control in Clinical Microbiology.—Panel Discussion, Moderator, Raymond Bartlett, Panel: H. E. Morton, H. Friedman, V. Pidcoe.
348	11-25-69	Eighth Annual Industry-Sponsored Guest Lecture: Transfer Factor and Cellular Immunity. —H. Sherwood Lawrence
CM12	12-15-69	Practical Methods of Differentiating Non-Fermenting Gram-Negative Rods.—Panel Discussion, Moderator, Eileen Randall, Panel: E. G. Scott, J. Cocklin.
349-427	1970s	Meetings 1970 to 1979—No. 349 through 427
349 & CM13	1-27-70	Immune Response of Patients with Bacterial Infection as an Aid to Diagnosis and Epidemiology.—Erwin Neter, MD
350	2-24-70	Australia Antigen and Hepatitis: Introduction.—Dr. Baruch Blumberg Purification, Chemical & Physical Properties of Australia Antigen. —Irving Millman, PhD Fluorescent Test for Au in Liver. —Veronica Coyne, MD Complement-Fixation Test for Au. —Scott Mazzur, PhD
CM14	3-16-70	New Techniques and Advances in the Laboratory Diagnosis of Haemophilus Infections.—Panel Discussion, Moderator, Viola Mae Young, Panel: K. R. Cundy, B. Delette, N. Huang

351	3-24-70	Graduate Student Program: Effect of Cyclophosamide on the Ontogeny of the Humoral Immune Response in Chickens.—Stephen Lerman, BS Synthesis of Macromolecules in Herpesvirus Infected Cells Deprived of Arginine. —George Mark, III, A.B. Role of Methionine in Replication of Type 5 Adenovirus.—Donald Pett, A.B. Evidence for the Fragmentation of High Molecular Weight Transforming DNA Upon Entry Into the Competent Recipient.—Jack Michalka, BS Effect of Sulfhydryl Reagents on the Ribosomes of B. subtilis.—Rajinder Ranu, B.V.Sc., Dr. Akira Kaji
CM15	5-11-70	Anaerobic Infection as Related to the Clinical Laboratory.—Panel Discussion, Moderator, Earle Spaulding, Panel: S. M. Finegold, R. W. Schaedler, E. Randall, E. G. Scott
352	5-26-70	Some Aspects of Teaching Microbiology: Undergraduate.—Wesley Hutchinson, PhD Medical Microbiology.—Amedeo Bondi, Jr., PhD Graduate School.—Albert Moat, PhD Post-Doctoral Fellows.—Earle Spaulding, PhD
353	9-26-70	Cellular Mechanisms of Immunity and Tolerance to Microbial Antigens: Welcome and Chairman—Fred Rapp, PhD Introduction—Herman Friedman, PhD Cellular Formation of Antibody During Immunity and Tolerance to E. coli Lipopolysaccharide Antigens in Mice. —Goran Moller Mechanism of Tolerance Induction to Salmonella Flagellin Antigens in Rats. —Gordon Ada, PhD The in vitro Induction of Tolerance to Salmonella Flagellin.—Erwin Diener, MD, PhD
CM16	10-5-70	Use of the Kirby-Bauer Antibiotic Susceptibility Test in Routine Clinical Microbiology.—(A film narrated by W. W. Kirby) followed by a talk by Theodore G. Anderson

354	10-27-70	Application of Gas Chromatography to Microbiology: Gas Chromatographic Instrumentation in Use.—Gerald Umbreit Important Techniques and Art of Gas Chromatography.—Walter Supina, PhD Virological Applications of Pyrolysis Gas Chromatography.—Earle Byrne, MD Computerization of Pyrolysis Gas Chromatography Results in Clinical Microbiology.—Richard Clark, BS, M.S.
CM17	11-16-70	Laboratory Diagnosis and Antibiotic Susceptibility Testing of The Systemic Mycotic Agents.—Panel Discussion, Moderator, Charlotte Campbell, Panel: S. Shadomy, E. Randall, V. Iralu. B. Cooper
355	11-24-70	The Ninth Annual Industry-Sponsored Guest Lecture: The Ecology of Infectious Disease.—Dr. Rene Dubos
CM18	12-14-70	Current Problems in Clinical Microbiology.—Panel Discussion, Moderator, Harry E. Morton, Panel: K. R. Cundy, E. G. Scott, R. R. Gutekunst
CM19	1-18-71	Practical Use of Computer Techniques in the Requesting and Reporting of Routine Clinical Microbiology.—Panel Discussion, Moderator, Richard B Clark, Panel: P. Greenhalgh, J. Stoller, V. M. Young
356	1-26-71	Phylogeny of the Immune Response and Immunoglobulins.—Michael Sigel, PhD
357	2-23-71	Studies on a Temperature Sensitive Mutant of E. coli With a Lesion in the Acylation of Lysophosphatidic Acid.—K. Hechemy, H. Goldfine Physical Separation of Genetic Markers and its Relationship to Their Recombination Index in Hemophilus influenzae Transformation.—J. Bendler, S. H. Goodgal Primary Clonal Antibody Production in vitro.—N. Klinman Synthesis of Adenovirus Capsid Proteins in vitro.—J. M. Wilhelm, H. S. Ginsburg
CM20	3-1-71	Recent Progress in the Laboratory Diagnosis of Non-gonococcal Venereal Diseases.—Panel Discussion, Moderator, Jay Satz, Panel: W. Ceglowski, T. J. Linna, L. Nicolas

358	3-23-71	Graduate Student Program: Introduction—James Prier Characteristics of a Feedback Resistant Phosphoglycerate Dehydrogenase.—Ilga Winicov The T Antigen Induced by SV40 Virus.—Bert Del Villano The Role of Arginine in the Replication of Herpes Virus.—George Mark Intracellular Localization of Sites of Ribosome Biosynthesis in "Protoplasts" of Strep. faecalis 9790.—George Roth
CM21	4-12-71	Practical Methods in the Laboratory Identification of Problem Gram Negative Bacilli.—Panel Discussion, Moderator, Eileen L. Randall, Panel: K. R. Cundy, R. E. Weaver
CM22	5-17-71	Recent Progress in the Laboratory Diagnosis of Gonorrhea.—Panel Discussion, Moderator, James E. Prier, Panel: L. Shapiro, T. G. Anderson, W. Cole, H. Beilstein
359	5-25-71	Awards Presented to the Following Individuals: Excellence in Teaching.—Dr. Joseph Gots Excellence in Graduate Studies.—George Mark Excellence in Research.—Drs. Gertrude & Werner Henle Presentation of a timely topic by Dr. W. Henle
CM23	10-4-71	New Applied Techniques in Anaerobic Clinical Microbiology.—Panel Discussion, Moderator, Kenneth R. Cundy, Panel: S. M. Finegold, E. L. Randall, V. L. Sutter
360	10-26-71	Tumor Viruses and Immunity: Chairman—Herman Friedman, PhD Leukemia Virus Induced Immunosuppression.—Walter Ceglowski Role of Interferon and Antibody in Virus Induced Leukemia.—Frederick Wheelock, MD Immune Response to Tumor Virus and Tumor Specific Antigens.—Paul Koldovsky, MD
CM24	11-15-71	Role of the Clinical Microbiology Laboratory in Environmental Microbiology.—Panel Discussion, Moderator, Herman Friedman, Panel: E. Ossman, E. Bernhardt, R. M. Swenson, Discussants, E. H. Spaulding, R. R. MacGregor

361	11-23-71	The Tenth Annual Industry-Sponsored Guest Lecture: Ecology and the Anaerobes.—Ralph Wolfe, PhD
CM25	12-13-71	Evaluations of Recently Introduced "Rapid" Methods in Clinical Microbiology.—Panel Discussion, Moderator, Theodore G. Anderson, Panel: G. Campbell, R. R. Gutekunst, R. R. Clark, H. Friedman
CM26	1-17-72	Practical Means of Identification of Yeast-like Fungi Commonly Encountered in Clinical Microbiology.—Panel Discussion, Moderator, Billy H. Cooper, Panel: V. Iralu, R. R. Marples, J. Cocklin
362	1-25-72	Functional Properties of the Streptococcal Cell Wall-Hutton D. Slade
363	2-22-72	Bacterial Vaccines—New Approaches: Chairman—Herman Friedman Cholera Vaccines: Immunization with Subunit Fractions.—Paul Actor, PhD The Nature of RNA-Rich Subcellular Immunogens From Salmonella.—Toby Eisenstein, PhD Immunogenic Ribosomal Preparations From Pneumococci.—Harriet C. W. Thompson Immunization Against Pneumococcal Infections.—Robert Austrian
CM27	3-6-72	Recent Advances in Automation of Blood Culturing and Antibiotic Susceptibility Testing in Clinical Microbiology.—Convener, Eileen Randall, Guest Speakers: H. D. Isenberg, J. R. Waters
364	3-28-72	Graduate Student Program: Moderator—James Prier Substructures of Capsids and Procapsids of Coxsackievirus B3.—Sara Beatrice Selection and Partial Characterization of Temperature Sensitive Mutants of Type 5 Adenovirus.—Marcia Ensminger The Kinetics of Antibody Production and Affinity in Mice Immune and Partially Tolerant to 2-4 Dinitrophenyl Determinant.—Allan Pickard The Identification of Bacteria From Plate Cultures by Non-Pyrolytic Gas Liquid Chromatography.—Thomas Wade

365	5-23-72	Awards Presented to the Following Individuals: Excellence in Teaching-Two Co-Recipients.—W. G. Hutchinson & E. Spaulding Excellence in Graduate Studies.—Maria Ensinger Excellence in Research.—B. Blumberg
CM28	10-16-71	Nosocomial Infections.—Discussants: R. R Gutekunst, M. A. Benarde, C. Hegh
366	10-30-72	Functions of the Central Office of the American Society for Microbiology.—Dr. Asger Langlykke
CM29	11-13-72	Laboratory Diagnosis of Malaria.—Moderator, Vichazelu Iralu, Lecturer and Discussant, Mae Melvin
367	11-27-72	Labor Relations—Laws Affecting Union and Management.—Charles Halpin,Jr.
CM30	12-4-72	Laboratory Diagnosis of Bacteremia.—Moderator, Eileen L. Randall, BD Vacutainer Blood Culture Tubes.—Kenneth Cundy A High Sucrose Containing Blood Culture Bottle.—Elvyn G. Scott A Radiometric Method of Detecting Bacteremia.—Eileen L. Randall
CM31	1-15-72	Cholera.—Moderator, Eileen L. Randall Pathogenesis and Clinical Aspects.—Nathaniel Pierce Laboratory Aspects.—Harry L. Smith Vibrio parahemolyticus.—Prabhavathi Fernandes
368	1-29-73	Mycoplasma and Mycoplasma-like Agents of Animal and Plant Disease.—Karl Maramorosch
CM32	2-19-23	Uncommon Gram-Negative Rods. Enterobacter agglomerans.—Richard Gutekunst Nosocomial Herellea Infections.—Fritz Blank Odd Ball Gram Negative Rods.—Eileen L. Randall
369	2-26-73	The Structure and Assembly of Membrane-enclosed Viruses.—Dr. Purnell Choppin, MD

CM33	4-23-73	Contamination of Biologicals and Other Materials; and Infections Acquired in Hospitals and by Laboratory Personnel.—Lecturer and Discussant, Gerard McGarrity, Moderator, Theodore Anderson
370	5-21-73	New Antibiotics for the Treatment of Infections Caused by Gram Negative Organisms.—Donald Kaye, MD
CM34	6-4-73	Laboratory Improvement Activities at the Center for Disease Control.—Dr. Charles Hall
371	10-29-73	Bacteremia and Disease.—Donald Kaye, MD New Antibiotics for the Treatment of Infections Caused by Gram Negative Organisms.—Donald Kaye, MD
372	11-26-73	The Clinical and Immunological Implications of Gonococcal Pili.—Thomas M. Buchanan, M. D.
373	1-28-74	Role of Bacterial Adherence in Infection and Immunity.—Ronald Gibbons
374	2-18-74	Isolation and Identification of Anaerobes in the Hospital Laboratory.—John Washington
375	3-4-74	The History of Identification of the Enterobacteriaceae.—Henry Isenberg, William Ewing
376	4-29-74	Potentiation of Cell-Mediated Immunity to Microbial and Tumor Associated Antigen: the Immono-potentiation effects of BCG and Selective Immunosuppressants.—George Mackanesse
377	5-31-74	Venezuela: Experiences and Encounters in a Primitive Jungle Laboratory.—Sister Regina Rowan
378	6-14-74	Sugarloaf Meeting Speaker Not Identified.
379	9-23-74	Regionalization of Hospital Clinical Laboratory Services.—Frank Young
380	10-15-74	Non-dermatophytic Mycology.—John Utz
381	11-26-74	Some Aspects of Antifungal Antibiotics.—Paul S. Hoeprich.
382	1-21-75	Cascade Control of Glutamine Synthetase Activity.—Earl R. Stadtman

383	2-17-75	Diagnostic Procedures for the Identification and Differentiation of Streptococci. —Richard R. Facklam, PhD
384	3-24-75	The Current Status of Venereal Disease in Philadelphia—Clinical and Laboratory Consideration.—Leslie Nicholas, MD, Henry Beilstein
385	5-23-75	Associaton of Herpesviruses with Veneral Disease and Cancer.—Dr. Fred Rapp
386	9-22-75	Foundation for Microbiology Lecture: Pulmonary Delayed Hypersensitivity: A Model for in vitro Mediator Production.—Quentin N. Myrvik, PhD
387	10-6-75	Anaerobic Bacteriology: The State of the Art.—Lillian V. Holdeman
388	11-17-75	Significance of Human (T strain) Mycoplasmas as Pathogens: The Role of the Diagnostic Laboratory.—Ruth Kundsin, PhD, Peter Brant
389	1-26-76	The Education and Training of a Microbiologist—A Panel Discussion: Sister Regina Rowan, Elizabeth Fowler, Dr. Albert Moat, Dr. Joseph Gots, Moderator—Norman Willett
390	2-23-76	The Mouse (and the Human?) Mammary Tumor Viruses.—Dr. Dan H. Moore
391	3-29-76	Microbiology in Industry: Moderator—Dr. Joseph F. Pagano Microbes and Antibiotics—The Discovery of Gentamicin.—Dr. Marvin Weinstein, Gerard Wagman Microbes as Chemists.—Dr. C. John DiCuollo, Dr. Joseph Valenta, Dr. George Greenspan Microbes in the Petroleum Industry. —Dr. Allen L. Haskin
392	4-19-76	First Annual Stuart Mudd Lecture: Lipid Specific Exotoxins.—Dr. Alan W. Bernheimer
393	5-21-76	Plagues and Pestilences Two Centuries Ago.—Dr. John Duffy
394	6-7-76	Tuberculosis—What's your Ziehl-Neelson Rating?—Dr. Russell Schaedler. MD, PhD

395	9-27-76	The History of the Swine Influenza Outbreak at Fort Dix and the Subsequent Immunization Program.—Dr. Ronald Altman
396	10-25-76	An Update on Hepatitis B Virus.—Dr. Irving Millman
397	11-29-76	Phagocytosis: Normal and Abnormal.—Dr. Anthony J. Sbarra
398	1-31-77	Experimental Models for Studying Pseudomonas aeruginosa Infections. —Dr. Richard S. Berk
399	2-24-77	General Perspectives and Utility of in vitro Assays.—Dr. Frederick de Serres Industrial Aspects.—Dr. David Brusick Mutagenicity as a Predictor of Carcinogenicity.—Dr. Mark Hite
400	3-28-77	Important Pathogenic Fungi Encountered in the Diagnostic Laboratory.—Dr. Fritz Blank
401	4-25-77	Second Annual Stuart Mudd Lecture: Australia Antigen and the Biology of Hepatitis B. Virus.—Dr. Baruch Blumberg
402	5-20-77	Legionnaire's Disease—The Evidence for Infection with a Chlamydial Agent. —Gary L. Lattimer, MD
403	6-13-77	Automation in Clinical Microbiology. —Richard Tilton, PhD
404	9-26-77	E. coli as an Agent of Intestinal Disease. —Dr. Harry Smith
405	10-17-77	Ultrastructural Investigations of Intracellular Killing of Human Patahogenic Fungi.—Billy H. Cooper, PhD
406	11-28-77	Bacterial Surface Recepters: Their Role in Virus Infections.—Dr. Manfred Bayer
407	12-19-77	Cheese—Its Art and Science.—Dr. George Somkuti
408	1-30-78	Viral Vaccines—Influence of Vaccine Technology on Epidemiological Strategy: Smallpox Eradication.—Dr. Ben Rubin Protection Against Rabies.—Dr. Ted Wiktor
409	2-27-78	Clinical Parasitology: Factual & Fictional. —Dr. John Thompson, Jr.

410	3-27-78	Epstein-Barr Virus Associated Malignancies: Burkitt's Lymphoma & Nasopharyngeal Carcinoma.—Werner Henle, Gertrude Henle
411	4-24-78	Third Annual Stuart Mudd Lecture: Diphtheria: Reflections on the Evolution of An Infectious Disease.—Dr. Alwin Pappenheimer, Jr.
412	5-22-78	Dietary Preferences, Cancer & Longevity.—Dr. Morris Ross
413	6-12-78	Hospital Infection and Control: Where Microbiology Meets the Patient.—Dr. Carl Walter
414	9-25-78	Human Babesiosis in the United States.—Dr. Franklin Neva
415	10-09-78	Microbiological Considerations in the Management of Common Genital Infections.—Dr. King Holmes
416	11-27-78	Properties of Hepatitis B and Related Viruses—A New Family of Oncogenic Agents?—Dr. Jesse Summers
417	12-11-78	Examples of Microbial Ecology: Wine & Cheese.—Dr. Wayne Umbreit
418	1-22-79	The Antibiotics of the 1970s and 1980s: Circumventing Bacterial Resistance Plasmids.—Christopher Martin, MD
419	2-26-79	Methanogenesis.—Ralph Wolfe, PhD
420	3-26-79	Infectious Dermatoses of the Dog.—Robert Schwartzman, V. MD
421	4-23-79	Fourth Annual Stuart Mudd Lecture: Bacterial Genetics and Disease: The Plasmid Connection.—Joseph Gots, PhD
422	5-21-79	The Diagnosis of Tuberculosis Infection by Skin Tests.—Merrill Chase, PhD
423	6-25-79	Protozoology.—John Thompson, Jr., PhD
424	9-24-79	Newer Drugs in the Treatment of Parasitic Diseases.—Brinton Miller, PhD
425	10-22-79	The Pathogenicity of the Enterobacteriaceae.—William Ewing, PhD
426	11-26-79	Modern Virology From Alpha to Omega— An Update on Research in Slow Virus Diseases.—Clarence Gibbs, PhD New Discoveries About Korean Hemorrhagic Fever.—Clarence Gibbs, PhD

427	12-17-79	The Microbial Flora of the Dento-Gingival Area: A Challenge to the Microbiologist.—Max Listgarten, D.D.S.
428-527	1980s	Meetings 1980 to 1989—No. 428 through 527
428	1-28-80	Viruses as Etiologic Agents of Juvenile Diabetes.—Richard Crowell
429	2-25-80	Human Tumor Antigens Detected by Hybridoma Antibodies.—Meenhard Herlyn, Dr. Med. Vet.
430	3-24-80	The Multiple Action of Staphylococcal Enterotoxins.—Merlin Bergdoll, PhD
431	4-28-80	Fifth Annual Stuart Mudd Lecture: Electron Microscopy Then and Now: Forty Years of Research on the Structure of Microorganisms.—Thomas Anderson, PhD
432	5-19-80	Aspects of Mutagenicity Testing.—Verne Ray, PhD
433	6-23-80	Helminthic Zoonoses, or Man's Best Friend Revisited.—Ruth Leventhal, PhD
434	9-22-80	What You Should Know About Animal Care Before Conducting Animal Experiments: Prevention of Animal Diseases.—Dr. Gordon Eaton
435	10-27-80	The Immunology of Malaria.—Dr. William Weidanz
436	11-24-80	Autologous Mixed Lymphocyte Interactions in Mice: Functional Aspects and Relation to Autoimmunity.—J. Bruce Smith, MD
437	12-15-80	The Structure and Regulated Expression of the Alpha-Fetoprotein and Albumin Genes in the Mouse.—Shirley Caldwell Tilghman, PhD
438	1-26-81	Laboratory Diagnosis of Viral Diseases: Reality or Fantasy.—Jay Satz
439	2-23-81	Slow Viruses: Animal Models and Human Diseases.—Neal Nathanson
440	3-23-81	Clinical Microbiology.—Eileen Randall, PhD
441	4-27-81	Hepatitis B Vaccine: Behind the Scenes of Development.—Irving Millman, PhD
442	5-18-81	Sixth Annual Stuart Mudd Lecture: New Approaches to the Study of Rabies Virus.—Hilary Koprowski

443	6-22-81	Legionella micdadei (Pittsburgh Pneumonia Agent): Investigational Approach and Discovery.—A. William Pasculle, Sc.D.
444	9-28-81	Random Thoughts on Pneumococcus and Pneumococcal Infection.—Robert Austrian, MD
445	10-26-81	Hepatitis B-Like Virus of Domestic Ducks.—William Mason, PhD
446	11-23-81	Interferons and Their Applications.—Paul E. Came, PhD
447	12-14-81	Parvovirus: The Disease and the Vaccine. —Peter Felsberg, V. MD
448	1-25-82	Anaerobes and Pelvic Inflammatory Disease.—Gale Hill, PhD
449	2-22-82	Determining and Reporting the MIC Options and Implications.—Paul Ellner, PhD
450	3-22-82	A Micromethod for the Analysis of Cryoglobulins via Laser Nephelometry: Evaluation and Comparison to Clq Binding Activity in Autoimmune Diseases in Pediatrics.—Lily Yang, PhD
451	4-26-82	Seventh Annual Stuart Mudd Lecture: The Genetics of Auto-immune Thyroiditis. —Noel Rose, PhD, MD
452	5-24-82	Oncogene Expression in Tumor Cells. —Susan Astrin, PhD
453	6-28-82	Opportunistic Fungal Infections in Patients with Neoplastic Diseases.—Joseph Aisner, MD
454	9-22-82	Legionella—The Disease, the Bacterium, and Its Nutritional Requirements. —John C. McKitrick, PhD
455	10-25-82	Role of Posttranslational Modifications to Glycoproteins in Enveloped Virus Replication.—Milton Schlesinger, PhD
456	11-22-82	What Robert Koch Would Have Loved to Know About the Tubercle Bacillus. —Mark Kaplan, MD
457	12-13-82	Wines and Cheeses: Discussion and Tasting.—Mr. Robert Perna
458	1-24-83	Unusual Aspects and Agents of Venereal Disease.—John Dooley, MD

459	2-28-83	RAS Oncogenes in Retroviruses and Human Tumors.—Edward Scolnick, MD
460	3-21-83	Heterogeneity of Macrophage Function.—Page Smith Morahan, PhD
461	4-25-83	Eighth Annual Stuart Mudd Lecture: Structure-Function Relationship Between the Bacterial Chromosome and Cell Membrane.—Moselio Schaechter, PhD
462	5-23-83	Acquired Immune Deficiency Syndrome: An Overview.—Wil Whittington
463	6-27-83	Monoclonal Antibodies: Development and Use in Diagnostic Systems.—Laurence McCarthy, PhD
464	9-26-83	The Epidemiology of Infant Botulism.—Sarah Long, MD
465	10-17-83	A Microbiologists' Guide to the Cephalosporin Antibiotics—Past, Present and Future.—Paul Actor, PhD
466	11-28-83	Progress in the Development of Vaccines Against Cholera, Typhoid Fever and Traveler's Diarrhea: An Overview.—Dr. Myron Levine
467	12-12-83	Anaerobes That Feed You.—Leonard Bull, PhD
468	1-23-84	AIDS in Infants and Children.—Anthony Minnefor, MD, F.A.A.P.
469	2-27-84	Can One Vaccinate Against the Human Cytomegalovirus?—Stanley Plotkin, MD
470	3-26-84	Unusual Microorganisms of Genital Disease: Chlamydia and Ureaplasma.—Sally Hipp, PhD
471	4-30-84	Ninth Annual Stuart Mudd Lecture Tracking Environmental Mutagens With Bacteria.—Philip Hartman
472	5-21-84	Immunological Mechanisms in Neurological Disease.—Byron Waksman, MD
473	6-25-84	Pathogenic Mechanisms of Diarrheal Disease.—Dr. Richard Hornick
474	9-24-84	Malaria in the Modern World.—Jean Bowdre, PhD
475	10-22-84	Genetics, Virulence and Immunity in Shigellosis.—Samuel Formal

476	11-26-84	Microbiology Grand Rounds: Aspergillosis—Case Presentation and Discussion of Organism.—Dr. George Talbot, MD, Helen Buckley
477	12-17-84	A History of the Eastern Pennsylvania Branch of the A.S.M.—Dr. Harry Morton, D.Sc.
478	1-28-85	Workshop on Hospital and Community Acquired Respiratory Infections.—Brian Murphy, MD
479	2-25-85	Special Poster Session Presented by Area Graduate Students and Branch Members.
480	3-25-85	Blood Cultures: New Technology and Old Controversies.—Donald Jungkind, PhD
481	4-29-85	Tenth Annual Stuart Mudd Lecture: Cell Biology of Human Aging.—Leonard Hayflick, PhD
482	5-20-85	Lyme Disease.—Jorge Benach, PhD
483	6-24-85	HTLV III and Influenza Virus—Two Viruses Fifty Years Apart.—Stephen Petteway, PhD, Horace Pettit, MD
484	9-30-85	Toxic-Shock Syndrome Toxin-1 and Other Toxins in TSS.—Patrick Schlievert, PhD
485	10-28-85	Spongiform Encephalopathies.—David Asher, MD
486	11-25-85	Virulence Factors of Pseudomonas aeruginosa.—Barbara Iglewski
487	12-16-85	And No Medicine or Any Other Defense Availed.—Dale Smith, PhD
488	1-27-86	Infectious Disease Rounds.—John Molavi, MD
489	2-24-86	Pathogenesis of Human Periodontal Disease: Actinobacillus actinomycetemcomitans vs. Host Defenses.—Bruce Shenker [for Norton Taichman,D.D.S., PhD]
490	3-31-86	New Developments in Hepatitis.—Robert Gerety, MD, PhD
491	4-28-86	Second Annual Branch Poster Session.—Area Graduate Students & Branch Members
492	5-19-86	Eleventh Annual Stuart Mudd Lecture: Host Control of Staphylococcus aureus in Focal Lesions.—Frank Kapral, PhD

493	6-16-86	The Microbiologist's Role in Reducing the Cost of Nosocomial Infections.—Bruce Hamory, MD
494	9-22-86	Laboratory Diagnosis of Virus Infections: Rapid and Not so Rapid.—Dr. Marilyn Menequs
495	10-27-86	Clinical Impact of Antimicrobial Susceptibility Testing.—Matthew Levison, M.D
496	11-24-86	Developing a Human Malaria Vaccine: Who Said It Would Be Easy?—Dr. James Young
497	12-15-86	Seeing is Believing?—Russell Maulitz, MD, PhD
498	1-26-87	Lymphokine Activated Killer Cells.—Joseph D. Irr, PhD
499	2-23-87	Microbiologic and Pharmacologic Aspects of Infection in the Neutropenic Cancer Patient.—George Drusano, MD, PhD
500	3-23-87	Fungal Chemotherapy: Novel Agents and Therapeutic Strategies.—Dr. George Kobayashi
501	4-27-87	Third Annual Branch Poster Session.—Area Graduate Students and Branch Members
502	5-18-87	Twelfth Annual Stuart Mudd Lecture: An Unusual Life Style of a Large DNA Virus and Other Stories.—Dr. Allan Granoff
503	6-22-87	Current Status and Changing Patterns of Parasitic Diseases.—Peter Schantz, V. MD, PhD
504	9-28-87	Listerosis.—Bennett Lorber, MD
505	10-26-87	Pathogenesis of Group A Streptococci.—Vincent Fischetti, PhD
506	11-23-87	Clinical and Laboratory Characteristics of Actinomycetales.—Geoffrey Land, PhD
507	12-14-87	500TH MEETING OF THE EASTERN PENNSYLVANIA BRANCH OF THE AMERICAN SOCIETY FOR MICROBIOLOGY: A Celebration.
508	1-25-88	Acanthamoeba Keratitis—Elisabeth Cohen, MD
509	2-29-88	Lymphokine Activated Killer Cells.—Joseph Irr, PhD

510	3-28-88	Haemophilus Influenzae Type B Vaccines.—John Robbins, MD, Rachel Schneerson, MD
511	4-11-88	Thirteenth Annual Stuart Mudd Lecture: Epitope Hunting in the Cholera/Coli Enterotoxin Forest.—Richard Finkelstein, PhD
512	5-23-88	Antibiotic Susceptibility Tests—Pitfalls and Potential: A mini-Symposium and Branch Poster Presentation Night.
513	6-20-88	Diarrheal Diseases in the 1980s.—Peter H. Gilligan, PhD
514	9-26-88	Antimicrobial Proteins From Human Saliva.—Daniel Malamud, PhD
515	10-24-88	Intimacy of the Syphilis Spirochete—Host Interaction.—Joel B. Baseman, PhD
516	11-28-88	Practical Impediments to the Diagnosis and Management of Systemic Mycosis: Challenges for Clinical Microbiology.—David J. Drutz, MD
517	12-19-88	Fleas, Ticks, Saliva, and Man.—Margaret Fisher, MD
518	1-30-89	Impact of the Diagnostic Virology Lab on the Management of Respiratory Infections.—Adamadia Deforest, PhD
519	2-27-89	A Novel Bioamplication Method for Diagnostic Testing.—Timothy Block, PhD
520	3-27-89	Strategies for Aids Therapeutics.—Martin Rosenberg, PhD
521	4-24-89	Fourteenth Annual Stuart Mudd Lecture: Protective and Autoimmune Epitopes of Streptococcal M Proteins.—Edwin H. Beachey, MD
522	5-22-89	Branch Poster Session night.
523	6-19-89	Applications of DNA Probes to the Study of Antimicrobial Resistance.—Fred Tenover, PhD
524	9-25-89	Laboratory Diagnosis of Mycoplasma and Legionella Infections.—David Smalley, PhD
525	10-30-89	Bacterial Ribotyping: A Tool for Epidemiology and Pathogenesis.—Terrence Stull, MD

526	11-20-89	The ABC'S of Streptococci: Alpha, Beta and the Confusing Others.—Dr. Roberta Carey
527	12-11-89	History of Wine Making in Australia. —Mr. Robert Whale
528-619	1990s	Meetings 1990 to 1999—No. 528 through 619
528	1-29-90	Molecular Interaction Between a Common Cold and Its Receptors.—Dr. Richard Colonno
529	2-26-90	Grand Rounds and Clinical Correlations— Cases from the Laboratory, Infectious Diseases and Pathology Departments of the Medical College of Pennsylvania and St. Christopher's Hospital for Children.
530	3-12-90	Coagulase Negative Staphylococcal Infections: Pathogenesis and Epidemiology.—Donald Goldman, MD
531	4-23-90	Fifteenth Annual Stuart Mudd Lecture: Salmonella: Understanding Pathogenicity and Development of Vaccine Strategies. —Roy Curtiss, III, PhD
532	6-25-90	New Antifungal Agents.—David J. Drutz, MD
533	9-24-90	The Epidemiology of Nosocomial Infections in Intensive Care Unit Patients. —Dr. William Jarvis
534	10-29-90	Molecular and Cellular Biology of Listeria Monocytogenes Pathogenesis. —Dr. Daniel Portnoy
535	11-19-90	Clinical Cases from The Hospital of the University of Pennsylvania and Temple University.
536	12-17-90	Wines of Spain.—Mr. Andrew Fruzzetti
537	1-28-91	The Gram Stain Revisited.—Charles Sanders
538	2-25-91	Antibiotic Proteins of Human Neutrophils.—Dr. John Spitznagel
539	3-25-91	The Pertussis Vaccine Controversy. —Alison Weiss, PhD
540	4-29-91	Sixteenth Annual Stuart Mudd Lecture: The Switch Between EBV Latency and Replication.—I. George Miller
541	5-20-91	Annual Poster Session Presentations.

542	6-24-91	Food Associated Parasites.—John H. Thompson, Jr., PhD
543	9-23-91	Biological Warfare: Historical Perspectives and Lessons Derived from the Gulf War. —James Poupard, PhD
544	10-28-91	In Honor of Dr. Richard Crowell: The Role of Receptors in Viral Infections.— Richard L. Crowell, PhD
545	11-25-91	Current Concepts in Mucosal Immunity and Vaccine Development.—Jerry R. McGhee, PhD
546	12-16-91	Death at an Early Age—Tuberculosis in Ancient Egypt.—Michael R. Zimmerman, MD, PhD
547	1-27-92	When is a 510K Not a Food Race? Ans: When it is an in vitro MedicalDevice.—Sharon L. Hansen, PhD
548	2-24-92	Vancomycin Resistant Enterococci—The MRSA of the 90'S.—Henry Fraimow, PhD
549	3-31-92	In Honor of Smith Hall and Dr. Harry Morton: Contemporary Issues in Microbiology. —Richard B. Thomson, Jr., PhD The History and Significance of Smith Hall.—Robert Kohler
550	4-27-92	Seventeenth Annual Stuart Mudd Lecture: Implications of Mycoplasma Infection in AIDS Progression.—Joel Baseman, PhD
551	6-22-92	Invasion or Invitation: The Role of Lipooligosaccharide in the Pathogenesis of Non-enteric Mucosal Pathogens. —Michael A. Apicella And The Annual Branch Poster Session.
552	9-21-92	Chlamydial Disease of the Female Genital Tract: Mechanisms of Pathogenesis. —Roger G. Rank, PhD
553	10-26-92	Mechanisms of Salmonella Survival Within Macrophage Phagolysosomes.—Samuel Miller, MD
554	11-23-92	Evaluation of Polymerase Chain Reaction for Detection of Chlamydia trachomatis. —Donald Jungkind—PhD
555	12-9-92	Microbiology and Infectious Diseases in the 21St Century Academic Medical Center—A Colloquium in Honor of Amedeo Bondi.

556	1-25-93	New Insights in Gastroenteritis (Or: The Viral Surprises of Pooh!) —Mary K. Estes, PhD
557	2-15-93	Macrophages and Mycobacterial Disease: Cofactors and Mechanisms of Resistance.—Bruce Zwilling, PhD
558	3-22-93	Microbiology in Today's Biologic/Biomedical Curriculum—Is It an Essential and Distinct Scientific Discipline? A Panel Discussion: Sharon L. Zablotney, PhD, Burton J. Landau, PhD, John Tudor, PhD, Pamela Tabery, PhD, Moderator, Norman Willett, PhD
559	4-26-93	Eighteenth Annual Stuart Mudd Lecture: Pathogenicity of Enterohemorrhagic E. coli (The Cause of the Jack-In-The-Box Outbreak).—Alison D. O'Brien, PhD
560	6-21-93	The Long Road to Standardized Antifungal Susceptibility Testing.—Robert A. Fromtling, PhD
561	9-27-93	Clinical Relevance—Patient Outcome and Specimen Rejection.—Franklin P. Koontz, PhD
562	10-25-93	Name that Organism! The Game of Puzzling Pediatric Pathogens.—Joseph M. Campos, PhD
563	11-22-93	Educating the Public about Animal Research and the Animal Rights Movement. —Adrian R. Morrison, D.V.M., PhD
564	12-18-93	"Abigail, A Negress" The Role and the Legacy of African Americans in the 1793 Philadelphia Yellow Epidemic. —Phil Lapsansky,
565	1-24-94	Pathogenic Mechanisms in AIDS. —Dr. Thomas M. Folks
566	2-28-94	Frogs, Sharks, and Men: New Sources of Antibiotics.—Michael Zasloff
567	3-21-94	Microbiology and the Undergraduate Science Curriculum—Are We Trying to Produce a Marketable Product?—Judy Kandel, PhD, Followed by A Panel Discussion: Paul Actor, PhD, John Tudor, PhD, Vijay Juneja, PhD, Moderator, Norman Willett, PhD

568	5-9-94	Nineteenth Annual Stuart Mudd Lecture: The Structures and Function of Gonococcal Iron-Utilization Receptors.—Philip Frederick Sparling, MD
569	6-20-94	Gene Therapy of Cystic Fibrosis.—John F. Englehardt, PhD
570	9-26-94	Will Insight into the Pathogenesis of Meningitis Lead to New Therapeutic Modalities?—Ralph van Furth, MD
571	10-24-94	Therapy and Prophylaxis of Respiratory Syncytial Virus Infection in Infants with a Humanized Monoclonal Antibody. —Susan Dillon, PhD
572	11-21-94	Susceptibility Testing of M. tuberculosis and other Mycobacteria in Clinical Laboratories.—Leonid B. Heifets, MD, Sc.D.
573	12-19-94	Mycobacteriology in the Century after Robert Koch.—Robert C. Good
574	1-23-95	The New Microbiologic Monsters: Opportunistic Fungi.—Dr. Michael G. Rinaldi
575	2-27-95	Emerging Coccidial and Microsporidial Infections.—Lynne S. Garcia
576	3-22-95	Microbial Diversity: Life in the Microcosmos.—Kenneth L. Anderson
577	4-24-95	Twentieth Annual Stuart Mudd Lecture: Regulation of Virulence by Cell to Cell Communication—A Language that Hurts.—Barbara H. Iglewski, PhD
578	6-19-95	Can Molecular Diagnostics Play a Role in Parasite Diagnosis?—Judith Weiss, PhD
579	9-11-95	Emerging Viruses: How Hot is the Zone? —Richard L. Hodinka, PhD
580	10-23-95	From the Basic Lab to the Clinical Bench, The Discovery of Human Granulocytic Ehrlichiosis.—J. Stephen Dumler, MD
581	12-12-95	From Bamboo Rats to Humans: The Emergence of *Penicillium marneffi* as an AIDS Indicator.—Chester R. Cooper, Jr., PhD
582	1-22-96	New Ways of Looking at Microbiology in the New Year—A Panel Discussion. —Raymond H. Kiefer, PhD, Steven D. Gaylor, Judith A. O'Donnell, MD

583	2-26-96	DNA Vaccines Against HIV Infection.—Bin Wang, PhD
584	3-25-96	The Internet for Research and Teaching: Why and How to Get Started.—Jean A. Douthwright, PhD
585	4-22-96	First Distinguished Branch Member Lecture in Honor of Harry Morton: In Pursuit of Sadam Hussein's Biological Arsenal: The Personal Account of a Biological Warfare Inspector in Iraq.—Raymond A. Zilinskas, PhD
586	5-6-96	Towards an Understanding of the Intracellular Lifestyle of Pathogenic *Neisseriae*.—Magdalene So. PhD
587	6-24-96	New and Emerging Yeast Pathogens.—Kevin Hazen, PhD
588	9-30-96	Antibiotic Resistant Bacteria: Are they a Concern in the Food Supply.—David Swerdlow, MD
589	10-28-96	Viral and Cellular Factors that Activate HSV-1 Gene Expression.—Priscilla A. Schaffer, PhD
590	11-25-96	Beyond the Streptococci: New Catalase-Negative Gram-Positive Cocci of Clinical Interest.—Kathryn L. Ruoff, PhD
591	12-16-96	*Campylobacter* Infection and the Guillain-Barre Syndrome.—Ban Mishu Allos, MD
592	1-27-97	Opioids Are Immunosuppressive and Induce Sepsis.—Toby K. Eisenstein, PhD
593	2-24-97	*Cryptococcus neoformans* Pathogenesis and New Approaches to Therapy.—Arturo Casadevall, MD, PhD
594	3-24-97	Challenges in Microbiology Education—Dr. Norm Willett, A Roundtable Discussion.
595	4-28-97	Second Distinguished Branch Member Lecture in Honor of Earle Spaulding: Old Technologies Detect New Species in a Routine Clinical Microbiology Laboratory.—Marie B. Coyle, PhD
596	5-19-97	Role of IL-12 in Experimental Leishmaniasis.—Phillip Scott, PhD
597	6-23-97	Parasites Associated with Exotic Foods.—Susan Novak, PhD

598	9-22-97	How Does Group A *Streptococcus* Eat Flesh?—James M Musser, MD
599	10-27-97	B-Lactamases: Extended Spectrum and Inhibitor Resistance.—David M. Shlaes, MD
600	11-17-97	*Helicobacter pylori* and the Pathogenesis of Upper Gastrointestinal Disorders.—Martin J. Blaser, MD
601	12-15-97	Are All Diseases Infectious?—Bennett Lorber, MD
602	1-26-98	Enteric Pathogen Exploitation of Host Cell Processes.—B. Brett Finlay,
603	2-23-98	Unwelcome Guests with Master Keys: How HIV Uses Chemokine Receptors for Infection.—Robert Doms, MD, PhD
604	3-23-98	Intimate Strangers: Unseen Life in Earth—Conveying Microbial Literacy to the Public.—Cynthia Needham, PhD
605	4-27-98	Third Distinguished Branch Member Lecture in Honor of Amedeo Bondi: Changing Antimicrobial Resistance in the '90s: The Hot Spots.—Clyde Thornsberry, PhD
606	6-22-98	Infectious Diseases with Skin Manifestations.—Guy Webster, MD
607	9-14-98	A Reason to Reflect and Celebrate—Our 600th Meeting.—Drs. James Poupard, Adamadia Deforest, Bruce Kleger, Robert Austrian and James Prier
608	10-26-98	Virulence Determinants of Uropathogens.—Harry L.T. Mobley, PhD
609	11-23-98	Glycopeptide Resistance in *Staphylococci*.—Fred C. Tenover, PhD
610	12-14-98	TB: An Historical Perspective at the End of the Millennium.—Patrick J. Brennan, MD
611	1-25-99	Impact of the Mucosal Immune System in Current Microbiology and Immunology.—Jiri Mestecky, MD
612	2-22-99	Cytokines and Innate and Adaptive Immune Responses to Intracellular Pathogens.—David M. Mosser, PhD
613	3-22-99	Ancient DNA, Microbial Diversity and Molecular Microbial Ecology: Learn by Doing.—Raul J. Cano. PhD

614	4-27-99	Fourth Distinguished Branch Member Lecture in Honor of Kenneth Cundy: Microbiology and Its Continuing Impact on Public Health.—Burton W. Wilcke, Jr., PhD
615	5-24-99	Sequencing Bacterial Genomes.—Karen A. Ketchum, PhD
616	9-13-99	Coordinate Regulation of Virulence Factors in *Vibro cholerae*.—Victor J. DiRita, PhD
617	10-25-99	Update on Campylobacter Infections. —Irving Nachamkin, Dr.P.H.
618	11-22-99	Molecular Mechanisms of Antimicrobial Resistance in *Helicobacter pylori*. —James Versalovic, MD, PhD
619	12-14-99	Can We Have a Talk—Bug to Bug?—Joseph M. Campos, PhD
620-702	2000s	Meetings 2000 to 2009—No. 620 through 702
620	1-24-00	DNA Vaccines for Malaria.—William O. Rogers, MD, PhD
621	2-21-00	Bubonic Plague as a Historical-Microbiological Problem.—Frederick A. Meier, MD
622	3-20-00	Contemporary Scientific Controversy and the Internet.—Marcia C. Linn, PhD
623	4-17-00	Fifth Distinguished Branch Member Lecture in Honor of Bruce Kleger: Introduction to Bioterrorism.—Andre Weltman, MD
624	5-15-00	Molecular Aspects of Human Respiratory Syncytial Virus Pathogenesis.—Dr. Gail Williams Wertz
625	6-19-00	New Strategies for Antimicrobial Susceptibility Testing.—Janet Fick Hindler
626	9-25-00	Epidemiological Studies on Drug Resistant Pneumococcal Pneumonia.—Joshua P. Metlay, MD, PhD
627	10-23-00	Blood is a Very Good Juice—Blood Cultures in the New Millennium.—Mel Weinstein, MD
628	11-20-00	How Lymphocytes Make Decisions.—Steven L. Reiner, MD
629	12-11-00	Role of the Modern Mycology Laboratory in the New Millennium.—Mahmoud A. Ghannoum, M.Sc., PhD

630	1-22-01	Oxygen-Dependent Innate Host Defenses in *Salmonella* Pathogenesis.—Ferric C. Fang, MD
631	2-26-01	Keeping *E. coli* 0157 Down on the Farm.—Michael P. Doyle, PhD
632	3-26-01	Sixth Distinguished Branch Member Lecture in Honor of Henry Beilstein: Antibiotic Resistance: Microbes on the Defense.—Stuart B. Levy, MD
633	4-23-01	Biologic Weapons: Responding to the Threat of Bioterrorism.—Ronald M. Atlas, PhD
634	5-14-01	Signs of Latency in Ongoing Viral Replication During Therapy with Anti-Retroviral Therapy.—Martin Markowitz, MD
635	6-25-01	Detection of Intestinal Protozoa Using Routine O&P and Rapid Immunoassay Methods: Recommendations Based on Clinical Relevance.—Lynne Garcia, M.S.
636	9-24-01	Animal Transmissible Spongiform Encephalopathy in the U.S.—Linda Detwiler, DVM.
637	10-22-01	Mechanism of Action and Clinical Impact of New Anti-fungal Agemts for *Candida albicans* and other Fungal Infections. —Annette Reboli
638	12-10-01	So You've Had Turista: What Next? A Review of Some Less Common Travelers' Illnesses.—Edward Johnson, MD
639	1-21-02	Bioterrorism—Bacteria as Bioterror Agents: Pathogenesis and Protection.—Karen Elkins, PhD,
640	2-25-02	Bacterial Vaginosis as a Risk Factor for Pregnancy, Complications and HIV Acquisition.—Sharon L. Hillier, PhD
641	3-25-02	Bacteria as Model Systems: Molecular Genetics, Biochemistry, Ecology & Evolution.—Amy Cheng Vollmer, PhD
642	4-22-02	Seventh Distinguished Branch Member Lecture in Honor of Norman Willett: Bugs, Gums and Heart Disease: Pathogenic Strategies of the Periodontal Pathogen *Porphyromonas gingivalis* in Endothelial Cell Inflammation.—Caroline Genco, PhD

643	6-24-02	Pathogenic Mechanisms of Gram-Positive Pathogens.—Bettina Buttaro
644	9-23-02	What's Eating You?—Molecular Pathogenesis of *Vibrio vulnificus*.—Paul A. Gulig, PhD
645	10-28-02	West Nile Virus: Basic Biology and Current Approaches to Therapy and Vaccines.—James Meegan, MD
646	12-16-02	Rapid Diagnosis and Detection of Antibiotic Resistance in Real-Time, and the Impact on Health Care.—Frank Cockerill, MD
647	1-27-03	Targets for Anti-Malaria Drugs.—Akhil Vaidya, PhD
648	2-24-03	A Postcard from Antarctica: Bacteria and Ozone Hole.—Robert Miller
649	3-24-03	The Third Golden Age of Microbiology—Coming Soon to Your Neighborhood.—Abigail A. Salyers, PhD
650	4-28-03	Eighth Distinguished Branch Member Lecture in Honor of Carl Abramson: Snakes, Sex, Sushi and Saunas.—Bennett Lorber, MD
651	6-16-03	Are We Really Ready for a Bioterrorism Event?—Richard Kellogg
652	9-22-03	SARS: Epidemiology, Diagnosis and Molecular Biology.—Teresa Peret
653	10-20-03	RNA Amplification Methods for Virus Detection.—Christina Ginocchio
654	12-15-03	Animal Zoonosis and Emerging Infections in Humans.—Corrie Brown
655	1-26-04	The Role of Enteric Bacteria as Provocateurs of Inflammatory Bowel Disease and as Facilitators of the Development of Oral Tolerance.—John J. Cebra, PhD
656	2-23-04	The Genomics Revolution from Microbes to Mammals.—George Weinstock, PhD
657	3-22-04	What Determines the Host-Specificity of *Salmonella*? An Active—Learning Seminar.—Stanley Malloy, PhD
658	4-26-04	Ninth Distinguished Branch Member Lecture in Honor of James Prier: Are the Feds Waiving Goodbye to the CLIA Regulations of Clinical Laboratory Testing?—Mark S. Birenbaum, PhD

659	6-21-04	Contributions of Molecular Technologies for the Diagnosis and Management of Infectious Diseases.—Jeanne Jordan, PhD
660	9-27-04	Shiga Toxins: Potent Poisons and Pathogenicity Determinants.—Alison O'Brien, PhD
661	10-25-04	Molecular Mechanisms of Bacterial Colonization.—Jeffery N. Weiser
662	12-13-04	Yellow Fever.—Alan Barrett, PhD
663	1-24-05	The Challenge of Antimicrobial Research. —David Payne, PhD, DSc.
664	2-28-05	Use of Host Gene Expression Responses to Identify Stage-Specific Progression of Illness from Exposure to Biothreat Agents.—Dr. Marti Jett
665	3-28-05	Is Microbiology in the 21st Century the Red Queens Race ?—Jennie Hunter-Cevera, PhD
666	4-25-05	Tenth Distinguished Branch Member Lecture in Honor of Josephine Bartola: A Perspective on Disease Control in Philadelphia.—Caroline C. Johnson, MD
667	6-20-05	Medical Mycology in the Twenty-first Century.—Mary Brandt, PhD
668	9-26-05	The Public Health Laboratory: Catalyst for Preparedness.—Michael Pentella, PhD
669	10-24-05	Koch's Postulates, Bioweapons, and Dinosaur Extinction: Connecting the Dots with the 'Damage-Response Framework'.—Arturo Casadevall, MD, PhD
670	12-12-05	Syntrophy Through Starvation: Do the General Principles of Ecology Apply to Microorganisms?—Hazel Barton, PhD The meeting was preceded by three presentations by area graduate students.
671	1-23-06	Characterization of the 1918 Pandemic Influenza Virus.—Christopher F. Basler, PhD
672	2-27-06	Vaccination for Persistent Viruses—Pipe Dream or Reality?—Marcia A. Blackman, PhD
673	3-27-06	Entwining Research and Teaching. —**Graham C. Walker**, PhD

674	5-8-06	Eleventh Distinguished Branch Member Lecture in Honor of Richard Crowell: The Importance and Challenge of Vaccine Development.—Emilio A. Emini, PhD
675	6-27-06	From Microscopy to DNA Analysis: CDC's Laboratory Diagnostic Approaches in Parasitology.—Alexandre J. DaSilva, PhD
676	9-25-06	Soooo, You Think You Know Clinical Microbiology—Pediatric Case Presentations.—Joseph M. Campos, PhD
677	10-23-06	Mapping the Global Impact of a High-Energy Metabolic Intermediate on a Signal Transduction Network.—Alan Wolfe, PhD
678	12-11-06	Staphylococcal Virulence and Its Regulation.—Dr. Richard Novick
679	1-22-07	Viruses as Gene Delivery Vehicles.—James M. Wilson, MD, PhD
680	2-26-07	Using Phage Lytic Enzymes to Control Pathogenic Bacteria.—Vincent A. Fischetti, PhD
681	3-26-07	Microbiology that Changed the World.—Kenneth L. Anderson, PhD
682	4-23-07	A Philadelphia Story, Cystic Fibrosis.—Peter Gilligan, PhD
683	6-25-07	The Global Threat of Drug Resistant Tuberculosis and Update on Recent Outbreaks of Interest in Pennsylvania.—Stephen Ostroff, MD
684	9-24-07	Fusariums, Contact Lenses & Fungal Keratitis: The Unsolved Puzzle.—Vishnu Chaturvedi, PhD
685	10-29-07	Acinetobacter: An Infectious Agent Emerges from an Ancient Lineage Revealing Sources of Bacteria Diversity.—Nicholas Ornston, PhD
686	11-26-07	Retrovirus Transmission and CNS Disease.—Brian Wigdahl, PhD
687	12-10-07	What is a Microbial Species: New Insights from Genomics.—Dr. James M. Tiedje
688	1-28-08	Community and Health-Care Associated MRSA Infections.—Jason Y. Kim, MD

689	2-25-08	Where Will Future Antibacterials Come from: Big Pharma Versus Biotech? —Clarence L. Young, MD, F.I.D.S.A.
690	3-24-08	Removing the Mask from Metabolism. —Patricia J. Kiley, PhD
691	4-28-08	Microbial CSI—Using DNA Fingerprinting to Identify the Source of Disease Outbreaks.—Joanne M. Bartkus, PhD
692	6-23-08	Academic Perspectives in Emerging Gram-Negative Resistance.—David M. Livermore, PhD
693	9-22-08	Brain to the Immune System: "Do Not Overact" Immunosuppressive Neuropeptides to the Rescue.—Doina Ganea, PhD
694	10-20-08	The Evolution of Sex in Fungi.—Joseph Heitman, MD, PhD
695	12-8-08	Antibiotic Resistant Enterococci: What Makes Good Commensals go Bad (and why are they Dragging *S. aureus* Down with them)?—Michael S. Gilmore, PhD The meeting was preceded by three graduate student presentations.
696	1-26-09	Gram-Negative Resistance Trends and the KPC Experience.—Alan T. Evangelista, PhD, (ABMM)
697	2-23-09	Nosocomial Fungal Infections: A Look at Emerging Pathogens and Future Threats.—Michael A. Pfaller, MD
698	3-16-09	Intersection of Research and Teaching: Creating Conditions That Lead to Change in a Learner's Brain.—Alix Darden, PhD
699	4-27-09	The Call of the Wild: Understanding AIDS by Natural SIV Hosts.—Guido Silvestri, MD
700	9-28-09	Horizontal Gene Transfer Among Pathogens During Chronic Infections Serves as a Counterpoint to the Host's Adaptive Immune System.—Garth Ehrlich, PhD
701	10-26-09	Genetics and Genomics of Antifungal Resistance.—Thomas Edlind

702	12-14-09	Functional Genomic Approaches to the Study of Malaria.—Manuel Llinas PhD. The regular program was preceded by four graduate student presentations: Chimeric PyMSP1/8 Vaccine Elicits Specific and Durable Protective Immune Response Against Lethal *Plasmodium yoelii*. —James Alaro. Nod 2-Dependent Responses to Colonization by Streptococcus pneumonia.—Kimberly Davis. Expression of Interfering RNAs from a Chimeric, HIV-1Tat-Inducible RNA Pol.—Viraj Sanghvi. Control of Growth of the Mother Cell During Twin Spore Formation by *Bacillus subtilis*.— Panagiotis Xenopoulos.

NAME INDEX FOR MEETINGS FROM 1936 THROURH 2009

Abramson, C. [312, 650]
Abramson, S. [173, 194, 234]
Actor, P. [363, 465, 567]
Ada, G. [353]
Adye, J. [300]
Ahmad, F. [314]
Aisner, J. [453]
Ajl, S. [287, 314]
Alaro, J. [702]
Albertson, J. [309, 314]
Alexander, J. [278]
Alexander-Jackson, E. [216]
Allen, E. [222]
Allen, J. [346]
Allen, R. [216]
Allison, M. [194]
Allos, M. [591]
Altman, R. [395]
Anderson, K. [576, 681]
Anderson, T. F. [148, 150, 155, 164, 168, 171, 185, 220, 431]

Anderson, T. G. [219, 254, 311, CM2, CM8, CM16, CM22 CM25, CM33]
Andrews, W. [287]
Apicella, M. [551]
Appleton, J. [151, 241]
Aronson, J. [142, 149, 199, 221]
Asher, D. [485]
Astrin, S. [452]
Austrian, R. [332, 337, 363, 444, 607, CM3]
Bacon, H. [139]
Bailin, M. [312]
Balczuk, N. [267]
Baldridge, R. [339]
Barclay, E. [305]
Bard, R. [257, 263]
Barksdale, L. [274]
Barnes, I. [339]
Barnes, K. [147]
Barnes, M. [147]

Barrett, A. [662]
Bartell P. [247, 296]
Bartkus, J. [691]
Bartlett, R. [CM11]
Bartola, J. [666]
Barton, H. [670]
Baseman, J. [515, 550]
Basler, C. [671]
Batson, H. [212]
Batson, O. [139]
Baumgarten, W. [220]
Bayer, M. [406]
Beachey, E. [521]
Beale, H. [228]
Beamer, W. [174]
Beatrice, S. [364]
Beilstein, H. [306, 384, 632, CM22]
Bekierkunst, A. [309]
Bell, J. [144]
Bellew, H. [146]
Benach, J. [482]
Benarde, M. [CM28]
Bendler, J. [357]
Benenson, A. [338]
Bergdoll, M. [430]
Berk, R. [398]
Berle, L. [171]
Bernhardt, E. [CM24]
Bernheimer, A. [343, 392]
Berry, L. Joe [249, 284, 297]
Beskid, G. [281]
Billen, D. [289]
Billingham, R. [325]
Birenbaum, M. [658]
Blackman, M. [672]
Blair, J. [166]
Blank, F. (Ph.D) [400]
Blank, F. [CM4, CM32]
Blaser, M. [600]

Blau, E. [243]
Blevins, A. [168]
Bliss, E. [249]
Block, T. [519]
Bloom B. [334]
Blumberg, B. [350, 365, 401]
Blumfeld, R. [131]
Blumstein, G. [154]
Blundell, G. [164, 174]
Boerner, F. [118, 119, 135, 143, 160, 191]
Boger, W. [198]
Boltjes, B. [195]
Bondi, A. [155, 162, 183, 186, 248, 264, 352, 555, 605]
Boroff, D. [295]
Bourdillon, J. [152]
Bowdre, J. [474]
Brachman, P. [261]
Brandt, M. [667]
Brant, P. [388]
Braun, A. [168]
Brennan, P. [610]
Brillaud, A. [305]
Briody, B. [240]
Broadhurst, J. [168]
Brodie, A. [225]
Brown, A. [246]
Brown, C. [654]
Brown, J. [257]
Brusick, D. [399]
Buchanan, T. [372]
Bucher, C. [129, 147, 159, 174]
Buck, T. [197]
Buckley E. [305]
Buckley, H. [476]
Bull, L. [467]
Burnet, F. [226]
Byrne, E. [354]

Calesnick, E. [198]
Calkins, H. [132]
Came, P. [446]
Campbell, C. [CM17]
Campbell, E. [155]
Campbell, G. [CM25]
Campos, J. [562, 619, 676]
Cannon, P. [172]
Cano, R. [613]
Cantor, A. [170]
Carey, R. [526]
Carp. R. [289, 296, 319]
Carpenter, C. [197]
Carr, M. [179]
Carski, T. [290]
Casadevall, A. [593, 669]
Case, E. [139]
Casman, E. [126, 183]
Cassel, W. [211, 240, 270]
Ceglowski, W. [347, 360, CM9, CM20]
Cerny, J. [346]
Chalian, W. [134]
Chambers, L. [119, 152, 157, 184]
Chanock, R. [323]
Chase, M. [253, 301, 422]
Chaturvedi, V. [684]
Cheronis, N. [242]
Choppin, P. [369]
Church, C. [126]
Ciminera, J. [173]
Clancy, C. [204, 213, 258]
Clark, J. [347]
Clark, R. [354, CM1, CM5, CM19, CM25]
Clawson, J. [184]
Cocklin, J. [CM12, CM26]
Cohen, E. [330, 508]
Cohen, H. [236]

Cohen, S. [180, 184, 214, 232]
Cole, W. [CM22]
Colonno, R. [528]
Conner, R. [302]
Cooper, B. [405, CM17, CM26]
Cooper, C. [581]
Copeland, J. [CM5]
Coriell, L. [344, 344]
Covert, S. [196]
Coy, N. [245]
Coyle, M. [595]
Coyne, V. [350]
Creamer, A. [246]
Creech, H. [216]
Crouch, R. [174]
Crowell, R. [288, 428, 544, 674]
Crumb, C. [187, 195]
Cundy, K. [614, CM14, CM18, CM21, CM23, CM30]
Curtiss, R. [531]
Czarnetzky, E. [119, 124, 127, 132]
Dannenberg, A. [234, 309]
Darden, A. [698]
DaSilva, A. [675]
Davis, J. [233, 242, 262]
Davis, K. [702]
DeCourcy, S. [297]
Defendi, V. [268, 307, 338]
Deforest, A. [331, 518, 607]
DeLamater, E. [222, 227]
Delette, B. [CM14]
Del Villano, B. [358]
Derrick, M. [235]
de Serres, F. [399]
DiCuollo, C. [391]
Diener, E. [353]
Dienes, L. [203]
Dietz, C. [183, 186]
Diller, I. [216]

Dillon, S. [571]
DiRita, V. [616]
Doms, R. [603]
Dooley, J. [458]
Douthwright, J. [584]
Doyle, M. [631]
Dozois, T. [146, 155]
Drake, M. [222]
Drusano, G. [499]
Drutz, D. [516, 532]
Dubos, R. [355]
Duffy, J. [393]
Dumler, J. [580]
Dutcher, R. [326]
Eagle, H. [143, 237, 321]
Earley, E. [259, 275]
Eaton, G. [434]
Edlind, T. [701]
Ehrich, W. [176, 255]
Ehrlich, G. [700]
Eiman, J. [140, 167, 179]
Eisenberg, G. [254]
Eisenstein, T. [345, 363, 592]
Ellis, A. [160]
Ellner, P. [449]
Emini, E. [674]
Emmart, E. [176]
Emmons, E. [239]
Engle, C. [300]
Englehardt, J. [569]
Engley, F. [180, 198]
Ensinger, M. [365]
Ensminger, M. [364]
Erf, L. [164]
Estes, M. [556]
Evangelista, A. [696]
Evans, G. [CM6]
Ewing, W. [375, 425]
Eyring, H. [168]

Ezekiel, D. [293]
Facklam, R. [383]
Fagan, R. [244, 261]
Fang, F. [630]
Farber, M. [228]
Fare, L. [338]
Favour C. [223]
Fazal, A. [300]
Fearing, M. [240]
Feinberg, R. [212]
Feldman, W. [263]
Felsberg, P. [447]
Felsen, J. [169]
Felton, H. [162, 179, 195]
Felton, L. [122]
Fenner, F. [266]
Fernandes, P. [CM31]
Fertig, J. [121]
Field, A. [340]
Finegold, S. [CM15, CM23]
Finger, I. [256]
Finkelstein, A. [139]
Finkelstein, R. [511]
Finlay, B. [602]
Firor, W. [134]
Fischetti, V. [505, 680]
Fisher, M. [517]
Fishman, M. [330]
Fitzpatrick, F. [163, 193, 213, 224]
Flick, J. [267]
Flippen, H. [264]
Flosdorf, E. [119, 123, 146, 155, 162]
Folks, T. [565]
Forbes, M. [273]
Force, E. [291]
Forster, C. [193]
Foster, J.W. [168]
Fowler, E. [178, 191. 224, 250, 389]
Fowler, H. [167]

Fowler, R. [140, 157, 179]
Fraimow, H. [548]
Fraimow, W. [CM10]
Frank, H. [282]
Frankel, J. [247, 275]
Free, B. [CM6]
Free, E. [277]
Freedman, H. [310]
Friedman, H. [259, 281, 294, 353, 360, 363, CM11, CM24, CM25]
Fritz, M. [312]
Fromtling, R. [560]
Fruzzetti, A. [536]
Fukuyama, T. [278]
Furcolow, M. [154]
Furmanski, P. [345]
Gaby, W. [219, 277, 281, 283]
Ganea, D. [693]
Garcia, L. [575, 635]
Gard, S. [279]
Gerety, R. [490]
Ghannoum, M. [629]
Gibbons, R. [373]
Gibbs, C. [426]
Gibson, E. [155]
Gilden, R. [319]
Gilead, Z. [319]
Gilligan, P. [513, 682]
Gilmore, M. [695]
Ginsberg, H. [288, 319]
Ginsburg, D. [209]
Ginsburg, H. [357]
Ginsburg, I. [295]
Girardi, A. [222, 247, 319]
Glenn, J. [149]
Gold, H. [133]
Goldfine, H. [357]
Goldman, D. [530]
Gollub, S. [295]

Good, R. [573]
Goode, W. [136]
Goodgal, S. [357]
Goodner, K. [221, 244]
Gots, J. [198, 215, 224, 225, 232, 248, 256, 268, 287, 300, 359, 389, 421]
Goulet, N. [291]
Graham, A. [293]
Grainger, T. [155]
Granoff, A. [502]
Green, J. [318]
Greenbaum, S. [151]
Greene, J. [331]
Greenhalgh, P. [CM19]
Greenspan, G. [391]
Gregory, J. [CM7]
Griffin, C. [290]
Grosebeck, M. [252]
Groupe, V. [274]
Gunter, J. [151]
Gunther, C. [132, 145, 180, 229]
Gutekunst, R. [CM7, CM18, CM25, CM28, CM32]
Haff, R. [340]
Halbert, S. [276]
Hall, C. [CM34]
Halpin, C. [367]
Hamory, B. [493]
Hampil, B. [141, 200, 221, 238]
Hansen, S. [547]
Harris, S. [195,200, 228, 240, 255, 276, 294, 325, 334]
Harris, T. [176, 219, 228, 240, 255, 276, 294, 295, 311, 325, 334]
Harrison, J. A. [178, 191, 246, 250, 224]
Hartman, P. [233, 471]
Haskins, A. [391]

Havas, H. [284, 303]
Havens, W. [147, 255]
Hayflick, L. [280, 323, 481]
Haymann, H. [298]
Hazen, K. [587]
Hechemy, K. [357]
Hegh, C. [CM28]
Heidelberger, M. [175]
Heifets, L. [572]
Heineman, H. [CM9]
Heitman, J. [694]
Henderson, H. [120, 222, 258]
Henle, G. [167, 174, 191, 195, 200, 222, 326, 359, 410]
Henle, W. [152, 157, 158, 167, 174, 191, 195, 200, 214, 222, 272, 326, 342, 359, 410]
Herlyn, M. [429]
Hewell, B. [154]
High, R. [192]
Hildreth, E. [333]
Hill, G. [448]
Hillier, J. [148, 216]
Hindler, J. [625]
Hipp, S. [470]
Hitchens, A. [135, 138]
Hite, M. [399]
Hoban, R. [164]
Hobby, G. [264]
Hodges, J. [174]
Hodinka, R. [579]
Hoeprich, P. [381]
Hogan, R. [306]
Holdeman, L. [387]
Holmes, K. [415]
Hoogerheide, J. [146]
Hopkins, W. [250]
Hornick, R. [473]
Horton, R. [273]

Hottle, G. [121]
Houser, E. [297]
Hraba, T. [346]
Huang, N. [CM4, CM14]
Huff, J. [220]
Hughes, D. [299]
Hummeler, K. [227, 290, 334]
Hunt, A. [215]
Hunt, W. [121, 236]
Hunter-Cevera, J. [665]
Hutchinson, W. [165, 218, 352, 365]
Hwang, M. [139]
Ichelson, R. [131]
Iglewski, B. [486, 577]
Ingalls, T. [327]
Ingham, J. [121]
Iralu, V. [CM17, CM26, CM29]
Irr, J. [498, 509]
Isenberg, H. [375, CM27]
Isquith, A. [314, 318, 324]
Israel, H. [156]
Jackson, Stanley [270]
Jackson, Sterling [271]
Jackson, W. [CM10]
Jakubowitch, R. [CM5]
Jarvis, W. [533]
Jasewicz, L. [252]
Jensen, E. [247, 291]
Jett, M. [664]
Johnson, C. [666]
Johnson, F. [146, 168]
Johnston, R. [CM4]
Jones, C. [135, 139, 143]
Jones, H. [164]
Jones, J. [193]
Jordan, J. [659]
Juneja, V. [567]
Jungkind, D. [480, 554]
Kalle, G. [287, 300]

Kalter, S. [183]
Kandel, J. [567]
Kaplan, Abram [267]
Kaplan, Albert [275, 293]
Kaplan, M. [456]
Kapral, F. [281, 317, 492]
Karush, F. [228]
Kast, C. [123]
Kaye, D. [370, 371]
Kearns, W. [168]
Kelner, A. [186]
Kelser, R. [197, 221]
Kereluk, K. [282]
Ketchum, K. [615]
Kiley, P. [690]
Kim, J. [688]
Kimball, A. [146]
Kimmelman, L. [187]
Kincov, J. [121]
Kleger, B. [607, 623]
Klein, M. [183, 187, 192, 200, 215, 229, 233, 244, 247, 259, 269, 275]
Kligman, A. [217, 229, 271]
Klinman, N. [318, 320, 357]
Knaysi, G. [194]
Knerr, W. [121]
Knight, R. [332, CM4, CM10]
Kobayashi, G. [500]
Kocholaty, W. [158, 167]
Koft, B. [215, 241]
Koh, W. [224]
Kohler, R. [549]
Koldovsky, P. [360]
Kolmer, J. [123, 160, 196, 239]
Koontz, F. [561]
Koprowski, H. [230, 442]
Korngold, L. [304]
Kozub, W. [CM6]

Kraemer, P. [312]
Kral, F. [265]
Kravis, L. [218, 227]
Kritchevsky, D. [289]
Kundsin, R. [388]
Kunkel, L. [125]
Lackman, D. [119, 124, 132, 145, 148]
Land, G. [506]
Landau, B. [558]
Landy, M. [310]
Langlykke, A. [366]
Langner, P. [140]
Lapsansky, P. [564]
Larkin, E. [326]
Lattimer, G. [402]
Lawrence, H. [348]
Lawrence, W. [318]
Leberman, P. [170, 211, 217]
LeBoy, P. [302]
Lecce, J. [245]
Lee, H. [184, 190]
Lee, M. [184]
Leikind, M. [138]
Lenhart, N. [297]
Lerman, S. [351]
Lesher, E. [289]
Lessner, J. [236]
Leszynski, W. [159]
Lev, M. [273]
Leventhal, R. [433]
Levine, A. [324]
Levine, M. [466]
Levine, S. [297]
Levinson, H. [242]
Levison, M. [495]
Levitt, N. [324]
Levy, R. [258]
Lichstein, H. [285]

Lief, F. [296]
Lincoln, I. [179]
Linn, M. [622]
Linna, T. [CM20]
Listgarten, M. [427]
Liu, O. [291]
Live, I. [144, 162, 265]
Livermore, D. [692]
Llinas, M. [702]
Lockwood, J. [137, 141]
Long, E. [120, 149]
Long, S. [464]
Lorber, B. [504, 601, 650]
Love, W. [121, 144]
Loveland, R. [205]
Loveless, M. [188]
Lucke, B. [123]
Ludlum, S. [193]
Lukens, M. [118, 135, 143, 160]
Luria, S. [322]
Lurie, M. [120, 127, 142, 149, 156, 163, 173, 194, 234, 258]
Lynch, E. [239]
Lynch, H. [137]
Lynn, R. [245]
Lytle, J. [240]
MacGregor, R. [CM24]
Mackanesse, G. [376]
MacLean, E. [168]
MacLeod, C. [260]
MacNeal, W. [130, 168]
Magasanick, B. [308]
Mahoney, J. [153]
Malamud, D. [514]
Malloy, S. [657]
Mandle, R. [297]
Manko, M [CM3]
Manson, L. [267, 289]
Maramorosch, K. [368]

Mark, G. [351, 358, 359]
Marples, R. [CM26]
Martin, Chris [418]
Martin, Collier [139]
Mason, W. [445]
Maulitz, R. [497]
Maurer, P. [334]
Mazzur, S. [350]
McAlack, R. [346]
McCarthy, L. [463]
McCaughan, J. [262]
McCracken, M. [165]
McCrea, J. [251]
McGarrity, G. [344, CM33]
McGhee, J. [545]
McGuinness, A. [124, 162, 195, 221]
McKenna, J. [294]
McKitrick, J. [454]
McLean, R. [284]
Meier, F. [621]
Melvin, M. [CM29]
Menequs, M. [494]
Menzel, A. [168]
Mestecky, J. [611]
Metlay, J. [626]
Meyer, H. [327]
Michalka, J. [351]
Micklin, E. [CM10]
Miley, G. [164]
Millar, A. [121]
Miller, A. [192, 198, 203, 225]
Miller, B. [424]
Miller, C. [220]
Miller, G. [347]
Miller, I. [540]
Miller, Robert [648]
Miller, Ruth [127, 211, 235]
Miller, S. [553]

Millman, I. [350, 396, 441]
Minnefor, A. [468]
Miovic, M. [345]
Moat, A. [239, 259, 270, 314, 352, 389]
Mobley, H. [608]
Mohan, R. [246, 298]
Molavi, J. [488]
Moller, G. [353]
Moon, R. [324]
Moore, D. [390]
Morahan, P. [460]
Morgan, I. [124]
Morrison, A. [563]
Morton, H. [119, 127, 141, 144, 150, 157, 163, 165, 167, 170, 180, 186, 211, 217, 245, 477, 549, 585 CM7, CM11, CM18]
Mosser, D. [612]
Mudd, S. [119, 124, 145, 148, 155, 165, 179, 182, 186, 190, 205, 213, 221, 222, 233, 242, 258, 262, 269, 283, 297]
Munder, R. [178]
Munoz, J. [227, 235, 238, 249]
Murphy, B. [478]
Musser, J. [598]
Myrvik, Q. [386]
Nachamkin, I. [617]
Natale, P. [259]
Nathanson, N. [439]
Nazerian, K. [326]
Needham, C. [604]
Neefe, J. [177]
Nelson, J. [125]
Nelson, W. [154, 156]
Neter, E. [349, CM13]
Neuman, R. [286]
Neva, F. [414]

Nicholas, L. [384]
Nichols, W. [307]
Nickel, L. [312]
Nicolas, L. [CM20]
Nightingale, F. [283]
North, L. [180]
Northrop, J. [125]
Novak, S. [597]
Novick, R. [678]
Nowotny, A. [310]
Obold, W. [171]
O'Brien, A. [559, 660]
O'Donnell, E. [339]
O'Donnell, J. [582]
Ogburn, C. [325]
O'Kane, D. [278]
Okono, C. [121]
O'Leary, D. [159]
Ornston, N. [685]
Oskay, J. [245]
Ossman, E. [CM24]
Ostroff, S. [683]
Ottolenghi, A. [295]
Pacis, M. [168]
Pagano, J. [391]
Palmer, C. [154]
Panos, C. [323]
Pappenheimer, A. [152, 411]
Pardee, A. [316]
Park, J. [273]
Parker, E. [168]
Parr, E. [142, 149]
Pasculle, A. [443]
Pasternak, V. [211]
Paucker, K. [251, 275]
Payne, D. [663]
Pedersen, K. [133]
Pelouze, P. [153]
Pennell, R. [190, 218]

Pentella, M. [668]
Peoples, D. [245]
Pepper, D. [124]
Perez, J. [200]
Perna, R. [457]
Petermann, M. [152]
Peterson, A. [282]
Pett, D. [351]
Petteway, S. [483]
Pettit, H. [119, 124, 483]
Pfaller, M. [697]
Phalle, M. [209]
Pickard, A. [364]
Pidcoe, V. [CM9, CM11]
Pierce, N. [CM31]
Pijper, A. [180, 205]
Pillsbury, D. [209]
Pinto, C. [340]
Pizer, L. [314]
Plotkin, S. [327, 469, CM3]
Polevitzky, K. [148, 241]
Pollikoff, R. [234]
Porges, N. [243, 252]
Portnoy, D. [534]
Poupard, J. [543, 607]
Preer, J. [204, 276, 302]
Price, A. [174, 202]
Prier, J. [286, 333, 358, 607, 658, CM3, CM9, CM22]
Quinn, J. [CM5]
Rabin, H. [259]
Raefler, J. [156]
Rake, A. [293]
Rake, G. [168]
Rakoff, A. [137]
Randall, E. [303, 440, CM2, CM6, CM12, CM15, CM17, CM21, CM23, CM27, CM30, CM31, CM32]

Rank, R. [552]
Rao, N. [238]
Rapp, F. [353, 385]
Ratcliff, H. [187]
Rauscher, F. [274]
Ravdin, I. [124]
Raven, C. [147]
Rawlings, W. [121]
Ray, V. [432]
Rebell, G. [209]
Redowitz, E. [140]
Reeves, H. [314]
Reich, P. [323]
Reimann, H. [126, 136, 147, 154, 174, 185, 202]
Reiner, S. [628]
Reinhold, J. [202]
Rettew, G. [177]
Reyniers, J. [210]
Rhodes, R. [262]
Rich, M. [307, 326]
Richards, O. [185]
Rights, F. [121]
Rigney, E. [171]
Riley, R. [CM9]
Rinaldi, M. [574]
Robbins, J. [510]
Rockborn, G. [265]
Rogers, M. [268]
Rogers, W. [620]
Ronkin, R. [302]
Roos, C. [134, 146]
Rosanoff, E. [225]
Rose, E. [221]
Rose, M. [151]
Rose, N. [451]
Rose, S. [127, 158, 202]
Rosebury, T. [160]
Rosenberg, M. [520]

Rosenberger, R. [127, 174]
Ross, M. [412]
Roth, G. [358]
Rothblat, G. [280]
Rothstein, E. [267]
Rous, P. [128]
Rowan, R. [377, 389]
Rowe, W. [335]
Rubin, B. [288, 325, 340, 408]
Rule, A. [123]
Ruoff, K. [590]
Sage, D. [173, 179]
Sall, T. [283]
Sallman, B. [206]
Sanders, C. [537]
Sanghvi, V. [702]
Santer, M. [287]
Santer, U, [263]
Satz, J. [318, 438, CM20]
Sawitz, W. [229]
Saylor, R. [142, 149]
Sbarra, A. [397]
Schaechter, M. [461]
Schaedler, R. [394, CM15]
Schaffer, P. [589]
Schantz, P. [503]
Scharff, M. [320]
Schatz, A. [242, 246, 271]
Schatz, V. [242]
Scheidy, S. [246]
Scherp, H. [123]
Schlesinger, M. [455]
Schlievert, P. [484]
Schlottman, D. [225]
Schneerson, R. [510]
Schorr, S. [215, 233]
Schreck, K. [277, 297, 317]
Schuchardt, L. [235, 238, 249]
Schultz, M. [140]

Schwartzman, R. [420]
Scolnick, E. [459]
Scott, D. [263, 289, 303]
Scott, E.G. [CM12, CM15, CM18, CM30, 254]
Scott, J. [144, 147, 164]
Scott, P. [596]
Scott, T. [150, 195]
Segall, S. [282]
Seibert, F. [120, 133, 149, 156, 176]
Seibles, T. [250]
Sevag, M. [132, 145, 152, 155, 196, 198, 206, 211, 215, 225, 235, 242]
Shadomy, S. [CM17]
Shaffer, F. [195]
Shapiro, L. [CM22]
Sharpless, G. [216]
Shaw, D. [123, 136]
Shelburne, M. [155]
Shenker, B. [489]
Shlaes, D. [559]
Shockman, G. [283, 298]
Shramm, J. [138]
Shtasel, T. [339]
Sigel, M. [195, 203, 218, 222, 227, 229 234, 356]
Silberman, R. [283]
Silvestri, G. [699]
Singer, A. [191]
Siu, R. [215]
Skeggs, H. [220, 232]
Sloan, B. [331]
Smalley, D. [524]
Smibert, R. [280]
Smith, A. [213]
Smith, D. [487]
Smith, Harry [243, 404, CM31]
Smith, Helene [324]

Smith, J. [436]
Smith, Louis [145]
Smith, L. W. [216]
Smith, P. [211, 217, 245, 259, 280]
Smith, W. [190, 199]
Smolens, J. [132, 152, 179]
Smythe, D. [284]
So, M. [586]
Sokol, F. [347]
Solowey, M. [198]
Somerson, N. [323]
Somkuti, G. [407]
Sommer, H. [158]
Sonstein, S. [345]
Sparling, P. [568]
Spaulding, E. [136, 156, 186, 219, 224, 229, 238, 254, 264, 277, 352, 365, 595, CM1, CM15, CM24]
Spitzer, J. [303, 311]
Spitznagel, J. [538]
Spizizen, J. [196, 220, 238]
Spotnitz, M. [332]
Sprague, J. [186]
Springer, G. [254, 267, 273]
Stadtman., E. [382]
Stavitsky, A. [184]
Steers, E. [196, 215]
Stelos, P. [320]
Stempen, H. [218]
Stern, K. [152]
Stevens, K. [255]
Stewart, G. [341]
Stinebring, W. [247, 274]
Stock, N. [326]
Stockmal, R. [345]
Stockton, J. [238]
Stokes, J. [123, 124, 136, 141, 158, 168, 222]

Stoller, J. [CM19]
Strawinski, R. [173, 178]
Strumia, M. [119, 126]
Stubbs, E. [142, 144]
Stull, T. [525]
Summers, J. [416]
Supina, W. [354]
Sussdorf, D. [294]
Sutter, V. [CM23]
Sweeney, F. [CM2, 261, 317]
Swenson, R. [CM24]
Swerdlow, D. [588]
Szybalski, W. [248]
Tabachnick, J. [236]
Tabery, P. [558]
Taichman, N. [489]
Takeya, K. [258]
Talbot, G. [476]
Taliaferro, W. [181]
Tashman, S. [215]
Taylor, I. [168]
Taylor, M. [318]
Teller, D. [287]
Teller, I. [192]
Tenover, F. [523, 609]
Thiele, E. [202, 218]
Thompson, H. [363]
Thompson, J. [409, 423, 542]
Thomson, R. [549]
Thornsberry, C. [605]
Tiedje, J. [687]
Tilghman, S. [437]
Tilton, R. [403]
Tint, H. [296]
Tiselius, A. [133]
Toennies, G. [268]
Tolchin, S. [236]
Topping, N. [233]
Trelawny, G. [246]

Tripodi, D. [324]
Tudor, J. [558, 567]
Tuft, L. [132]
Tumilowicz, J. [326]
Tytell, A. [286]
Umbriet, G. [354]
Umbreit, W. W. [292]
Umbreit, Wayne. [417]
Unger, J. [291]
Utz, J. [380]
Valenta, J. [391]
Vali, N. [283]
Van Furth, R. [570]
Van Slyke, C. [153]
Velicer, L. [331]
Versalovic, J. [618]
Verwey,W. [163,173, 178, 179, 192, 202, 203, 206, 225, 235, 238,240,257]
Vogt, A. [145, 149]
Wade, T. [364]
Wagman, G. [391]
Wagner, R. [251, 328]
Waksman, B. [329, 472]
Waksman, S. [161]
Walker, G. [673]
Walker, L. [179]
Wallis, A. [188]
Walter, C. [413]
Wang, B. [583]
Warren, G. [178]
Washington, J. [374]
Wasserman, A. [236, 250]
Waters, J. [CM27]
Watson, D. [336]
Weaver, R.E. [CM21]
Webster, G. [606]
Weed, L. [220]
Wegner, W. [314]

Weidanz, W. [435]
Weinstein, Marvin [391]
Weinstein, Mel [627]
Weintraub, S. [121]
Weiser, J. [661]
Weiss, A. [539]
Weiss, C. [186, 236, 262]
Weiss, J. [578]
Weissman, S. [323]
Wells, D. [145, 154]
Wells, W. [131, 142, 145, 160, 171, 187, 195]
Weltman, A. [623]
Werner, G. [261]
Wertz, G. [624]
West, M. [236, 240]
Whale, R. [527]
Wheelock. F. [360]
White, P. [168]
Whittington, W. [462]
Wigdahl, B. [686]
Wiktor, T. [333, 408]
Wilcke, B. [614]
Wilhelm, J. [357]
Willard, C. [195]
Willett, N. [389, 558, 567, 594, 642]
Williams, N. [204, 241, 243, 250, 271]
Willits, C. [282]
Wilmer, D. [203]
Wilson, D. B. [330]
Wilson, D. W. [220]
Wilson, J. [679]
Wilson, P. [313]
Winicov, I. [358]
Winterscheid, L. [222]
Wintersteiner, O. [168]
Wise, R. [252, 261, 264]
Wohl, M. [202]

Wolfe, A. [677]
Wolfe, F. [139]
Wolfe, R. [361, 419]
Wolman, I. [123]
Wong, S. [286]
Wood, T. [256]
Woodward, C. [168]
Woolley, D. [189]
Wright, L. [197, 220, 241]
Wuerthele-Caspe, V. [216]
Xenopoulos, P. [702]
Yang, L. [450]
Young, C. [689]
Young, F. [379]
Young, J. [496]

Young, V. [CM19]
Zablotney, S. [558]
Zajac, I. [283, 312]
Zappasodi, P. [156, 234, 258]
Zasloff, M. [566]
Zeldow, B. J. [250]
Zellat, J. [275]
Zilinskas, R. [585]
Zillessen, F. [121]
Zilliken, F. [278]
Zimmerman, M. [546]
Zinder, N. [315]
Zintis, H. [209]
Zittle, C. [137, 152]
Zwilling, B. [557]

SUBJECT INDEX FOR MEETINGS FROM 1936 THROURH 2009

Antimicrobial Agents & Chemotherapy: **general**—[124, 141, 161, 168, 184, 186, 189, 196, 238, 250, 270, 271, 283, 287, 293, 332, 370, 371, 391, 418, 514, 566, 571, 645]; anti-HIV [520, 634]; fungal [257, 500, 532, 637]; mycobacterial [557, 572, 683]; parasitic [424, 647]; viral [300, 571, 645]; **drugs**—actinomycin [331], aureomycin [215], cephalosporins [465], glycopeptides, [609], neomycin [213], nitrofurans [225], penatin [167], penicillin [177, 183, 192, 225, 237, 250, 273,], promin [163], proteins [538], stilbamidine [240], streptomycin [186, 187, 190, 192, 211, 225, 236, 249], sulfa [132, 137, 155, 163, 168, 173, 178, 183, 196, 215, 262], terramycin [217]; betalactamases [183, 599, 696]; synergy-antagonism: [233, 264]

Antimicrobial Susceptibility Testing: **general**—[259, 306, 311, 495, 512, 625]; automation [403, 547]; disk [254]; fungal [560]; mic [449]; mycobacterial [502]

Antiseptics and Disinfectants: [127, 159, 168, 180, 277]

American Society for Microbiology (Society of American Bacteriologists): [134, 165, 366]

Awards: [359, 365]

Bacteriology: acid fast bacilli [120, 138, 216, 257, 456]; *Acinetobacter* [685]; *Actinobacillus* [489]; Actinomycetes [506]; adherance [373]; anaerobes [367, 374, 387, 448, 468]; *Bacillus* [194, 224, 242, 281, 293, 295, 298, 702]; *Brucella* [154, 157, 161]; *Campylobacter* [591, 617]; *C. diphtheriae* [119, 242, 274]; chlamydia [552]; *Clostridium* [145, 262, 278, 295, 324]; *Cornynebacterium* [245]; dental [151, 642]; *E. coli* [173, 215, 224, 236, 263, 284, 287, 289, 298, 303, 318, 324, 345, 346, 353, 357, 404, 511, 559, 609, 631]; *Ehrlichia* [580]; enteric bacilli [212, 238, 375, 602, 660]; *Enterobacteriaceae* [425]; enterococci [278, 358, 548, 655, 695]; Friedlander's bacterium [146]; gram positive cocci [590, 643]; *H. influenzae* [121, 179, 312, 357, 510]; *H. parainfluenzae* [312]; *H. parapertussis* [162]; *H. pertussis* [146, 155, 162, 179, 235, 238, 257]; *Helicobacter* [600, 618]; intestinal flora [273]; lactobacilli [171, 215, 243]; *Legionella* [402, 443, 454, 524]; *Leptospira* [147, 246]; *Listeria* [504, 534]; luminous [178]; lymphogranuloma agent [234]; meningococci [121]; microaerophiles [158]; micrococci [154]; Mima-Herella [303]; mitrochondria [233]; mycobacteria [187, 258, 332, 610]; *Neisseria* [586]; gonococci [153, 170, 197, 372, 568]; oral flora [241, 250, 271, 427]; paracolon bacilli [195, 351]; pneumococci [122, 145, 165, 198, 225, 332, 363, 444, 626]; *Porphyromonas* [642]; *Pseudomonas* [277, 398, 486]; psittacosis agent [234]; rickettsia [140]; *Salmonella* [198, 250, 256, 312, 345, 353, 363, 531, 553, 554, 630, 657]; *Serratia* [216, 250, 263, 324]; *Shigella* [475]; slow lactose fermenters [140]; spirochetes [160, 227, 515]; staphylococci [126, 166, 196, 261, 277, 281, 283, 297, 317, 339, 345, 430, 484, 492, 530, 678, 688, 695]; streptococci; [119, 124, 132, 137, 152, 154, 155, 180, 228, 276, 295, 298, 303, 312, 323, 362, 383, 505, 521, 526, 590, 598, 702]; *Streptomyces* [246]; symbiosis; [151]; typhoid bacillus; [132, 165,180]; UTI Pathogens [608]; venereal disease [384]; *Vibrio* [263, 511, 616, 644]; *Yersinia* [270, 393, 621]

Biochemistry: [133, 137, 145, 171, 172, 190, 193, 196, 215, 232, 235, 236, 241, 242, 249, 259, 262, 278, 281, 283, 285, 289, 292, 293, 298, 300, 305, 308, 311, 313, 314, 316, 318, 322, 323, 324, 330, 338, 339, 345, 350, 351, 357, 358, 382, 419, 437, 455, 461, 471, 528, 534, 538, 551, 568, 592, 613, 641, 652, 675, 677, 680, 690, 693]

Biological Warfare & Bioterrorism: [543, 585, 623, 633, 639, 651, 664, 668, 669]

Clinical Microbiology: [337, 354, 379, 403, 440, 480, 549, 561, 562, 595, 601, 619, 676]

Diagnostic Microbiology: [229, 348, 494, 518, 659]

Diseases: AIDS [462, 468, 550, 565, 581, 640, 686, 699, 702]; anthrax [261]; aster yellow [125]; bacteremia (septicemia) [179, 371]; Bang disease [147]; black leg aggression [144]; botulism [464]; brucellosis [265, 274]; catarrh [125]; cancer (tumors) [168, 216, 252, 268, 284, 303, 376, 399, 412, 432, 452, 453, 471, 499]; chlamydia [470, 552]; cholecystitis [121]; cholera [185, 346, 363, 616]; chronic [700]; clostridial [121]; CNS [636, 686]; coryza [125]; cystis fibrosis [569, 682]; diabetes [428]; diarrheal [513]; diphtheria [411]; dental [462, 489]; emerging infections [654]; eye [186]; fungal [149]; genital [147, 552]; gonorrheal [153, 170, 306]; Gullain Barre syndrome [591]; hepatitis [490]; influenza [141, 144, 147, 164, 395]; intestinal (dysentery) [169, 174, 187, 195, 404, 556, 559, 600, 638, 655]; keratitis [508]; laryngitis [179]; legionaire's [402, 443]; leptospirosis [147]; leprosy [120]; leukemia [119, 326, 347]; listeriosis [504]; lyme [482]; lymphoma [342]; lymphopathia venereum [139]; melioidosis [332]; meningitis [121, 167, 570]; mycobacterial [557, 573]; mycosis [516]; ophthalmia [144]; paratyphoid [121]; pelvic inflamatory [448]; pertusiss [249]; pneumonia [702]; rabies [147]; respiratory tract [337, 444, 478, 626]; rheumatic fever [131, 140, 228]; rickettsial [141]; rocky mountain spotted fever [233]; skin [420, 606]; spotted fever [163]; stomatitis [150]; syphilis [135, 143, 160, 191, 239, 515]; swine influenza [144, 147, 164]; tetanus [134]; toxic shock syndrome [484]; tuberculosis [120, 127, 142, 149, 156, 163, 173, 176, 190, 194, 224, 236, 258, 309, 332, 394, 422, 546, 683]; tularemia [144, 145]; typhoid [121, 140]; typhus [193, 213]; urinary tract [311, 608]; vaginitis (vaginosis) [137, 640]; venereal [170, 384, 385, 458]; viral (general) [141]; whooping cough [162, 179]; yellow fever [564, 662]

Educational: [121, 159, 165, 190, 224, 260, 352, 377, 389, 555, 558, 563, 567, 582, 584, 594, 604, 622, 657, 673, 698]

Electron Microscopy & Ultrastructure: [148, 150, 155, 157, 164, 165, 171, 185, 194, 213, 245, 258, 334, 347, 405, 431]

Epidemiology: [261, 317, 327, 349, 525, 530, 533, 626, 652]

Food Microbiology: [171, 198, 282, 542, 588, 597, 650]

Genetics—Genomics: [126, 256, 323, 338, 351, 356, 357, 421, 461, 475, 511, 521, 523, 525, 569, 589, 615, 641, 656, 664, 677, 679, 687, 692, 694, 700, 701, 702]

Grand Rounds & Clinical Cases: [488, 529, 535]

History: [134, 138, 139, 144, 375, 393, 431, 441, 456, 477, 483, 497, 546, 549, 564, 573, 607, 610, 621, 662, 671, 681]

Hospitals & Infection Control: [277, 379, 478]

Immunology:—**general** [119, 120, 121, 123, 124, 126, 127, 132, 134, 137, 140, 142, 143, 145, 146, 147, 152, 156, 162, 163, 165, 175, 176, 188, 211, 216, 227, 228, 230, 234, 235, 238, 240, 245, 247, 249, 255, 256, 265, 276, 288, 294, 301, 303, 305, 311, 312, 317, 318, 319, 320, 324, 325, 329, 330, 331, 334, 336, 338, 339, 340, 345, 346, 347, 348, 349, 350, 351, 353, 356, 357, 358, 360, 364, 372, 376, 381, 386, 397, 406, 429, 435, 436, 437, 450, 451, 460, 474, 492, 498, 509, 521, 544, 545, 551, 553, 557, 592, 596, 602, 603, 611, 612, 628, 630, 635, 693, 702]; allergy [140, 154, 253]; BCG [142, 376]; toxins (toxoids & antitoxins) [121, 126, 133, 152, 155, 252, 284, 287, 310, 312, 318, 331, 339, 343, 345, 392, 430, 511, 660]; tuberculin; [133, 154, 156, 176, 223]

Interferon: [328, 340, 360, 446]

Laboratory Regulations: [305, 658]

Medical Microbiology: [269]

Memorials: Bergey [127]; Rosenberger [174]; Morton & Smith Hall [549]

Miscellaneous: aging [481]; Antartica [648]; blood groups [254]; cell biology [481, 577]; cell surface [279]; cheese [407, 457]; colonization [661]; dairy [243]; diet [412]; evolution [641]; future of microbiology [649, 665]; germ free animals [210, 273, 280]; industrial microbiology [391]; labor relations [367]; laboratory animals [262, 434, 563]; microbial diversity [576, 614]; microbial ecology [355, 361, 417, 641, 670]; ozone hole [648]; sex [650]; wine [457, 527, 536]

Mycology:—**general**—[154, 159, 167, 168, 216, 217, 271, 380, 381, 400, 405, 453, 516, 560, 629, 667, 694]; *Aspergillus* [476]; *Blastomyces* [240]; chemotherapy [257, 500, 532, 637, 701]; *Coccidioides* [149, 575]; *Cryptococcous* [593]; dermatomycosis/dermatosis [265, 420]; drug resistance [701]; *Fusarium*

[684]; *Gliocladium* [196]; histoplasmin; [192]; keratitis [684]; Microsporidia [575]; nosocomial [697]; yeast [211, 239, 256, 259, 270, 587, 637]; *Penicillium* [581]

Mycoplasmas (PPLP, L-Forms, & Ureaplasmas): [211, 217, 245, 259, 280, 312, 323, 368, 388, 470, 524, 550]

Nosocomial Infections: [493, 533, 697]

Opportunistic Infections: [574]

Parasitology:—**general**—[181, 409, 424, 433, 503, 517, 542, 578, 597, 650, 675]; *Acanthamoeba* [508]; *Babesia* [414]; *Leishmania* [596]; malaria [435, 474, 496, 620, 647, 702]; myxomatosis [514]; protozoa [178, 191, 256, 276, 302, 423, 635]

Phase Contrast Microscopy: [185, 224, 227]

Public Health: [306, 478, 614, 666, 668, 669]

Resistance to Antimicrobials: [198, 233, 248, 249, 418, 487, 523, 548, 588, 605, 609, 618, 626, 632, 646, 663, 683, 692, 695, 696]

Serology: agglutination tests [147, 165, 180, 228]; complement fixation [135, 234, 350]; darkfield [165]; leprosy [120]; salmonella [250]; syphilis [191, 239]; wasserman test [118, 160]

Techniques: air sampling (airborne studies) [145, 154, 158, 160, 171, 173, 187, 195, 344]; anaerobic culture [136, 151, 167]; animal studies [163]; agar cup plate [127]; bacterial counts [140]; bio amplification [519]; biochemical tests [170]; blood culture [480, 627]; DNA fingerprinting [691]; electrical impulse [236]; electrophoresis [156, 304]; filtration (filters) [127, 192]; fluorescence-luminescence [146, 156, 290, 295, 306, 346, 347, 350]; frei test [139]; gas (liquid) chromatography [354, 364]; gram stain [537]; hand washing [157]; in vitro assays [399]; laminar air flow [344]; laser nephelometry [450]; lyophilisation [118]; media [132, 138, 153, 183, 197, 246]; molecular diagnostics [578]; monoclonal assay [463, 571]; PCR [554]; PPD [149]; radiation [123]; replicate plating [250]; RNA amplification [653]; slide test [145]; sonication [119]; staining [211, 213, 217]; sterilization [129]; turbidity [243]; UV therapy (irradiation) [164, 173, 181]; ventilation [131]

Tissue Culture: [247, 280, 286, 312, 321, 324, 347]

Vaccines: anti-viral [408, 672]; Bang disease [147]; BCG [142]; cholera [363, 466]; CMV [469]; *Haemophilus* [510]; hepatitis [441]; HIV [583]; influenza [195, 272, 296]; malaria [496, 620, 702]; parvovirus [447]; pertusiss [195, 539]; polio [257, 296]; rabies [333]; rubella [327]; Salmonella [531]; smallpox [296, 408]; travelers diarrhea [466]; typhoid [466]; typhus [184]; veterinary [244]; west nile [645]

Virology: **general**—[171, 197, 214, 226, 232, 244, 289, 293, 338, 354, 369, 406, 438, 455, 502, 544, 679]; adenovirus [261, 288, 318, 319, 324, 331, 335, 340, 351, 357, 363]; animal viruses [152]; B Virus [286]; bacteriophage [125, 130, 164, 168, 180, 196, 224, 238, 315, 324, 680]; bovine [275]; Burkitt's lymphoma [326, 342]; cancer (and tumor viruses) 123, 128, 216, 307, 319, 360, 385, 390, 410, 416, 459]; canine distemper [265]; CMV; [489]; coxsackie [227, 288, 312, 324, 364]; dengue [247]; detection [653]; diabetes (& viruses) [428]; diarrheal [473]; EB virus [342, 410, 540]; emerging [579]; encephalitis [485]; enteric [171, 291, 556]; hemorrhagic fever [426]; hepatitis [177, 222, 396, 401, 416, 445, 490]; herpes virus [275, 300, 331, 351, 358, 385, 589]; retroviruses (HIV, HTLV); [459, 487, 603, 634, 640, 686, 702]; inclusion bodies [168]; influenza [123, 152, 157, 164, 167, 174, 191, 195, 214, 240, 251, 266, 272, 275, 286, 291, 296, 395, 483, 671]; lukemia [307, 360]; myxomatosis [514, 266]; neurological [472]; parotitis [195]; parvovirus [447]; poliomyelitis [123, 151, 193, 247, 257, 259, 267, 296, 312, 324]; polyoma [307]; pseudo rabies [275, 293]; rabies [230, 261, 333, 347, 442]; rapid diagnosis [494, 518]; respiratory [136]; rhinovirus [230]; rhinitis [340]; rouse sarcoma [274]; RSV [571, 625]; rubella [327]; SARS [652]; slow [426, 439]; smallpox [296, 338, 408]; spongiform encephalopathy [485, 636]; SV40 [319, 335, 358]; swine influenza [164, 395]; vaccinia [270]; west nile [645]

Appendix IV

The Clinical Microbiology Section of The Eastern Pennsylvania Branch: 1967 To 1973

Starting in the 1960s, increasing numbers of Branch members identified themselves as Clinical Microbiologists. In 1967, the Clinical Microbiology members decided that the Branch was not adequately serving their needs and they decided to split from the Branch. It took great effort on the part of all parties involved to keep this new Clinical Microbiology Group as a functioning part of the Branch. This effort paid off, since the compromise was to form the Clinical Microbiology Section, run by an independent steering committee, but still part of the Branch. The goal of the Clinical Microbiology Section was to serve the needs of all Clinical Microbiologists in the Philadelphia area. This decision resulted in major changes in the scope of activities offered to Philadelphia Microbiologists.

Starting in 1967, a series of Clinical Microbiology Section meetings were held on a different evening, independent of the regular Branch meetings, and with a separate numbering system. In the early 1970s, it became apparent that attendance at the Clinical Microbiology Section meetings was increasing significantly, and there was a significant decrease in attendance at the regular Branch meetings. In 1973, there was agreement that the Branch would be restructured to accommodate the Clinical Microbiology members, and that there would be one executive committee to direct all activities. This resulted in the elimination of the Clinical Microbiology Section and restoration of a unified Branch.

The following is information on the thirty-four Clinical Microbiology Section meetings held between 1967 and 1973. These meetings were scheduled on Monday evenings so not to interfere with the regular Tuesday Branch meeting. After the June 1973 meeting, the Branch resumed the practice of just one set of regular meetings, however, these meetings were switched to Monday evenings, rather than the usual Tuesday.

The Thirty-Four Clinical Microbiology Section Meetings: 1967-1973

1967

CM1—September 25: Training in Clinical Microbiology. Earle Spaulding, Convener—Richard Clark

CM2—December 4: Microbial Aspects in the Diagnosis and Management of Urinary Tract Infections. A Panel Discussion, Moderator, Theodore Anderson, Panel: A. Rubin, F. J. Sweeney, L. Karafin and E. Randall

1968

CM3—February 5: (Joint meeting, regular branch meeting #337) Acute Respiratory Infections and Its Laboratory Management. A Panel Discussion, Moderator, James E. Prier, Panel: R. Austrian, M. Manko, and S. A. Plotkin

CM4—April 8: Chronic Respiratory Disease and Its Laboratory Management. A Panel Discussion, Moderator, Ralph Knight, Panel: N. Huang, R. Johnston, and F. Blank

CM5—September 16: Application of Fluorescent Microscopy in the Clinical Laboratory. A Panel Discussion, Moderator, Richard Clark, Panel: J. Copeland, J. Quinn and R.A. Jakubowitch

CM6—November 4: New Techniques and Advances in Blood Culturing. A Panel Discussion, Moderator, Eileen Randall, Panel: G. Evans, B. Free and W. Kozub

1969

CM7—January 13: Mycoplasma. Moderator, Harry E. Morton: Epidemiology of *Mycoplasm pneumoniae* Infections in Marine Recruits. Richard R. Gutekunst Mycoplasm of the Uterine Cervix. James E. Gregory Isolation and Identification of Mycoplasma. Harry E. Morton

CM8—March 17: Universal Standards for Antimicrobial Susceptibility Testing. Theodore Anderson

CM9—May 12: Role of the Clinical Laboratory in the Diagnosis of Viral and Rickettsial Disease. Moderator, James E. Prier:
Technics for Isolation and Identification of Viruses. Randolph Riley
Technics for Serologic Identification of Virus Infection. Walter Ceglowski

Virus Diagnostic Services in the Hospital Laboratory. Herbert Heineman

Public Health Laboratories as Reference Agencies in Virus Diagnosis. Vern Pidcoe

CM10—September 29: Advances in the Isolation and Identification of Mycobacteria. A Panel Discussion, Moderator, Ralph Knight, Panel: E. Micklin, W. Jackson and W. Fraimow

CM11—November 3: Practical Methods of Quality Control in Clinical Microbiology. A Panel Discussion, Moderator, Raymond Bartlett, Panel: H. E. Morton, H. Friedman and V. Pidcoe.

CM12—December 15: Practical Methods of Differentiating Non-Fermenting Gram-Negative Rods. A Panel Discussion, Moderator, Eileen Randall, Panel: E. G. Scott and J. Cocklin.

1970

CM13—January: (Joint meeting, regular branch meeting #349) Immune Response of Patients with Bacterial Infection as an Aid to Diagnosis and Epidemiology. Erwin Neter

CM14—March 16: New Techniques and Advances in the Laboratory Diagnosis of Haemophilus Infections. A Panel Discussion, Moderator, Viola Mae Young, Panel: K. R. Cundy, B. Delette and N. Huang

CM15—May 11: Anaerobic Infection as Related to the Clinical Laboratory. A Panel Discussion, Moderator, Earle Spaulding, Panel: S. M. Finegold, R. W. Schaedler, E. Randall and E. G. Scott

CM16—October 5: Use of the Kirby-Bauer Antibiotic Susceptibility Test in Routine Clinical Microbiology. A film narrated by W. W. Kirby) followed with a talk by Theodore G. Anderson

CM17—November 16: Laboratory Diagnosis and Antibiotic Susceptibility Testing of The Systemic Mycotic Agents. A Panel Discussion, Moderator, Charlotte Campbell, Panel: S. Shadomy, E. Randall, V. Iralu and B. Cooper

CM18—December 14: Current Problems in Clinical Microbiology. A Panel Discussion, Moderator, Harry E. Morton, Panel: K. R. Cundy, E. G. Scott and R. R. Gutekunst

1971

CM19—January 18: Practical Use of Computer Techniques in the Requesting and Reporting of Routine Clinical Microbiology. A Panel

Discussion, Moderator, Richard B Clark, Panel: P. Greenhalgh, J. Stoller and V. M. Young

CM20—March 1: Recent Progress in the Laboratory Diagnosis of Non-gonococcal Venereal Diseases. A Panel Discussion, Moderator, Jay Satz, Panel: W. Ceglowski, T. J. Linna and L. Nicolas

CM21—April 12: Practical Methods in the Laboratory Identification of Problem Gram Negative Bacilli. A Panel Discussion, Moderator, Eileen L. Randall, Panel: K. R. Cundy, and R. E. Weaver

CM22—May17: Recent Progress in the Laboratory Diagnosis of Gonorrhea. A Panel Discussion, Moderator, James E. Prier, Panel: L. Shapiro, T. G. Anderson, W. Cole and H. Beilstein

CM23—October 4: New Applied Techniques in Anaerobic Clinical Microbiology. A Panel Discussion, Moderator, Kenneth R. Cundy, Panel: S. M. Finegold, E. L. Randall and V. L. Sutter

CM24—November 15: Role of the Clinical Microbiology Laboratory in Environmental Microbiology. A Panel Discussion, Moderator, Herman Friedman, Panel: E. Ossman, E. Bernhardt and R. M. Swenson. Discussants: E. H. Spaulding and R. R. MacGregor

CM25—December 13: Evaluations of Recently Introduced "Rapid" Methods in Clinical Microbiology. A Panel Discussion, Moderator, Theodore G. Anderson, Panel: G. Campbell, R. M. Gutekunst, R. R. Clark and H. Friedman

1972

CM26—January 17: Practical Means of Identification of Yeast-like Fungi Commonly Encountered in Clinical Microbiology. A Panel Discussion, Moderator, Billy H. Cooper, A Panel: V. Iralu, R. R. Marples and J. Cocklin

CM27—March 6: Recent Advances in Automation of Blood Culturing and Antibiotic Susceptibility Testing in Clinical Microbiology. Convener, Eileen Randall, Guest Speakers: H. D. Isenberg and J. R. Waters

CM 28—October 16: Nosocomial Infections. Discussants: R. R Gutekunst, M. A. Benarde and C. Hegh

CM29—November 13: Laboratory Diagnosis of Malaria. Moderator, Vichazelu Iralu, Lecturer and Discussant, Mae Melvin

CM30—December 4: Laboratory Diagnosis of Bacteremia. Moderator, Eileen L. Randall:
BD Vacutainer Blood Culture Tubes. Kenneth Cundy
A High Sucrose Containing Blood Culture Bottle. Elvyn G. Scott
A Radiometric Method of Detecting Bacteremia. Eileen L. Randall

1973

CM31—January 15: Cholera. Moderator, Eileen L. Randall
  Pathogenesis and Clinical Aspects. Nathaniel Pierce
  Laboratory Aspects. Harry L. Smith
  Vibrio parahemolyticus. Prabhavathi Fernandes
CM32—February 19: Uncommon Gram-Negative Rods.
  Enterobacter agglomerans. Richard Gutekunst
  Nosocomial Herellea Infections. Fritz Blank
  Odd Ball Gram Negative Rods. Eileen L. Randall
CM33—April 23: Contamination of Biologicals and Other Materials; and Infections Acquired in Hospital and by Laboratory Personnel. Moderator, Theodore Anderson. Lecturer and Discussant, Gerard McGarrity
CM34—June 4: Laboratory Improvement Activities at the Center for Disease Control.—Dr. Charles Hall

Appendix V

Eastern Pennsylvania Branch Sponsored Symposia 1969 To 2009

The first Branch sponsored symposium was held 6-7 November 1969, at the Holiday Inn on City Line Ave. James Prier was chairman with Jay Satz and Richard Gutekunst serving as co-chairs. It was called the "Northeast Regional Conference on Rubella." More than two hundred people attended this first symposium. The symposium was such a success that it was decided to repeat it each November and over the next several years it became an established tradition to have the Annual Branch Sponsored Symposium each November. As one looks through the material on the symposium series in the Branch Archives, the name of the series and number assigned to symposia show several variations. One of the most common designations was the "XXth Annual November Clinical Microbiology Symposium." The complication of a numbering system started in 1972 with the initiation of the first Branch Sponsored Basic Science Symposium entitled "Virus Tumorigenesis and Immunogenesis." There have been eight symposia designated as Basic Science Symposia held since 1972. Initially, they were held in either April or June, but in 1986 it was held in November, in place of what had become known as the Annual November Clinical Microbiology Symposium, which was not held that year. Also in 1994, another year without a November symposium, a Basic Science Symposium was held in December. Therefore, of the 45 Branch Sponsored Symposia, there are essentially eight Basic Science Symposia and 37 Clinical Microbiology November Symposia. It should be noted that in several of the Branch sponsored November Symposia there was considerable effort to combine Basic Science and Clinical Microbiology topics, further complicating the numbering system. The only years since 1969 without a November Symposium (except for the one in December) were 1981 and 2007. In 1981, after several planning meetings, it was determined that the caliber of speakers needed to present on the chosen topic were not available, and the difficult decision was made to cancel the meeting for that year. In 2007, there was difficulty in deciding on a topic and finding a chairperson. In addition, the National Laboratory Training Network, which had been providing assistance starting in 2001, notified the Branch that they could no longer provide their support. These factors lead to the decision not to go forward with a symposium that year.

It is interesting to note the issue of using a numbering system for this series of symposia. There are books and brochures collected by various symposium committee members who, after reviewing the attached list, will try to discover why their particular symposium is listed as the twelfth or twenty-second when these numbers may not match those on the list that follows. The reason for inconsistencies is related to how the Basic Science Symposia were counted. Initially there were two independent numbering systems, however, as the symposia evolved into a combination of clinical and basic science subject matter, two independent numbering systems were abandoned.

It should be noted that work on many of the symposia did not end with the close of the meeting. In many cases the symposium resulted in a published book of the proceedings. This process often took at least one year, or more, to complete.

Titles of the 45 Branch Sponsored Symposia: 1969 to 2009

1—1969 Nov. Northeast Regional Conference on Rubella
2—1970 Nov. Symposium on Immunoglobulins
3—1971 Nov. Symposium on Australia Antigen
4—1972 date? Virus Tumorigenesis and Immunogenesis
5—1972 Nov. Symposium on Opportunistic Pathogens
6—1973 Nov. Symposium on Quality Control in Microbiology
7—1974 Nov. Modern Methods in Medical Microbiology: Systems and Trends
8—1975 April Tumor Virus Infections and Immunity
9—1975 Nov. The Clinical Laboratory as an Aid in Chemotherapy of Infectious Disease
10—1976 Nov. Infection Control in Health Care Facilities: Microbiological Surveillance
11—1977 June Infection, Immunity, and Genetics
12—1977 Nov. Immunoserology in the Diagnosis of Infectious Disease
13—1978 April Microbial Infection and Autoimmunity
14—1978 Nov. Diagnosis of Viral Infections: The Role of the Clinical Laboratory

15—1979 Nov. Clinical Parasitology: Current Concepts in a Changing Science
16—1980 Nov. New Horizons in Medical Mycology
17—1981 June Host Defenses to Intracellular Pathogens
18—1982 Nov. Hepatitis B: The Virus, the Disease and the Vaccine
19—1983 Nov. Infections in the Compromised Host: Laboratory Diagnosis and Treatment
20—1984 Nov. Urogenital Infections: New Developments in Laboratory Diagnosis and Treatment
21—1985 Nov. Clinical Implications of Antimicrobial Resistance: Mechanisms, Testing Problems and Epidemiology
22—1986 Nov. Host Defenses and Immunomodulation to Intracellular Pathogens
23—1987 Nov. Rapid Methods in Clinical Microbiology: Present Status and Future Trends
24—1988 Nov. Infection Control: Dilemmas and Practical Solutions
25—1989 Nov. Emerging Developments in Infectious Diseases: Challenges for the '90s
26—1990 April Second Conference on Candida and Candidiasis: Biology, Pathogenesis and Management
27—1990 Nov. Innovations in Antiviral Development and the Detection of Virus Infections
28—1991 Nov. Antimicrobial Susceptibility Testing: Critical Issues for the '90s
29—1992 Nov. The Migration of Infectious Diseases: Five Hundred Years After Columbus
30—1993 Nov. Antimicrobial Resistance—A Crisis in Health Care
31—1994 Dec. Vaccines: Preventive Strategies for the Twenty-first Century
32—1995 Nov. Resistance, Epidemiology, Laboratory Methods: The Gram-Positive Perspective
33—1996 Nov. Diagnosis of Infectious Diseases Using Molecular Methods: Impact on the Laboratory and Patient Care
34—1997 Nov. Chronic Infectious Diseases: Mechanisms, Diagnosis and Treatment
35—1998 Nov. Clinically Relevant Microbiology in the Era of Managed Care
36—1999 Nov. New Technologies Driving Microbiology into the Twenty-first Century: Applying Genomics, Microarrays &

             Combinatorial Chemistries to Drug Discovery, Bacterial
             Pathogenesis and Vaccine Development
37—2000 Nov. Chronic Viral Hepatitis: Advances in Laboratory Testing,
             Therapy and Prevention
38—2001 Nov. Clinical Microbiology Update 2001
39—2002 Nov. Infections of the Central Nervous System
40—2003 Nov. Clinical Mycology Update: 2003
41—2004 Nov. Antimicrobial Resistance & Emerging Infections: A Public
             Health Perspective
42—2005 Nov. STDs and Other Genital Tract Infections: Current Status
             and Future Trends
43—2006 Nov. Emerging Community and Healthcare-Associated Infectious
             Diseases
44—2008 Nov. Challenging Issues in Antimicrobial Resistance
45—2009 Nov. Molecular Testing in Clinical Microbiology: Advancements,
             Challenges, Future Directions

The Branch Archives contain a collection of material from each of these symposia (except for one Basic Science symposium), with some variation in the amount of material collected each year. Some interesting facts concerning registration fees, cost of hotel rooms, and location of the symposia are presented. By looking at this full list, one can get an accurate idea of how this series of Branch symposia evolved over the years.

When one looks at the titles of these 45 symposia, the variety of subjects is impressive. In addition to the actual symposium, many of these also resulted in a book of collected papers. The first symposium proceedings was published by Charles C. Thomas; then starting in 1971, most were published by the University Park Press (through 1978), followed by Plenum Press, starting in 1981. Because the Symposium Committee worked so well together, the work involved in producing the book was often far more labor intensive than organizing the actual symposium. The work on the book extended the role of the chairperson for at least a full year. During this time the chairperson had to prod authors for manuscripts and then edit the final version for inclusion in the book. If one compares the symposium date with the publication date, it can be noted that many of the books took two years to result in a final publication. Over twenty-two books were published from the Branch Symposium Series. Unfortunately, except for the early works, the publisher was reluctant to provide the Branch with many extra free copies. At the current time, the Branch Archives only contains twelve of the published books. (Those with a [*] following the publication). It should be noted that these publications, especially in the early years, generated a significant amount of royalties for the Branch. As the years progressed, and external competition grew, there was a decrease in the amount of the royalties. However, the book royalties, combined with the money generated by commercial sponsorship for this very popular symposium series, provided the Branch with a financial base that enabled it to experiment with many innovative projects that would not have been possible without this financial base.

The following is a listing that contains some of the more interesting symposium details. The list documents the variations of the registration fee (starting at $25) as well as how the cost for a single room ($27 at the Marriott Hotel in 1974) increased considerably over the years.

EASTERN PENNSYLVANIA BRANCH ASM MASTER LIST OF SYMPOSIA: 1969 TO 2009

1) 1969 Nov. 6-7 Northeast Regional Conference on Rubella
   Chairman: James E. Prier, Co-Chairs: J. E. Satz & Richard Gutekunst
   Registration $25, Location: Holiday Inn (City Line Ave.)
   Publication: 1973, Ed., H. Friedman. Charles C. Thomas Inc.

2) 1970 Nov. 12-13 Symposium on Immunoglobulins
   Chairman: Jay E. Satz, Program Chairman: H. M. Rawnsley
   Registration $35, Location: Marriott Hotel (City Line Ave.)
   Not published.

3) 1971 Nov. 8-9 Symposium on Australia Antigen
   Chairman: Jay E. Satz, Program Chairman: Baruch S. Blumberg
   Registration $35, Location: Marriott Hotel (City Line Ave.)
   Publication: 1973, Prier, J., and H. Friedman. Australian Antigen. University Park Press, Baltimore, MD.*

4) 1972 Virus Tumorigenesis and Immunogenesis (First Basic Science Symposium.)
   Chairman: Herman Friedman
   Publication: 1974, Eds., Ceglowski, W. and H. Friedman. Academic Press.
   (Information from July 1974 Newsletter summary.)

5) 1972 Nov. 2-3 Symposium on Opportunistic Pathogens
   Chairman: Jay E. Satz, Program Chairman: Earle H. Spaulding
   Registration $35, Location: Marriott Hotel (City Line Ave.)
   Publicatiom: 1974, Prier, J., and H. Friedman. Opportunistic Pathogens.
   University Park Press, Baltimore, MD.*

6) 1973 Nov. 15-16 Symposium on Quality Control in Microbiology
   Chairman: Jay E. Satz, Program Chairman: Richard R. Gutekunst
   Registration $35, Location: Marriott Hotel (City Line Ave.)
   Publication: 1975, Prier, J., J. Bartola and H. Friedman. Quality Control in Microbiology.
   University Park Press, Baltimore, MD.*

7) 1974 Nov. 14-15 Modern Methods in Medical Microbiology: Systems and Trends
   Chairman: Jay E. Satz, Program Chairman: Vern Pidcoe
   Registration $35, Location: Marriott Hotel (City Line Ave.) Single room $27
   Publication: 1976, Prier, J., J. Bartola and H. Friedman. Modern Methods in Medical Microbiology: Systems and Trends. University Park Press, Baltimore, MD.*

8) 1975 April 24-25 Tumor Virus Infections and Immunity
   Chairmen: Herman Friedman and Richard Crowell
   Registration $35, Location: Marriott Hotel (City Line Ave.)
   Publication: 1976, Crowell, R., H. Friedman and J. E. Prier. Tumor Virus Infections and Immunity. University Park Press, Baltimore, MD.*

9) 1975 Nov. 20-21 The Clinical Laboratory as an Aid in Chemotherapy of Infectious Disease
   Chairman: Jay E. Satz, Program Chairman: Amedeo Bondi
   Registration $50, Location: Hilton Hotel (34th Street) Single room $24
   Publication: 1977, Bondi, A., J. Bartola, and J. Prier. The Clinical Laboratory as an Aid in Chemotherapy of Infectious Disease. University Park Press, Baltimore, MD.*

10) 1976 Nov. 11-12 Infection Control in Health Care Facilities: Microbiological Surveillance
    Chairman: Jay E. Satz, Co-Chairs: William Ball & Kenneth Cundy
    Registration $50, Location: Hilton Hotel (34th Street) Single room $30
    Publication: 1977, Eds., Cundy, K. and W. Ball. University Park Press, Baltimore, MD.

11) 1977 June 6-7 Infection, Immunity, and Genetics
   Chairman: Herman Friedman, Co-Chairman: James Prier
   Registration $48, Location: Marriott Motor Hotel (City Line Ave.)
   Publication: 1978, Friedman, H., T. Linna and J. Prier. Infection, Immunity, and Genetics.
   University Park Press, Baltimore, MD.*

12) 1977 Nov. 10-11 Immunoserology in the Diagnosis of Infectious Disease
   Chairman: Jay E. Satz, Co-Chairs: T. Juhani Linna and H. Friedman
   Registration $50, Location: Hilton Hotel (Thirty-fourth Street) Single room $30
   Publication: 1979, Eds., Friedman, H., T. J. Linna and J. Prier. University Park Press, Baltimore, MD.

13) 1978 April 17-18 Microbial Infections and Autoimmunity
   Co-Chairmen: Herman Friedman and T. Juhani Linna
   Registration $40, Location: Marriott Motor Hotel (City Line Ave.)
   Published?

14) 1978 Nov. 9-10 Diagnosis of Viral Infections: The Role of the Clinical Laboratory
   Chairman: Robert R. Strauss, Co-Chairs: Steven Specter, David A. Lennette and Kenneth Thompson
   Registration $50, Location: Hilton Hotel (Thirty-fourth Street)
   Publication: 1979, Eds., Lennette, D. A., S. Spector and K. Thompson. University Park Press, Baltimore, MD.

15) 1979 Nov. 8-9 Clinical Parasitology: Current Concepts in a Changing Science
   Chairman: Robert R. Strauss, Co-Chairs: William Ball and C. Iralu
   Registration $75, Location: Hilton Hotel (Thirty-fourth Street)
   Publication: 1981, Ball, W. and V. Iralu. Immunological and Serological Aspects of Clinical Parasitology. Published by Eastern Pennsylvania Branch of ASM, Philadelphia, PA.*

16) 1980 Nov. 13-14 New Horizons in Medical Mycology
   Chairman: Richard L. Crowell, Program Chairperson: Helen Buckley
   Location: Marriott Motor Hotel (City Line Ave.)
   Published ?

17) 1981 June 10-12 Host Defenses to Intracellular Pathogens
Chairpersons: Toby K. Eisenstein and Herman Friedman
Registration $90, Location: Franklin Plaza Hotel
Publication: 1983, Eisenstein, T., P. Actor and H. Friedman. Host Defenses to Intracellular Pathogens. Plenum Press, New York, NY.*

18) 1982 Nov. 11-12 Hepatitis B: The Virus, the Disease and the Vaccine
Co-Chairs: Baruch S. Blumberg and Irving Millman
Registration $90, Location: Franklin Plaza Hotel, Single room $74
Publication: 1984, Millman, I., T. Eisenstein and B. Blumberg. Hepatitus B: The Virus, the Disease, and the Vaccine. Plenum Press, New York, N.Y.*

19) 1983 Nov. 17-18 Infections in the Compromised Host: Laboratory Diagnosis and Treatment
Chairman: Paul Actor, Co-Chairs: Alan Evangelista and James Poupard
Registration $95, Location: Franklin Plaza Hotel, Single room $74
Publication: 1986, Actor, P., A. Evangelista, J. Poupard and E. Hinks. Infections in the Compromised Host: Laboratory Diagnosis and Treatment. Plenum Press, New York, NY.*

20) 1984 Nov. 15-16 Urogenital Infections: New Developments in Laboratory Diagnosis and Treatment
Chairman: Amedeo Bondi, Co-Chairs: Joseph M. Campos and Donald Stieritz
Registration $110, Location: Franklin Plaza Hotel, Single room $76
Publcation: 19xx, Eds., Bondi, A., J. Campos and D. Stierietz. Plenum Press, New York, NY.

21) 1985 Nov. 21-22 Clinical Implications of Antimicrobial Resistance: Mechanisms, Testing Problems and Epidemiology
Chairman: Paul Actor, Co-Chairs: Eileen T. Hinks and Gerald D. Shockman
Registration $105, Location: Franklin Plaza Hotel
Publication: 19xx?, Eds., Actor, P., E. Hinks and G. Shockman. Plenum Press, New York, NY.

22) 1986 Nov. 19-21 Host Defenses and Immunomodulation to Intracellular Pathogens
Conveners: Toby K. Eisenstein, Ward E. Bullock and Nabil Hanna

Registration $150, Location: Adam's Mark Hotel (City Line Ave.) Single room $75
Publication: 19xx?, Eds., Eisenstein, T., W. Bullock and N. Hanna. Plenum Press, New York, NY.

23) 1987 Nov. 12-13 Rapid Methods in Clinical Microbiology: Present Status and Future Trends Chairman: Bruce Kleger, Co-Chairs: Eileen Hinks and Donald Jungkind
Registration $150, Location: Adam's Mark Hotel (City Line Ave.) Single room $78
Publication: 1989, Eds., Kleger, B., D. Jungkind, E. Hinks, and L.Miller Pleneum Press, New York, NY.

24) 1988 Nov. 3-4 Infection Control: Dilemmas and Practical Solutions
Chairman: Kenneth R. Cundy, Co-Chairs: Eileen T. Hinks and Bruce Kleger
Registration $150, Location: Adam's Mark Hotel (City Line Ave.) Single room $88

25) 1989 Nov. 9-10 Emerging Developments in Infectious Diseases: Challenges for the 90's
Chairman: Bruce Kleger, Co-Chairs: Olarae Giger and Donald Jungkind
Registration $150, Location: Adam's Mark Hotel (City Line Ave.) Single room $90

26) 1990 April 1-4 Second Conference On Candida and Candidiasis: Biology, Pathogenesis and Management
Chairperson: Helen R. Buckley
Registration $250, Location: Adam's Mark Hotel (City Line Ave.) Single room $94

27) 1990 Nov. 15-16 Innovations in Antiviral Development and the Detection of Virus Infections
Chairman: Timothy M. Block, Co-Chairs: Richard Crowell and Donald J. Jungkind
Registration $195, Location: Adam's Mark Hotel (City Line Ave.) Single room $94
Publication: 1992, Eds., Block, T., D. Jungkind, R. Crowell, M. Denison and L. Walsh. Pleneum Press, New York, NY.

28) 1991 Nov. 14-15 Antimicrobial Susceptibility Testing: Critical Issues for the 90's
Chairman: James A. Poupard, Co-Chairs: Bruce Kleger and Lori Walsh
Registration $195, Location: Adam's Mark Hotel (City Line Ave.) Single room $108
Publication: 1994, Poupard, J., L. R. Walsh and B. Kleger. Antimicrobial Susceptibility Testing. Plenum Press, New York, NY.*

29) 1992 Nov. 12-13 The Migration of Infectious Diseases: Five Hundred Years After Columbus
Chairman: Alan T. Evangelista, Co-Chairs: Carl Abramson and Bruce Kleger
Registration $195, Location: Adam's Mark Hotel (City Line Ave.) Single room $99

30) 1993 Nov. 11-13 Antimicrobial Resistance—A Crisis in Health Care
Chairman: Donald Jungkind, Co-Chairs: Gary Calandra and Henry Fraimow
Registration $215, Location: Adam's Mark Hotel (City Line Ave.)
Publication: 1995, Eds., Jungkind, D., J. Mortensen, H. Fraimow and G.Calandra. Pleneum Press, New York, NY.

31) 1994 Dec. 8-9 Vaccines: Preventive Strategies for the 21st Century
Chairperson: Toby K. Eisenstein, Co-Chairs: Robert Austrian and Jorg Eichberg
Registration $225, Location: Adam's Mark Hotel (City Line Ave.) Single room $90

32) 1995 Nov. 16-17 Resistance, Epidemiology, Laboratory Methods: The Gram-Positive Perspective
Chairperson: Linda A. Miller, Co-Chairs: Kathleen G, Beavis and Joel E. Mortensen
Registration $225, Location: Adam's Mark Hotel (City Line Ave.) Single room $99

33) 1996 Nov. 14-15 Diagnosis of Infectious Diseases Using Molecular Methods: Impact on the Laboratory and Patient Care
Chairman: Donald Jungkind, Co-Chairmen: Alan Evangelista & David Persing
Registration $225, Location: Adam's Mark Hotel (City Line Ave.) Single room $99

34) 1997 Nov. 6-7 Chronic Infectious Diseases: Mechanisms, Diagnosis and Treatment
Chairman: Ian A. Critchley, Co-Chair: Paul Actor
Registration $225, Location: Adam's Mark Hotel (City Line Ave.) Single room $101

35) 1998 Nov. 19-20 Clinically Relevant Microbiology in the Era of Managed Care
Chair: Kathleen Gleason Beavis, Co-Chair: Karen Carroll and Margaret M. Yungbluth
Registration $225, Location: Adam's Mark Hotel (City Line Ave.) Single room $113

36) 1999 Nov. 4-5 New Technologies Driving Microbiology into the Twenty-first Century: Applying Genomics, Microarrays & Combinatorial Chemistries to Drug Discovery, Bacterial Pathogenesis and Vaccine Development
Chairman: David M. Mosser, Co-Chair: Anna Feldman-Rosen
Registration $225, Location: Adam's Mark Hotel (City Line Ave.) Single room $105

37) 2000 Nov. 9-10 Chronic Viral Hepatitis: Advances in Laboratory Testing, Therapy and Prevention
Co-Chairs: Timothy M. Block, Mark A, Feitelson and Donald Jungkind
Registration $225, Location: Adam's Mark Hotel (City Line Ave.) Single room $107

38) 2001 Nov. 15-16 Clinical Microbiology Update 2001
Co-Chairs: Irving Nachamkin and Donald Stieritz
Registration $225, Location: Adam's Mark Hotel (City Line Ave.) Single room $110
First joint symposium with the National Laboratory Training Network assistance.

39) 2002 Nov. 14-15 Infections of the Central Nervous System
Co-Chairs: Laura J. Chandler and Olarae Giger
Registration $225, Location: Adam's Mark Hotel (City Line Ave.) Single room $110

40) 2003 Nov. 6-7 Clinical Mycology Update: 2003
Chairman: Donald D. Stieritz, Co-Chair: Karin L. McGowan
Registration $195, Location: Biomedical Research Building, University of Pennsylvania

41) 2004 Nov. 9-10 Antimicrobial Resistance & Emerging Infections: A Public Health Perspective
Co-Chairs: Irving Nachamkin and James Poupard
Registration $195, Location: Biomedical Research Building, University of Pennsylvania

42) 2005 Nov. 8-9 STDs and Other Genital Tract Infections: Current Status and Future Trends
Co-Chairs: Richard Hodinka and Donald Jungkind
Registration $195, Location: Biomedical Research Building, University of Pennsylvania

43) 2006 Nov. 8 Emerging Community and Healthcare-Associated Infectious Diseases
Co-Chairs: Linda A. Miller and Ebbing Lautenbach
Registration $125, Location: Biomedical Research Building, University of Pennsylvania

44) 2008 Nov. 14 Challenging Issues in Antimicrobial Resistance
Co-Chairs: Linda A. Miller and Anna Feldman-Rosen
Location: Biomedical Research Building, University of Pennsylvania

45) 2009 Nov. 12 Molecular Testing in Clinical Microbiology: Advancements, Challenges, Future Directions
Chair: Zi-Xuan Wang, Co-Chairs: Donald Jungkind and Laura Chandler
Registration $150, Location: Dorrance H. Hamilton Building, Thomas Jefferson University

* Book in Branch Archive Collection.

List of Chair and Co-chair of Symposia: 1969 to 2009

1) 1969 Nov.   Chair: James E. Prier, Co-Chairs: J. E. Satz & Richard Gutekunst
2) 1970 Nov.   Chair: Jay E. Satz, Program Chair: H. M. Rawnsley
3) 1971 Nov.   Chair: Jay E. Satz, Program Chairman: Baruch S. Blumberg
4) 1972 ?   Chair: Herman Friedman
5) 1972 Nov.   Chair: Jay E. Satz, Program Chairman: Earle H. Spaulding
6) 1973 Nov.   Chair: Jay E. Satz, Program Chairman: Richard R. Gutekunst
7) 1974 Nov.   Chair: Jay E. Satz, Program Chairman: Vern Pidcoe
8) 1975 April Chairs: Herman Friedman and Richard Crowell
9) 1975 Nov.   Chair: Jay E. Satz, Program Chairman: Amadeo Bondi
10) 1976 Nov. Chair: Jay E. Satz, Co-Chairs: William Ball & Kenneth Cundy
11) 1977 June Chair: Herman Friedman, Co-Chair: James Prier
12) 1977 Nov. Chair: Jay E. Satz, Co-Chairs: T. Juhani Linna & H. Friedman
13) 1978 April Co-Chairs: Herman Friedman and T. Juhani Linna
14) 1978 Nov. Chair: Robert R. Strauss, Co-Chairs: S. Specter, D. E. Lennett & K. Thompson
15) 1979 Nov. Chair: Robert R. Strauss, Co-Chairs: William Ball & C. Iralu
16) 1980 Nov. Chair: Richard L. Crowell, Program Chairperson: Helen Buckley
17) 1981 June Chair: Toby K. Eisenstein, Co-Chair: Herman Friedman
18) 1982 Nov. Co-Chairs: Baruch S. Blumberg and Irving Millman
19) 1983 Nov. Chair: Paul Actor, Co-Chairs: Alan Evangelista & James Poupard
20) 1984 Nov. Chair: Amedeo Bondi, Co-Chairs: Joseph M. Campos & Donald D. Stieritz
21) 1985 Nov. Chair: Paul Actor, Co-Chairs: Eileen T. Hinks & Gerald D. Shockman
22) 1986 Nov. Coveners: Toby K. Eisenstein, Ward E. Bullock & Nabil Hanna
23) 1987 Nov. Chair: Bruce Kleger, Co-Chairs: Eileen Hinks & Donald Jungkind
24) 1988 Nov. Chair: Kenneth R. Cundy, Co-Chairs: Eileen T. Hinks & Bruce Kleger
25) 1989 Nov. Chair: Bruce Kleger, Co-Chairs: Olarae Giger & Donald Jungkind
26) 1990 April Chair: Helen R. Buckley
27) 1990 Nov. Chair: Timothy M. Block, Co-Chairs: Richard Crowell & Donald J. Jungkind
28) 1991 Nov. Chair: James A. Poupard, Co-Chairs: Bruce Kleger & Lori Walsh
29) 1992 Nov. Chair: Alan T. Evangelista, Co-Chairs: Carl Abramson and Bruce Kleger
30) 1993 Nov. Chair: Donald Jungkind, Co-Chairmen: Gary Calandra & Henry Fraimow

31) 1994 Dec. Chair: Toby K. Eisenstein, Co-Chairmen: Robert Austrian & Jorg Eichberg
32) 1995 Nov. Chair: Linda A. Miller, Co-Chairs: Kathleen G, Beavis & Joel E. Mortensen
33) 1996 Nov. Chair: Donald Jungkind, Co-Chairs: Alan Evangelista & David Persing
34) 1997 Nov. Chair: Ian A. Critchley, Co-Chair: Paul Actor
35) 1998 Nov. Chair: Kathleen Beavis, Co-Chairs: Karen Carroll & Margaret M. Yungbluth
36) 1999 Nov. Chair: David M. Mosser, Co-Chair: Anna Feldman-Rosen
37) 2000 Nov. Co-Chairs: Tim Block, Mark Feitelson and Donald Jungkind
38) 2001 Nov. Co-Chairs: Irving Nachamkin and Donald Stieritz
39) 2002 Nov. Co-Chairs: Laura J. Chandler and Olarae Giger
40) 2003 Nov. Chair: Donald D. Stieritz, Co-Chair: Karin L. McGowan
41) 2004 Nov. Co-Chairs: Irving Nachamkin and James Poupard
42) 2005 Nov. Co-Chairs: Richard Hodinka and Donald Jungkind
43) 2006 Nov. Co-Chairs: Linda A. Miller and Ebbing Lautenbach
44) 2008 Nov. Co-Chairs: Linda A. Miller and Anna Feldman-Rosen
45) 2009 Nov. Chair: Zi-Xuan Wang, Co-Chairs: Donald Jungkind and Laura Chandler

Committee Members Listed by Year of Service 1969 to 2009

A year in brackets [xx] designates that the person served as a symposium chairperson or as co-chair for that year.

Note some members with two listings for one year, such as "90, 90", indicates that it was a year when the Branch had both a Clinical and Basic Science Symposium, and the person served on both committees.

Abramson, Carl 82, 83, 86, 87, 88, 89, 90, 90, 91 [92] 03
Actor, Paul 81 [83] 84 [85] 86, 89, 90, 90, 91, 93, 94 [97] 99, 00, 01
Aries, Kathleen 88
Austrian, Robert [94]
Axler, David 05, 06
Badger, Alison 86
Ball, William [76] 77, 78 [79]
Barbagallo, Stephanie 06
Bartola, Josephine 70, 71, 72, 73, 74, 75, 76, 77, 78, 79, 80, 81, 82, 83, 84, 85, 86, 87, 88, 89, 90, 90, 91, 93

Beavis, Kathleen [95] 96, 97 [98] 99
Beilstein, Henry 90, 92, 93, 94, 95, 96, 97, 98, 99, 00, 01, 02, 03
Berman, Richard 03
Block, Timothy [90] [00]
Blumberg, Baruch [71] [82]
Bondi, Amedeo [75] [84]
Brillinger, Beth 00
Bromke, Bruno 00, 01, 02
Buckley, Helen [80] [90] 83
Bullock, Ward [86]
Burdash, Nick 88, 89
Buttaro, Bettina 06, 09
Calandra, Gary [93]
Calderone, Richard 90
Campbell, George 89
Campos, Joseph [84]
Carroll, Karen [98]
Carthey, Blitz 00
Ceglowski, Walter 70, 80, 81, 83, 84, 86, 87
Cerwinka, Paul 92, 95, 96, 98, 99, 00, 01, 03, 04
Chaffin, LaJean 90
Chandler, Laura [02] 03, 04, 08 [09]
Collins, Frank 81
Critchley, Ian [97] 98, 99, 00
Crowell, Richard 74, [75] [80] [81] [90]
Cundy, Kenneth 74, 75, [76], 77, 78, 82, 83, 84, 85, 86, 87 [88] 89, 90, 90,
    91, 92, 93, 94, 95, 96, 97, 98, 99, 00, 01, 03, 02, 04, 05, 06
Deforest, Ada 90
Denison, Mark 90
Edelstein, Paul 88
Edwards, John 90
Eichberg, Jorg [94]
Eisenstein, Toby 77, 78, [81] 82, 83, 84, 85 [86] [94]
Escott, Shoolah 01, 03, 04, 05, 06
Evangelista, Alan [83] 84, 87, 89, 90, 91 [92] 95 [96] 97, 98
Feitelson, Mark [00]
Fraimow, Henry [93] 97
Frey, Carrie 90
Friedman, Harvey 90

Friedman, Herman 70, 71 [72] 74 [75] 75, [77] [78] [81]
Giger, Olarae 88 [89] 91, 92, 96, 97, 98, 99, 00 [02] 03, 04, 05, 06, 08, 09
Gorman, Jessica 90
Gots, Joseph 80
Grappel, Sarah 81, 90
Gutekunst, Richard [69] 70, 71, 72 [73] 74, 75, 76
Haller, Gary 87, 88
Halstead, Diane 89, 91, 92
Hanna, Nabil [86]
Hartwig, Kareen 99, 00, 01
Hinks, Eileen 83, 84 [85] 86 [87] [88]
Hodinka, Richard [05]
Howard, Dexter 90
Iacocca, Victor 87
Iralu, Charles [79]
Johnson, Caroline 93
Jungkind, Donald 80 [87] 88 [89] [90] [93] [96] 97, 98, 99 [00] 01, 02, 03 [05] [09]
Kleger, Bruce 86 [87] [88] [89] 90, 90 [91] [92] 93, 94, 96
Klein, Morton 82
Korzeniowski, Denise 03, 04, 05, 06
Lautenbach, Ebbing [06]
Lennette, David [78]
Levison, Matthew 97
Linna, Juhani 71, 72, 73, 74, 75, 76 [77] 77, 78 [78]
Maier, Edward 72
Mandle, Robert 77, 78
Manson, Lionel 78
Marble, Amy 95
McGowan, Karin 89 [03]
McHale, Barbara 93, 94, 95, 96, 97, 98, 99, 00
Miller, Linda 83, 84, 85, 86, 87, 88, 89, 91, 92, 93 [95] 97, 04 [06] [08]
Millman, Irving [82]
Mitchell, Robert 70
Mortensen, Joel 91, 92, 93, 94 [95] 96
Mosser, David [99]
Nachamkin, Irving 87, 92, 93, 94, 95, 96, 97, 98, 99, 00 [01] 02, 03 [04] 05, 06, 08, 09
Neimeister, Ron 87

Nisbet, Louis 83
Pagano, Joseph 79, 80, 81, 87
Perkins, Kathryn 83, 85
Persing, David [96]
Phillips, Michael 78
Pidcoe, Vern 70, 71, 72, 73, [74] 75, 76
Poupard, James [83] 85, 86, 87, 88. 89 [91] 92, 95, 00 [04]
Prier, James [69] 70, 71, 72, 73, 74, 75 [77] 87, 91
Raichle, Linda 93
Rawnsley, Howard [70] 71, 72, 73, 74, 75
Reboli, Annette 93
Rest, Richard 97, 98, 99
Rittenhouse, Stephen 91, 95
Rosen, Anna-Feldman 88, 89, 90, 90, 91, 92, 93, 94, 95, 96, 97, 98 [99] 00, 01, 02 03, 04, 05, 06 [08]
Russel, Lynn 92, 94, 95
Saluk, Paul 81
Satz, Jay [69][70] [71] [72] [73] [74] [75] [76] [77]
Shockman, Gerald [85] 93
Smith, Harry 77, 78, 79
Spaulding, Earle [72]
Specter, Steven [78]
Stieritz, Donald 77, 78, 79, 80, [84] 85, 86, 88, 89, 91, 92, 93, 95, 96, 97, 98, 99, 00 [01] 02 [03] 04, 05, 06, 08
Strauss, Robert [78] [79] 80, 81
Szymczak, Betsy 03, 04, 05, 06
Talbot, George 88
Tashchner, Robert 71
Tassoni, Ernest 76
Thompson, Kenneth 77, 78 [78]
Utrup, Linda 92
Walsh, Lori 89, 90 [91] 92, 95
Wang, Zi-Xuan [09]
Warren, Nancy 04
Wheelock, Frederick 78
Willett, Norman 72, 73, 75, 76, 82, 88, 91, 92
Wood, Craig 93
Youngbluth, Margaret [98]
Zajac, Barbara 84

Appendix VI

Eastern Pennsylvania Branch Sponsored Workshops: 1969 to 2009

The First Workshop was originally sponsored by the Branch Clinical Microbiology Section. It was held July 8-11 on "Fluorescent Antibody Techniques in the Clinical Laboratory." The registration fee was $25 and the workshop was held at the Pennsylvania Department of Health Laboratories, 2100 West Girard Avenue. The Branch conducted at least one workshop every year from 1969 to 2007. The goal was to conduct two workshops each year to serve the needs of Clinical Microbiologists in the Philadelphia area. The program temporarily was suspended in 2008 due to staffing and space availability.

#	DATE	YEAR	WORKSHOP
1	July 8-11	1969	Florescent Antibody Techniques in the Clinical Laboratory
2	October 21-24	1969	Advanced Fluorescent Antibody
3	March 3-6	1970	Advanced Enteric Bacteriology
4	November 17-20	1970	Systemic Mycosis
5	April 13-16	1971	Non-Fermenting Gram-Negative Rods
6	October 5-8	1971	Identification of Anaerobes
7	June 6-9	1972	The Isolation and Identification of Mycobacteria
8	November 14-17	1972	Parasitology
9	June 5-7	1973	Hospital Epidemiology: Hospital Environmental Surveillance and Quality Control in Clinical Microbiology Surveillance
10	Oct. 30-Nov.1	1973	Bacterimia and Blood Culturing
11	March 5-8	1974	The Identification of the Enterobacteriaceae
12	October 15-18	1974	Mycology
13	June 10-13	1975	Identification of Unusual Gram Negative Rods from Clinical Specimens

14	October 6-9	1975	Anaerobic Bacteriology: The VPI Method
15	June 7-10	1976	Isolation and Identification of Mycobacteria
16	November 3-5	1976	General Immunology
17	June 14-16	1977	Antimicrobial Susceptibility and Assay Methods
18	October 8	1977	Chlamydia Infections Old and New Facets of the Disease
19	October 18-21	1977	Isolation & Identification of Dermatophytes and Opportunistic Fungi
20	March 18	1978	An Overview of Diagnostic Virology
21	June 14-16	1978	Hospital Infection Control-Microbiology and Epidemiology
22	October 10-13	1978	Sexually Transmitted Diseases
23	June 26-28	1979	Clinical Protozoology
24	January 26	1980	Recent Advances in MIC Technology
25	June 24-26	1980	Clinical Helminthology
26	December 6	1980	Blood Cultures: Update and Current Controversies
27	June 23-25	1981	"New" Pathogens
28	January 30	1982	Clinical Anaerobic Bacteriology
29	June 29-July 1	1982	Medical Mycology
30	February 26	1983	Sexually Transmitted Diseases
31	June 28-30	1983	Rapid Serologic Methods
32	January 28	1984	Problems in Pediatric Microbiology
33	June 26-28	1984	Recent Advances in Laboratory Diagnosis of Diarrheal Disease
34	February 2	1985	Hospital and Community Acquired Respiratory Infections
35	May 21-23	1985	Skin and Diseases with Skin Manifestations
36	November 2	1985	Infections of the Central Nervous System
37	June 17-19	1986	Nosocomial Infections and Infection Control
38	November 1	1986	Antimicrobial Susceptibility Testing Current Methods and Clinical Impact
39	June 23-25	1987	Recent Advances in Parasitic Diseases
40	June 20-23	1988	Recent Advances in Gastrointestinal Tract Diseases
41	June 19-22	1989	Nonculture Methods for the Diagnosis of Infectious Diseases
42	October 14	1989	Tips for Teaching Clinical Microbiology at the Bench
43	June 25-28	1990	Mycology 101: Identification Methods
44	October 13	1990	New Perspectives On AIDS
45	June 24-27	1991	Current Concepts in Clinical Parasitology

46	October 16	1991	Medical Mycobacteriology in the 1990's
47	June 23-25	1992	Recent Advances in Genital Tract Infections
48	October 21	1992	Blood Cultures & Blood Borne Diseases
49	June 22-24	1993	Laboratory Diagnosis of Fungal Infections
50	October 7	1993	Molecular Diagnosis in Clinical Microbiology
51	June 21-23	1994	Laboratory Diagnosis of Respiratory Infections
52	June 20-22	1995	Clinical Parasitology—1995
53	October 12	1995	New & Reclassified Organisms of Clinical Importance
54	June 25-27	1996	What's New With Yeasts?
55	June 24-26	1997	Food & Water Borne Diseases of North America
56	June 23-25	1998	Infectious Diseases with Skin Manifestations—Diagnosis & Pathogenesis
57	June 22-23	1999	Practical Clinical Parasitology: Fundamentals, Review and Future Perspectives
58	June 20-21	2000	A New Era of Susceptibility Testing: Challenges and Methods
59	June 26-27	2001	Clinically Relevant, Rapid Microbiology
60	June 25	2002	Strategies for the Identification of Gram-Positive Bacteria in the Clinical Laboratory
61	June 17	2003	Bioterrorism and the Clinical Laboratory: Practical Information Needed in Every Microbiology Lab
62	June 21-23	2004	Introduction to Molecular Diagnostic Techniques
63	June 21-22	2005	Medical Mycology; A Refresher Course for Laboratorians
64	June 26-27	2006	Food and Waterborne Intestinal Parasites
65	June 26	2007	TB: Challenges for the Laboratory
		2008	No Workshop was held.
		2009	No workshop was held.

Appendix VII

Philadelphia Infection and Immunity Forum 1992 to 2009

A new tradition was initiated on 10 April 1992. This year marked the start of a new annual series designed to get graduate students and faculty members involved with the Branch. It was presented as an interdisciplinary forum for clinical and basic research scientists in the Philadelphia and the Delaware Valley areas interested in microbial pathogenesis and host response. Richard Rest, of Hahnemann University, was in charge of the committee that initiated the first Philadelphia Infection and Immunity Forum that was held at Thomas Jefferson University. This first Forum included internationally known speakers, a poster session, two Topical Sessions and informal discussions. This first meeting established the format for future Forums which became an annual event that continues to the present time. Once a Branch student chapter was established, organizing this annual forum became included as part of their activities. They organized this event while working closely with an Executive Committee member.

In addition to Dr. Rest as chairperson, the following individuals were on the first organizing committee: Toby Eisenstein, Richard Johnston, Joel Mortensen, Irving Nachamkin, Daniel Portnoy and Norton Taichman.

Summary of Forums: 1992 through 2009

1992 April 10, Thomas Jefferson University
Organizer and Chairperson: Richard Rest
Format: speakers, posters, topical discussions and exhibits

Speakers:
Barry Bloom, Albert Einstein School of Medicine, "Revisiting and Being Revisited by Mycobacteria"

John Mekalanos, Harvard Medical School, "New Insights in Cholera Pathogenesis and their Implications in Vaccine Development" [Dr. Mekalanos had to cancel at the last minute due to illness, Dr. Rest replaced him on the program.]

Richard Rest, Hahnemann University School of Medicine, "The Pathogenesis of Gonorrhea: Adhesins, Invasins and Life at Mucosal Surfaces"

Topical Discussions Groups: Microbial Virulence Factors, AIDS and Related Infections, Vaccines, Host Responses to Infection

1993 April 14, Penn Tower Hotel, at the University of Pennsylvania
Chairperson: Richard Rest
Format: speakers, posters, topical discussions and exhibits

Speakers:
Staffan J. Normark, Washington University, St. Louis, "Phagocytic Processing of Defined Epitopes Expressed in Bacteria for Class I and II MHC Presentation to T Cells"
Robert Lehrer, University of California, Los Angeles, "Defensins and Related Antimicrobial Peptides of Leukocytes"

Topical Discussion Groups: Microbial Virulence Factors, AIDS and Related Infections, Vaccines, Host Responses to Infection

1994 April 15, Penn Tower Hotel, at the University of Pennsylvania
Chairperson: Richard Rest
Format: speakers, student oral presentations, Philadelphia area research highlights

Speakers:
Roy Curtiss, III, Washington University, St. Louis, "The Life of Salmonella: Steps Towards Pathogenesis"
Victor Nussenzweig, New York University Medical Center, "How do Malaria Sporozoites Enter Hepatocytes?"

Philadelphia Research Highlights:
Carole A. Long, Hahnemann University, "Cellular Immunity to Malaria"
Charles Bevins, University of Pennsylvania, "Antimicrobial Peptides as Effectors of Mucosal Host Defense"

Jeffrey Weiser, University of Pennsylvania, "The Wily Ways of Parasites, Phenotypic Variations in Encapsulated Respiratory Pathogens"

1995 [There is no material deposited in the Branch Archives and no listing in either the Executive Committee minutes or Branch Newsletter for this Forum. This may have been due to the activities in the Branch to form a formal student chapter, and the forum this year, most likely was held as an informal meeting.]

1996 April 12, Penn Tower Hotel, at the University of Pennsylvania
Chairperson: Richard Rest
Format: speakers, posters, panel discussion, and exhibits

Speakers:
Ralph Isberg, Tufts University School of Medicine, "Intracellular Targeting of *Legionella pneumophila* Within Macrophages"
Daniel Portnoy, University of Pennsylvania School of Medicine, "Pathogenic Mechanisms of *Listeria monocytogenes*"
Robert Modlin, UCLA School of Medicine, "T-cell and Cytokine Responses in Mycobacterial Infection"

1997 April 11, SmithKline Beecham Pharmaceuticals, Collegeville, PA
Coordinator: Padmini Salgame
Format: speakers, posters and exhibits

Speakers:
Alan Sher, National Institute of Allergy and Infectious Diseases, "Required and Redundant Mechanisms of Host Resistance to Opportunistic Infections"
Diane Griffin, Johns Hopkins University, "The Many Ways that Measles Affects the Immune System: Mechanisms and Mysteries"
William R. Jacobs, Howard Hughes Medical Institute, "Tuberculosis Control: Beyond Isoniazid and BCG"

1998 April 17, SmithKline Beecham Pharmaceuticals, Collegeville, PA
Coordinator: Padmini Salgame
Format: speakers and posters

Speakers:
Keith Joiner, Yale University School of Medicine, "Secretion in *Toxoplasma gondii*: Unusual Routes, Unusual Destinations"

Rafi Ahmed, Emory University, "Immune Memory to Viruses"
Jorge Galan, SUNY at Stonybrook, "Modulation of the Host-Cell Actin Cytoskeleton by *Salmonella typhimurium*"

1999 April 16, SugarLoaf Conference Center, Germantown Pike
Coordinator: Padmini Salgame
Format: speakers, posters and exhibits

Speakers:
Eric G. Pamer, Yale University School of Medicine, "T Cell Responses to *Listeria monocytogenes*"
Eric J. Brown, Washington University School of Medicine, "Integrins and Integrin-associated Protein (CD47) in Host Defense and Immunity"

Local Presenters:
Yvonne Paterson, David Roos and Howard Goldfine, all from the University of Pennsylvania, and Padmini Salgame, Temple University

2000 May 4, SugarLoaf Conference Center, Germantown Pike
Coordinator: Padmini Salgame
Format: speakers, posters and exhibits

Speakers:
Douglas T. Golenbock, Boston University School of Medicine, "Toll Receptors: Guardians of the Innate Immune System"
Virginia L. Miller, Washington University School of Medicine, "*Yersinia enterocolitica*: A Model Organism for the Genetic Analysis of Virulence"
Philip A. Scott, University of Pennsylvania School of Veterinary Medicine, "Host and Parasite Factors in the Control of Cutaneous Leishmaniasis"

Student and Postdoc Oral Presentations:
Trish Darrah, and Somia Perdow Hickman, both from Temple University; John Po, MCP Hahnemann University; and Sarah Blass, University of Pennsylvania School of Veterinary Medicine

2001 May 4, SugarLoaf Conference Center, Germantown Pike
Coordinator: Betinna Buttaro
Format: speakers, posters and exhibits

Speakers:
Barbara H. Iglewski, University of Rochester, "Quorum Sensing Regulates Virulence and Antibiotic Sensitivity of *Pseudomonas aeruginosa*"
David G Russell, Cornell University, "Mycobacterium: Intracellular Infection by an Indigestible Microbe"

Local Speakers:
Thomas J. Rogers, Temple University, "Crosstalk Among G-Protein Coupled Receptors and Their Ligands"
Christopher A. Hunter, University of Pennsylvania, "Understanding the Role of NF-kB in Innate and Adaptive Immunity to Infection"
Susan R. Weiss, University of Pennsylvania, "Viral Determinants of Murine Coronavirus Pathogenesis"

2002 May 31, MCP-Hahnemann, Queen Lane Campus
Coordinator: Bettina Buttaro, Session Chairs: Padmini Salgame, Richard Rest, Marjan van der Woude, Toby K. Eisenstein and Akhil Vaidya
Format: speakers and posters

Speakers:
William R. Bishai, Johns Hopkins University, "Inciting Immunopathology: Is Tuberculosis More of an Autoimmune Disease Than We Thought?"
T.V. Rajan, University of Connecticut, "Host Defense Against a Metazoan Dwelling Parasite"

Session Speakers:
James Burns, MCP Hahnemann, "Malaria Vaccines: Immunization with Blood-Stage Antigen Combinations"
Ming H. Yuk, University of Pennsylvania, "Molecular Functions of the Bordetella Type III Secretion System"
Padmini Salgame, Temple University, "Functional Consequences of Differential Cytokine Expression Pattern in *Mycobacterium tuberculosis* Infected Dendritic Cells and Macrophages"

2003 May 30, Drexel University, Queen Lane Campus
Co-chairpersons: Bettina Buttaro and Richard Rest, Session Chairs: Padmini Salgame, and Toby Eisenstein
Theme: The Role of the Microbiologist in Biodefense
Format: speakers, posters and exhibits

Speakers:
Stephen A. Morse, CDC Bioterrorism Preparedness and Response Program, "The Role of the Microbiologist in Biodefense"
Richard F. Rest, Drexel University, "Anthrolysin O—A New Anthrax Toxin?"
Stuart N. Isaacs, University of Pennsylvania, "Smallpox Bioterrorism—Poxvirus Immune Evasion Proteins as Therapeutic Targets"
Robert W. Doms, University of Pennsylvania, "Closing the Door on HIV Infection"
Tanja Popovic, Epidemiologic Investigations Laboratory, Meningitis and Special Pathogens Branch, CDC, "Journal of a CDC Laboratorian: Daily Encounters with *Bacillus anthracis*"

2004 June 10, Drexel University, Queen Lane Campus
Co-chairpersons: Bettina Buttaro and Richard Rest, Session Chairs: Richard Rest and Toby Eisenstein
Format: speakers, posters, and exhibits

Speakers:
Brian Wigdahl, Drexel University, "HIV CNS Disease: Signature Sequences, Transactivators, and Bone Marrow Infection"
Fred C. Krebs, Drexel University, "Rational Design and Development of Polybiguanides as Safe and Effective Anti-HIV-1 Microbicides"
Bettina Buttaro, Temple University, "Persistence of *Streptococcus pyogenes*"
John D. McKinney, The Rockefeller University, "Hard Target: Persistence Mechanisms in Mycobacterium"
Rick L. Tarleton, University of Georgia, "CD8+ T Cells and Immune Control of *Trypanosoma cruzi* Infection"

2005 June 24, Drexel University, Queen Lane Campus
Co-chairpersons: Richard Rest and James Burns
Format: speakers, student platform presentations, and posters

Speakers:
Michele S. Swanson, University of Michigan, "Pht is Where it's at: Amino Acids Signal Legionella to Differentiate in Macrophages"
Edward Pearce, University of Pennsylvania, "Shaping the Immune Response to Pathogens: Role of Dendritic Cells".

2006 June 23, Drexel University, Queen Lane Campus
Chairperson: Simon Knight
Format: speakers, student platform presentations, posters and exhibits

Speakers:
Ralph Isberg, Tufts University, "Manipulation of Host Secretory Traffic by an Intracellular Pathogen"
David Artis, University of Pennsylvania, "Regulation of Immunity to Pathogens in the Gastrointestinal Tract"
Eleanor Riley, London School of Hygiene and Tropical Medicine, "Innate Immunity to Malaria"

Student and Postdoc Speakers:
Shannon L. Morgan, Temple University "*Streptococcus pyogenes* Persistence in Eukaryotic Co-culture"
Nazzy Pakpour, University of Pennsylvania School of Veterinary Medicine, "Central memory CD4+ T cells Require IL-12 to Become the Effector Cells that Maintain Immunity to *Leishmania major*"
Mathilde A. Poussin, University of Pennsylvania, "It takes More Than Listeriolysin O to Permeabilize a Listeria phagosome"
John Paul-Vermitsky, Drexel University, "Pdr1 Regulates Multidrug Resistance in *Candida glabrata*: Gene Disruption and Genome-wide Expression Studies"

2007 May 11, Drexel University, Queen Lane Campus
Chairperson: Simon Knight
Format: speakers, student platform presentations, posters and exhibits

Speakers:
Douglas Golenbock, University of Massachusetts, "TLR Activation and its Relationship to Human Disease"
John Gunn, Ohio State University, "Something About Mary That we Never Knew: Typhoid, Biofilms and Gallstones"
Gary H. Cohen, University of Pennsylvania, "Entry of *Herpes simplex* virus into Mammalian Cells"

Student and Postdoc Speakers:
Heather J. Painter, Drexel University, "Specific Role of Mitochondrial Electron Transport Chain in Blood-stage *Plasmodium falciparum*"

Anirban Roy, Thomas Jefferson University, "Failure to Open the Blood-brain Barrier and Deliver Immune Effectors to the CNS Tissue is Responsible for the Lethal Outcome of Silver-haired Bat Rabies Virus Infections"

Christine Bucks, Drexel University, "Distinct Death Pathways Mediate Quantity and Quality Deficiency of Chronic Antigen Stimulated Virus Specific CD8+ T cells" (Elsie Mosser was scheduled to present, but due to an emergency was replaced by C. Bucks.)

Estela M. Galvan, University of Pennsylvania School of Veterinary Medicine, "Development of a Pneumonic Plague Model for delta-*pgm Yersinia pestis* strains"

2008 May 9, Drexel University, Queen Lane Campus
Chairperson: Simon Knight
Format: speakers, student platform presentations, posters and exhibits

Speakers:
James Kaper, University of Maryland School of Medicine, "In vivo Gene Expression of *Vibrio cholerae* in a Human IVET (in vivo expression technology) Study"

Sara Cherry, University of Pennsylvania School of Dental Medicine, "Using High-Throughput RNAi Strategies to Identify Host Factors That Control Viral Replication and Innate Immunity"

Michael Apicella, University of Iowa, "The Role of Sialic Acid in the Survival of Nontypeable *Haemophilus influenzae* in the Human Host"

Student and Postdoc Speakers:
Salvador Almagro-Moreno, University of Delaware, "Functional and Transcriptional Analysis of Sialic Acid Degradation Gene Cluster Encoded on the Vibrio Pathogenicity Island-2 Region in Pathogenic *Vibrio cholerae* isolates"

Mariana E. Bernui, Drexel University, "N-acetycysteine (NCA) and STIMAL (liposome-encapsulated NAC) Impair Germination of *Bacillus anthracis* Spores and Increase the Ability of Macrophages to Limit Anthrax Growth in vitro"

Zhi Liu, University of Pennsylvania School of Verterinary Medicine, "Mucosal Penetration Primes *Vibrio cholera* for Host Colonization by Repressing Quorum Sensing"

Sharron L. Manuel, Drexel University, "Priming of the Tax-specific Cytotoxic T Lymphocyte Response Through Dentritic Cells: in vitro and in vivo"

2009 May 8, University of Pennsylvania School of Medicine Auditorium, Currie Blvd.
Chairperson: Simon Knight
Format: speakers, student platform presentations, posters and exhibits

Speakers:
Fidel P. Zavala, Johns Hopkins University, "Development of Memory CD8+ T Cell Subsets Against Liver Stages of Malaria Parasite"
Andreas J. Baulmer, University of California, Davis, "Salmonella: Life in the Inflamed Intestine"
Michele Kutzler, Drexel University, "Improved DNA Vaccine Strategies for the Induction of Mucosal Immunity"

Student and Postdoc Speakers:
Mary K. Collins, New York University School of Medicine, "Stromal Cell Differentiation Leads to Dendritic Cell Entrapment at the Maternal/fetal Interface"
Mitali Purohit, Drexel University, "Characterization of a Unique, Naturally Occurring Immunomodulatory Lipopeptide: 1-peptidyl-2-3-diacylglyceride (PDAG)"
Swati Thorat, Drexel University, "Influence of *Plasmodium yoelii* Macrophage Migration Inhibitory Factor on the Course of Blood Stage malaria"
Meera Nair, University of Pennsylvania, "Alternatively Activated Macrophage-derived RELMa Limits Helminth-induced Type 2 lung Inflammation"

Appendix VIII

The Stuart Mudd Memorial Lecture Series 1976 to 1995

On 6 May 1975, the Branch lost one of its most prominent members, Dr. Stuart Mudd, emeritus professors of Microbiology at the University of Pennsylvania. Dr. Mudd was a former president of the ASM as well as a former Branch president. A memorial service was held at the College of Physicians of Philadelphia on 30 May 1975. Branch member Dr. Joseph Gots, Professor and Acting Chairman of the Department of Microbiology of the University of Pennsylvania School of Medicine, spoke at this service. The official Branch representative was Dr. Richard Crowell, Professor of Microbiology and Immunology of Hahnemann Medical College and President of the Eastern Pennsylvania Branch. He announced that the Branch had decided to establish an Annual Stuart Mudd Memorial Lectureship. He stated that, "As a former president of our Branch, a distinguished microbiologist and a truly fine person, we trust that this memorial will serve to pay tribute to Dr. Stuart Mudd, who was a great inspiration to us all."

The Stuart Mudd lecture became an established Branch tradition each spring. A committee was formed and charged with selecting a speaker who was either a former student, a colleague or someone who made significant contributions to work initiated by Dr. Mudd. These annual lectures became both a scientific and social highlight of the year for the Branch. Mrs. Mudd was usually in attendance, accompanied by different family members. The Branch made certain that Mrs. Mudd was provided with an orchid corsage that was selected each year by Dr. Robert Mandle of Thomas Jefferson University. A formal brochure was produced each year and all speakers were awarded a plaque containing an engraved photograph of Dr. Mudd. This was followed by a special dinner which was always well attended.

Although the meetings were always a success, as the years progressed, the choice of lecturers had a natural evolution form an emphasis on former students to colleagues to speakers relating to Dr. Mudd's research.

It was decided that the supply of top caliber candidates available for this lectureship, that met the strict criteria that had been established by the selection committee, was becoming exhausted and that it was time to begin a new tradition that would recognize other prominent Branch members who made significant contributions to the Branch which Dr. Mudd had so enthusiastically supported. Therefore, the last Stuart Mudd Lecture was held on 24 April 1995, and the following year the Distinguished Branch Lectureship was initiated.

The Branch Archives has a rich collection of material relating to this lecture series. The Stuart Mudd Lecture collection demonstrates how important Dr. Mudd was to this Branch and how many of his students and colleagues carried on his work. The twenty selected speakers and their topics follow:

1976 LIPID SPECIFIC EXOTOXINS.
      Alan W. Bernheimer, PhD
1977 AUSTRALIA ANTIGEN AND THE BIOLOGY OF HEPATITIS B VIRUS. Baruch S. Blumberg, MD, PhD
1978 DIPHTHERIA: REFLECTIONS ON THE EVOLUTION OF AN INFECTIOUS DISEASE. Alwin M. Pappenheimer, Jr., PhD
1979 BACTERIAL GENETICS AND DISEASE: THE PLASMID CONNECTION.
      Joseph S. Gots, PhD
1980 ELECTRON MICROSOPY THEN AND NOW: FORTY YEARS OF RESEARCH ON THE STRUCTURE OF MICROORGANISMS.
      Thomas Foxen Anderson, PhD
1981 NEW APPROACHES TO THE STUDY OF RABIES VACCINE.
      Hilary Koprowski, MD [He was unable to give the lecture in person.]
1982 THE GENETICS OF AUTOIMMUNE THYROIDITIS.
      Noel Rose, PhD, MD
1983 STRUCTURE FUNCTION RELATIONSHIP BETWEEN THE BACTERIAL CHROMOSOME AND CELL MEMBRANE.
      Moselio Schaechter, PhD
1984 TRACKING ENVIRONMENTAL MUTAGENS WITH BACTERIA.
      Philip E. Hartman, PhD
1985 CELL BIOLOGY OF HUMAN AGING. Leonard Hayflick, PhD
1986 HOST CONTROL OF STAPHYLOCOCCUS AUREUS IN FOCAL LESIONS. Frank A. Kapral, PhD

1987 AN UNUSUAL LIFE CYCLE OF A LARGE DNA VIRUS.
Allan Granoff, PhD
1988 EPITOPE HUNTING IN THE CHOLERA/COLI ENTEROTOXIN FOREST. Richard Finklestein, PhD
1989 PROTECTIVE AND AUTOIMMUNE EPITOPES OF STREPTOCOCCUS-M PROTEINS. Edwin H. Beachey, MD
1990 SALMONELLA: UNDERSTANDING PATHOGENICITY AND DEVELOPMENT OF VACCINE STRATEGIES. Roy Curtiss III, PhD
1991 THE SWITCH BETWEEN EBV LATENCY AND REPLICATION. I. George Miller, Jr., MD
1992 IMPLICATIONS OF MYCOPLASMA INFECTION IN AIDS PROGRESSION.
Joel Baseman, PhD
1993 PATHOGENICITY OF ENTEROHEMORRHAGIC E. COLI (THE CAUSE OF THE JACK-IN-THE-BOX OUTBREAK). Alison D. O'Brien, PhD
1994 THE STRUCTURES AND FUNCTION OF GONOCOCCAL IRON UTILIZATION RECEPTORS. Philip Frederick Sparling, MD
1995 REGULATION OF VIRULENCE BY CELL TO CELL COMMUNICATION-A LANGUAGE THAT HURTS. Barbara Hotham Iglewski, PhD

Appendix IX

The Distinguished Branch Member Lectureship 1996 To 2006

After twenty years (1976-1995) of the Stuart Mudd Lecture Series it was decided, at a Branch Executive Committee meeting, that it was time for a change. There was a strong desire to maintain the general format that worked so well for the Stuart Mudd Lectures. This involved a lecture by a recognized speaker, followed by a dinner at a nearby restaurant for more informal discussions. It was decided that this new lecture series should be used to recognize Branch members who made special contributions to the success of the Branch over several years. The first Distinguished Branch Member Lecture was held on 22 April 1996, at Thomas Jefferson University, with Dr. Harry E. Morton chosen as the first Honoree. Like Dr. Morton, each Honoree who followed, was chosen for their various contributions to the Branch, and the lecturers associated with these events were chosen to reflect some aspect of the work or interests of the Honorees. Like the Stuart Mudd Lecture Series, the Distinguished Branch Lectures became one of the social, as well as a scientific, highlight of the Branch academic year. It gave an opportunity for newer members to not only celebrate the contributions of the more senior members; it gave all members an opportunity to get to know the Honoree better. This was accomplished by providing a biographical sketch of the Honoree, as well as providing a time for more informal remarks on the Honoree at the dinner that followed. The following is a list of the Honorees, starting in 1996 and continuing to the end of this series in 2006. As can be seen in the list below, a review of the lecturers and their subject matter not only reflects contemporary developments in microbiology, but reflects the broad range of interests of our Branch Honorees.

BRANCH HONOREES

1996 Harry E. Morton, Sc. D.
1997 Earle H. Spaulding, PhD
1998 Amedeo Bondi, PhD
1999 Kenneth R. Cundy, PhD
2000 Bruce Kleger, Dr. P.H.
2001 Henry R. Beilstein, PhD
2002 Norman P. Willett, PhD
2003 Carl Abramson, PhD
2004 James E. Prier, PhD
2005 Josephine Bartola, JD
2006 Richard L. Crowell, PhD

Lecturers and Lecture Titles 1996 through 2006

April 22, 1996 Raymond A. Zilinskas, PhD [Honoree: Harry E. Morton, Sc.D.] In Pursuit of Sadam Hussein's Biological Arsenal: The Personal Account of a Biological Warfare Inspector in Iraq.

April 28, 1997 Marie B. Coyle, PhD [Honoree: Earle H. Spaulding, PhD] Old Technologies Detect New Species in a Routine Clinical Microbiology Laboratory.

April 27, 1998 Clyde Thornsberry, PhD [Honoree: Amedeo Bondi, PhD] Changing Antimicrobial Resistance in the 90's: The Hot Spots.

April 27, 1999 Burton W. Wilcke, Jr., PhD [Honoree: Kenneth R. Cundy, PhD] Microbiology and Its Continuing Impact on Public Health.

April 17, 2000 Andre Weltman, MD [Honoree: Bruce Kleger, Dr. P.H.] Introduction to Bioterrorism.

March 26, 2001 Stuart B. Levy, MD [Honoree: Henry R. Beilstein, PhD] Antibiotic Resistance: Microbes on the Defense.

`April 22, 2002 Caroline Genco, PhD [Honoree: Norman P. Willett, PhD] Bugs, Gums and Heart Disease: Pathogenic Strategies of the Periodontal Pathogen *Porphyromonas gingivalis* in Endothelial Cell Inflammation.

April 28, 2003 Bennett Lorber, MD [Honoree: Carl Abramson, PhD] Snakes, Sex, Sushi and Saunas.

April 26, 2004 Mark S. Birenbaum, PhD [Honoree: James E. Prier, PhD] Are the Feds Waiving Goodbye to the CLIA Regulations of Clinical Laboratory Testing?

April 25, 2005 Caroline C. Johnson, MD [Honoree: Josephine Bartola, JD] A Perspective on Disease Control in Philadelphia.

May 8, 2006 Emilio A. Emini, PhD [Honoree: Richard L. Crowell, PhD] The Importance and Challenge of Vaccine Development.

Appendix X

The Annual Industry Sponsored Guest Lectureship Series: 1962-1971

In 1962, funding was received from seven industrial sponsors, each contributing $50, to initiate a guest lecturer series, the first lecture in this series was held in November of 1962. This lecture series continued each year from 1962 through 1971. This was the first project that relied on pharmaceutical companies to support Branch activities. Pharmaceutical companies played a significant role in funding many additional Branch projects over the years.

27 November 1962: First Annual Industry Sponsored Guest Lectureship
Experiments on the Role of Heredity in Experimental Sensitization.
Merril W. Chase, PhD, Rockefeller Institute for Medical Research, New York, N.Y.

26 November 1963: Second Annual Industry Sponsored Guest Lecturship
Enzyme Induction and Catabolite Repression.
Boris Magasanick, PhD, Department of Biology, Mass. Institute of Technology, Cambridge, MS.

24 November 1964: Third Annual Industry Sponsored Guest Lecturship
Growth of an RNA Bacteriophage.
Norton D. Zinder, PhD, Rockefeller Institute for Medical Research, New York, N.Y.

30 November 1965: Fourth Annual Industry Sponsored Guest Lectureship
Cellular Interaction and Metabolic Controls in Cultured Human Cells.
Harry Eagle, MD, Albert Einstein College of Medicine, Yeshiva University.

22 November 1966: Fifth Annual Industry Sponsored Guest Lectureship
The Interferons.

Robert Wagner, MD, Prof. of Microbiology, Johns Hopkins Univ. School of Medicine, Baltimore, MD.

28 November 1967: Sixth Annual Industry Sponsored Guest Lecturship
Adenovirus SV-40 Hybrid Viruses.
Wallace Rowe, MD, Laboratory of Viral Diseases, Institute of Allergy and Infectious Diseases, National Institute of Health, Bethesda, MD.

29 October 1968: Seventh Annual Industry Sponsored Guest Lecturship
A Re-Examination of the Germ Theory of Disease.
Gordon Stewart, MD, The School of Public Health, Univ. of North Carolina, Chapel Hill, N. C.

25 November 1969: Eighth Annual Industry Sponsored Guest Lectureship
Transfer Factor and Cellular Immunity.
H. Sherwood Lawrence, MD, Section of Immunology, New York University Medical School, New York, NY.

24 November 1970: Ninth Annual Industry Sponsored Guest Lectureship
Rene Dubos, MD, The Rockefeller University, New York, NY.

23 November 1971: Tenth Annual Industry Sponsored Guest Lectureship

Ralph S. Wolfe, PhD, Department of Microbiology, University of Illinois.

Appendix XI

Celebration of the 500th and 600th Branch Meetings: 1987 and 1998

The 500th Branch Meeting Celebration 15 December 1987

On 15 December 1987, the Branch took the occasion of the 500th Branch meeting to celebrate sixty-seven years of continuous activity. The celebration, at the University of Pennsylvania, started with an afternoon of presentations on the Branch, microbiology in general and history related events that occurred during each of the last seven decades.

The presentations were followed by a reception and a spectacular dinner attended by 164 guests at the Faculty Club of the University of Pennsylvania. In addition to Dr. Morman's lecture, following dinner, several long term Branch members were honored. Honorees included Drs. Harry Morton, Ruth Miller, Earle Spaulding, Amedeo Bondi and Morton Klein. A publication, "A History of the Eastern Pennsylvania Branch of the ASM: 1920 to 1987" edited by Drs. Linda A. Miller, Harry E. Morton and James A. Poupard, was introduced and distributed to those in attendance. This publication became a valuable resource as a collection of Branch history up to that time.

PROGRAM FOR THE 500TH BRANCH MEETING CELEBRATION
December 15, 1987

1:30 WELCOME
      James Prier, Philadelphia College of Osteopathic Medicine
1:35 INTRODUCTION
      James Poupard, The Medical College of Pennsylvania
1:45 THE EARLY YEARS 1920-1932
      James A. Poupard, The Medical College of Pennsylvania, and Harry E. Morton, The University of Pennsylvania

2:45 THE SECOND DECADE—1933-1939
Henry Beilstein, Hahneman University; PA College of Podiatric Medicine; Beaver College
3:10 THE WAR YEARS—INFECTIOUS DISEASE AND MICROBIOLOGY
Carl Abramson, Pennsylvania College of Podiatric Medicine
3:30 POST-WAR MICROBIOLOGY—1951-1970
Paul Actor, SmithKline and French Laboratories
3:5 THE ROLE OF THE BRANCHES IN THE AMERICAN SOCIETY FOR MICROBIOLOGY
Donald Shay, American Society for Microbiology, Archives Committee Center for the History for Microbiology, University of Maryland-Baltimore County, Catonsville, Maryland
4:10 MICROBIOLOGY TODAY AND PERSPECTIVE FOR TOMORROW
Nick Burdash, Philadelphia College of Osteopathic Medicine

Dinner Speaker: BACTERIOLOGY AT THE TURN OF THE CENTURY
Edward Morman, the Institute of the History of Medicine Baltimore, Maryland

The 600th Branch Meeting Celebration 14 September 1998

On 14 September 1998, the Branch celebrated the 600th meeting with an afternoon session at Thomas Jefferson University entitled "A Reason to Celebrate Our 600th Meeting." This involved five Philadelphia Microbiologists "Looking Back" at the last six hundred meetings and the microbiology represented at these meetings. The meeting was followed by a celebration dinner at Girasole Restaurant on Locust Street. The following topics were presented:

An Overview and Introduction
James A. Poupard, PhD, SmithKline Beecham Pharmaceuticals
A Look Back at Virology
Bruce Kleger, Dr. P. H., Bureau of Laboratories, Pennsylvania Department of Health
A Look Back at Mycobacterium
Adamadia Deforest, PhD, St. Christopher's Hospital for Children and Allegheny University of the Health Sciences

A Look Back at *Streptococcus pneumoniae*
Robert Austrian, MD, Professor and Chairman Emeritus, Department of Research Medicine, University of Pennsylvania School of Medicine
Searching for Dr. Arrowsmith
James E. Prier, PhD, Philadelphia College of Osteopathic Medicine

Introduction to a Special Branch Membership Project, "100 Years of Philadelphia Microbiology" James A. Poupard, SmithKline Beecham Pharmaceuticals.

Appendix XII

History of The Branch Newsletter: 1937 To 2009

The Branch Newsletter has gone through many variations throughout the years since 1937. It had its start as an annual edition, which was a key element in documenting the history of the Branch. During the early years of our history, when there were very scant minutes and no executive committee, the annual newsletter became the first document that provided a hint at what was going on beyond just a series of regular meetings. We are indebted to Dr. Harry Morton for provide the Branch with the first set of these valuable documents.

First Series: 1937-1948

The First Series of newsletters were produced from 1937 to 1948, starting when Dr. Morton became Branch Secretary/Treasurer. He decided to document our history in the form of an Annual Newsletter. He referred to the first edition in 1937 as Volume 1, Issue 1. With few exceptions, he continued the process each year through 1948.

1937 Volume 1 September—The first Branch newsletter.
1938 Volume 2 November
1939 (No volume was produced.)
1940 Volume 3 October
1941 Volume 4 April—Reported on wartime preparations.
1942 Volume 5 October
1943 (No newsletter, most likely due to WW II staffing issues.)
1944 Volume 6 January—This edition contains interesting comments on the Branch activities during the WW II years.
1945 Volume 7 January
1946 Volume 8 January
1947 No newsletter produced.
1948 Volume 9 May

Second or Sporadic Series: 1962-1974

When the First Series ended in 1948, there was a long period without any attempt to produce a Newsletter until 1962, when Dr. Carl Clancy, of Pennsylvania Hospital, was charged with producing a newsletter. Then newsletters were produced sporadically over the next twelve years.

1962 Dr. Carl Clancy, editor.
1966 Spring Dr. Leonard Zubrzyck, editor.
1967 Volume 14 December Dr. Leonard Zubrzyck, editor. He designated this as Volume Fourteen; however, there is no indication of how the number 14 was chosen.
1970 Volume 15 June Albert Moat, editor with Carl Abramson and Jay Satz as co-editors.
1971 Volume 16 June—Dr. Albert Moat, editor.
1972 and 1973, no newsletters were produced.
1974 Volume 17 July—Dr. Albert Moat editor.

The Continuous Series: 1977 to 2009

After an absence of three years, a new series of newsletters began in November (Volume 18 Number 1) of 1977. This was a monthly series (ten per year) and for several years served as the meeting notice for the regular monthly meeting. In the later years the number of issues was reduced to four each year. This series continued until 2009.

1977 to 1979, editor, Steven Specter
1979 to 1980, editor, Linda Creeden
1980 to 1983, editor, Margaret Cook and Albert Giovenella
1983 to 1978, editor, Albert Giovenella
1987 to 1988, editors, Allan Truant and Don Jungkind
1988 (Feb. to April) editors and publishers, Alan Truant, Donald Jungkind and Harry Smith
1988 (Sept) to 1999, editor, Anna Feldman-Rosen and publisher, Lori Walsh
1999 (Sept.) to 2009, editor, Julie Conaron

A significant decision was made in 2009 to discontinue the newsletter and replace it with a web-based communication system in 2010. Many factors need to be considered in the design of an adequate replacement for the newsletter. An ad hoc committee was formed to investigate possible new methods of communication. Recommendations from this committee are expected in 2010.

INDEX

Numerals within parentheses refer to meeting numbers.0

A

Abbott, Alexander Crever, 3, 32, 34-35, 37-38, 43-47, 49-50, 56-57, 61, 64, 72-73, 75-76, 149, 181, 193-201, 205-6, 230
  Hygiene of Transmissible Diseases, The, 46, 50, 200, 230
  Principles of Bacteriology, The, 46, 193
Abington Memorial Hospital, 81, 85, 186
Abramson, Carl (312, 650), 110, 121, 137, 153, 155, 170, 242-43, 283, 314, 351, 354, 375-76, 380, 383
Abramson, Samuel (173, 194, 234), 258, 262, 269
Abstracts of Bacteriology, 61
Academy of Natural Sciences of Philadelphia, 10
Academy of Sciences of Paris, 18
acid-fast bacilli, 246, 274
Acinetobacter, 165, 316
acquired immunodeficiency syndrome (AIDS), 136, 148, 153, 165, 302, 307, 309, 317, 360, 363, 373
Actinobacillus, 136
actinomycetes, 136
actinomycin, 107, 286
Actor, Paul (363, 465, 567), 137-41, 152, 159, 218, 238-39, 242, 244, 294, 302, 308, 349, 352, 354-55, 380

Ada, Gordon (353), 291
Adams, John, 45
Adam's Mark Hotel, 142, 144, 154-60, 168-69, 350-52
adenovirus, 106, 120, 284-87, 291, 294, 378
Adye, J. (300), 281
agar, 79, 260, 277
agar cup plate, 248
agglutination tests, 252, 260, 284
aging, 281, 372
Ahmad, Fazal (314), 283
Ahmed, Rafi, 365
air-sampling, 85, 96, 187
Ajl, Samuel J. (287, 314), 107, 243, 283
Alaro, James (702), 318
Albert Einstein Medical Center, 95, 103, 105, 109, 119, 121, 127, 130, 155, 173, 185, 214, 223, 239-41, 243, 362
Albertson, John N., Jr. (309, 314), 283
Albin O. Kuhn Library, 218
Alexander, James (278), 277
Alexander-Jackson, E. (216), 266
Allen, Emma G. (222), 267
Allen, Jerry (346), 289
Allen, R. (216), 266
Allergy, 98, 250, 254, 261, 268, 273, 364, 378
Allison, Marvin J. (194), 262
Almagro-Moreno, Salvador, 369
almshouses, 9, 42

385

Blockley, 41
  See also Blockley Hospital
  Philadelphia, 28
American Association of Clinical Chemists, 109
American Medical Association, 17, 42, 192, 229
American Philosophical Society, 10
American Society for Microbiology (ASM), 1, 46, 48-50, 59-60, 64-66, 73-74, 76-77, 81, 90-93, 116-18, 153-54, 181-82, 200-205, 208-13, 229-30, 232-33
  founding of, 46
  Archives Committee of, 79, 137, 163, 380
  general meetings of, 61, 73-74, 84, 101-2, 108, 116, 125, 162-63, 184, 201, 224
  and its Center for the History of Microbiology, 143
  See also Eastern Pennsylvania Branch of the American Society for Microbiology (ASM)
American Vertebrate Paleontology, 15
anaerobes, 84, 87, 106, 120, 124, 136, 294, 296, 301-2, 359
anaerobic culture, 87, 93, 129, 140, 297, 340, 360
Anaerobic Infection, 123, 291, 339
Anderson, Kenneth L. (576, 681), 309, 316
Anderson, Theodore G. (219, 254, 311, CM2, CM8, CM16, CM22 CM25, CM33), 97, 103, 156, 220, 241, 267, 273, 283, 287, 291, 294, 296, 338-40
Anderson, Thomas Foxen (148, 150, 155, 164, 168, 171, 185, 220, 431), 92, 139, 154, 219, 252-54, 256-58, 260, 372
Andrews, Willard (287), 279
animal studies, 18, 30, 66, 190

Annual Industry Sponsored Guest Lecture. *See under* Eastern Pennsylvania Branch of the American Society for Microbiology (ASM)
Annual Infection and Immunity Forums. *See under* Eastern Pennsylvania Branch of the American Society for Microbiology (ASM)
Annual Newsletter, 77-78, 84, 87, 89-90, 109, 207
anthrax, 59, 275, 367, 369
antifungal agent, 99, 156, 274, 313, 317, 361
antimicrobial agents, 5, 96, 110
Antimicrobial Resistance, 136, 141, 148, 156, 159, 166, 171, 173, 187, 305, 311-12, 344-45, 349, 351, 353, 375
antimicrobials, 5, 85, 95-96, 106, 110, 147, 165, 178, 187, 304, 312
antimicrobial susceptibility testing, 96, 106-7, 112, 142, 148, 154, 168, 187, 220, 338, 344, 351, 360
antitoxins, 44-45, 55, 177, 196, 247
antituberculosis, 20, 62
Apicella, Michael A. (551), 369
Appleton, Joseph L. T., Jr. (151, 241), 75, 205, 237-38, 253, 271
Armed Forces. *See* U.S. Army
Aronson, Joseph D. (142, 149, 199, 221), 73, 75-76, 205, 237-38, 251, 253, 267
Artis, David, 368
Asher, David (485), 303
Aspergillus flavus, 257
aster yellows, 248
aureomycin, 96, 266
Australia Antigen, 120, 124, 130, 290, 298, 343, 346, 372
Austrian, Robert (332, 337, 363, 444, 607, CM3), 151, 153, 156-57,

172, 223, 286-87, 311, 338, 351, 355, 381
  on pneumococcal vaccine, 153, 157, 172, 223
 awards
  Excellence in Graduate Studies, 124-25, 295
  Excellence in Research, 124-25, 295
  Excellence in Teaching, 125
  Smith Kline & French Award for Excellence. *See* Smith Kline & French Award for Excellence
Axler, David, 166-67, 223, 238, 240, 242, 244

B

Bacillus anthracis, 16, 369
bacillus Calmette-Guérin (BCG), 251, 296, 364
Bacon, Harry E. (139), 250
bacteremia, 125, 295-96, 340
bacterial colonization, 315, 318, 369
bacterial counts, 250
"Bacteria of Wounds and Skin Stitches, The" (Ghriskey and Robb), 51
Bacteriological Reviews, 61
bacteriology, xvii, 1-5, 7, 25-34, 36-40, 43-47, 57-58, 61-62, 67, 71, 92, 106, 181-82, 186-203, 227-28, 230-32
  establishment, 11, 13
  first-generation bacteriologists, 16-17, 19, 21-22
  institutions, 66-67
  second-generation bacteriologists, 49-52, 54-55, 62, 64-66
bacteriophage, 85, 96, 98, 106, 165, 248, 256-57, 260, 263, 268, 270, 284, 377
Bailey, W. Robert, 127
Bailin, Marilyn (312), 283

Balczuk, N. C. (267), 275
Baldridge, Robert (339), 287
Ball, Michael V., 38, 45-46, 49-50, 56-57, 63, 130, 132, 155, 167, 191, 221, 230, 347-48, 354
  Essentials of Bacteriology, 46, 50, 193, 230
Ball, William, 130, 132, 155, 167, 221, 347-48, 354
Baltimore City College, 49
Bang's disease, 252
Barclay, E. S. (305), 282
Bard, Raymond C. (257, 262), 99, 275
Barksdale, Lane (274), 277
Barnes, Kathryn (147), 252
Barnes, M. F. (147), 252
Bartell, Pasquale (247, 296), 272, 280
Bartlett, Raymond (CM11), 290, 339
Bartola, Josephine (666), xiii, 121, 127-29, 138, 155, 171, 173, 224, 241, 347, 375-76
Barton, Hazel (670), 315
Baseman, Joel B. (515, 550), 373
Batson, H. C. (212), 265
Batson, Oscar V. (139), 250
Baulmer, Andreas J., 370
Baumgarten, W. (220), 267
Bayer, Manfred (406), 298
Beachey, Edwin H. (521), 305, 373
Beale, Henry D. (228), 269
Beamer, William D. (174), 259
Beatrice, Sara (364), 294
Beaver College, 137, 380
Beavis, Kathleen Gleason, 152, 158, 160, 241, 351-52, 355
"Beginning of Bacteriology in Philadelphia, The" (McFarland), 14, 48, 207, 228, 230, 232
Behring, Emil von, 12, 50
Beilstein, Henry R. (306, 384, 632, CM22), 121, 137-38, 168, 170, 217, 222, 238-39, 242, 244, 282, 293, 297, 313, 340, 375, 380

Bekierkunst, Adam (309), 282
Bell, J. F. (144), 251
Bellevue-Stratford Hotel, 129, 209
Bellew, H. P. (146), 252
Benarde, M. A. (CM28), 295, 340
Bendler, J. (357), 292
Benenson, Abram (338), 287
Bennett, Joan, 164
Beregy Trust, 164, 224
Bergey, David Hendricks, 3, 34-35, 49-50, 56-57, 64-65, 68, 72-78, 149, 164, 181-82, 195-98, 200-201, 203-7, 224, 231, 237-38
   Bergey's Manual of Determinative Bacteriology, 50, 65, 68, 149, 204, 231
   Principles of Hygiene, The, 50
Bergey's Manual of Determinative Bacteriology (Bergey), 50, 65, 68, 149, 204, 231
Berk, Richard S. (398), 298
Berle, Lilliam C. (171), 258
Berlin, 14, 20-22, 36, 38, 40, 49-50, 191, 200
Bernhardt, E. (CM24), 293, 340
Bernheimer, Alan W. (343, 392), 130, 297, 372
Bernui, Mariana E., 369
Berry, L. Joe (249, 284, 297), 107, 122, 213, 237, 239, 278, 281
Beskid, George (281), 278
betalactamases, 85
Bevins, Charles, 363
bichloride of mercury. *See* Laplace's solution
Billingham, R. E. (325), 285
Billings, John Shaw, 32-35, 49-50, 193-97
biochemistry, 5, 85, 95, 106, 109, 119, 136, 147, 165, 186, 269, 279, 281, 285, 313
biological warfare, 148, 153, 166, 307, 310, 375. *See also* anthrax

bioterrorism, 148, 165-66, 180, 187
Birenbaum, Mark S. (658), 170, 314, 376
Bishai, William R., 366
Biven, David, 39, 199
Black Death. *See* plague
Blank, Fritz C. (CM4, CM32), 127-28, 288, 295, 338
Blau, Eva (243), 271
Bleckman, Diane, 127
Blevins, Anne (168), 257
Bliss, Eleanor (249), 272
Block, Timothy M. (519), 154, 168, 350, 352, 354-55
Blockley Hospital, 9, 17, 19, 24-26, 41-42, 54, 181, 190-91. *See also* Philadelphia General Hospital
blood culturing, 107, 112, 288, 294-95, 338, 340, 359
   automation of, 125, 294, 298, 340
Bloom, Barry R. (334), 155, 287, 362
Blumberg, Baruch S. (350, 365, 401), 124-25, 130, 140, 290, 346, 349, 354, 372
Blumfeld, Ruth (131), 249
Blumstein, G. I. (154), 254
Blundell, George P. (164, 174), 256, 259
Boerner, Frederick (118, 119, 135, 143, 160, 191), 86, 243, 246, 249, 251, 255, 261
Boger, William P. (198), 263
Boltjes, B. H. (195), 262
Bolton, B. Meade, 43-44, 196, 199
Bond, Thomas, 10
Bondi, Amedeo, Jr. (155, 162, 183, 186, 248, 264, 352, 555, 605), 86, 97, 129, 137, 141-42, 155, 159, 171, 211, 241-43, 256, 260, 347, 349, 354, 375
Boroff, Daniel (295), 280
Boston Society of Natural History, 15
Boston University, 158, 365

botulism, 302
bovine diseases, 22, 53, 272, 285
Bowdre, Jean (474), 302
Brachman, Phillip (261), 275
Brant, Peter (388), 297
Braun, Armin C. (168), 257
Brennan, Patrick J. (610), 159
Brillaud, Andre (305), 282
Briody, Bernard A. (240), 97, 243, 271
Broadhurst, Jean (168), 257
Brodie, Arnold F. (225), 268
Bromke, Bruno, 169
Brown, Albert (246), 272
Brown, Claude P., 72-76, 81, 204, 206, 210-11, 241
Brown, Eric J., 365
Brown, John H. (257), 274
Brucella abortus, 254-56
brucellosis, 85, 275, 277
Brusick, David (399), 298
Bryn Mawr College, 105, 107, 110, 122, 185-86, 213, 239-40
Bryn Mawr Hospital, 81, 85, 122, 127, 135, 138-39, 186, 242-44
Buchanan, Thomas M. (372), 296
Bucher, Carl J. (129, 147, 159, 174), 71, 73, 75-78, 83, 92, 206-7, 210, 233, 237-38, 242, 252, 255, 258
Buck, T. C., Jr. (197), 263
Buckley, Eleanor (305), 282
Buckley, Helen R. (476), 139, 154, 168, 221, 303, 348, 350, 354
Bucks, Christine, 369
Buecscher, Georganne K., 171, 223
Bug Club. *See* Microbiological Club
Bull, Leonard (467), 302
Bullock, Ward E., 142, 349-50, 354
Burdash, Nick, 137, 139, 153, 242-43, 380
Bureau of Health of Philadelphia. *See* Philadelphia Public Health Department
Burkitt's lymphoma, 106, 285, 288, 299

Burnet, Frank M. (226), 98, 268
Burns, James, 172, 366-67
Buttaro, Bettina, 166-71, 223, 238, 240, 242, 244, 365-67
B virus, 279
Byrne, Earle (354), 292

C

Calandra, Gary, 156, 351, 354
Calesnick, Eleanor J. (198), 263
Calkins, H. E. (132), 249
Calmette, Albert, 20, 62
Campbell, Charlotte (CM17), 292, 339
Campbell, E. P. (155), 254
Campbell, G. (CM25), 294
Campbell, Milton, 44
Campos, Joseph M. (562, 619, 676), 141, 312, 349, 354
Campylobacter, 148
cancer, 16, 96, 99, 106, 248, 297. *See also* tumors
candidiasis, 148, 154, 344, 350
canine heartworm, 16, 188
Cannon, Paul R. (172), 89
Cantor, A. (170), 258
Cappel Laboratories, 108
Carp, Richard (289, 296, 319), 279-80, 284
Carpenter, Charles M. (197), 263
Carr, M. (179), 259
Carroll, Karen, 160, 352, 355
Case, Eugene (139), 250
Casman, Ezra P. (126, 183), 248, 260
Cassel, William A. (211, 240, 270), 265, 271, 276
cauliflower mosaic virus (CMV), 136, 148
C. diphtheriae, 271
Ceglowski, Walter (347, 360, CM9, CM20), 126, 139, 242-43, 289-90, 292-93, 340, 346
cell biology, 372. *See also* aging

Center for Disease Control, 127, 296, 341
Center for the History for Microbiology, 137, 380
Cerny, Jan (346), 289
Cerwinka, Paul L., 152-53, 167, 220, 238-39, 241, 243-44
Chalian, William (134), 249
Chambers, Leslie A. (119, 152, 157, 184), 246, 253, 255, 260, 281
Chandler, Laura J., 166-67, 169, 174, 224, 238, 240, 244, 352-53, 355
Chanock, R. M. (323), 285
Charache, Patricia, 164, 224
Charity Hospital, 20
Charles C. Thomas Inc., 113, 346
Chase, Merrill W. (253, 301, 422), 98, 109, 299, 377
chemotherapy, 67, 98-99, 110, 120, 129, 248, 251, 260, 270, 274, 343, 347
Cheronis, Nicholas D. (242), 271
Cherry, Sara, 369
Children's Hospital of Philadelphia, 81, 85, 95, 103, 105, 119, 124, 147, 165, 185
Children's Seashore Home, 51
chlamydia, 130, 147, 302, 360
cholecystitis, 247
cholera, 17, 27, 42, 60, 127, 192, 256, 260, 289, 294-95, 302, 312, 363, 369, 373
Chronicles of the Society of American Bacteriologist (Cohen), 46, 230, 232-33
Church, Charles F. (126), 248
Ciminera, J. L. (173), 258
Civil War, 16, 29, 59, 188
Clancy, Carl F. (204, 213, 258), 102, 107, 109, 156, 220, 242, 264, 274, 383
Clark, Jefferson H., 75-76, 205, 237-38
Clark, Junius (347), 290

Clark, Paul F., 11, 227-28, 230
Pioneer Microbiologists of America, 11, 227-28, 230
Clark, Richard (354, CM1, CM5, CM19, CM25), 103, 118, 288, 292, 294, 338, 340
Claude Brown Clinical Laboratory, 81
Clawson, Jean R. (184), 260
Cleveland, Grover, 17
clinical parasitology (*see also* parasitology), 120, 132, 154, 158, 160, 298, 344, 348, 360-61
coagulase-negative staphylococcus (CNS), 367, 369
Coccidioides immitis, 253
Cocklin, J. (CM12, CM26), 290, 294, 339-40
Cohen, Barnett, 46, 61, 230, 232-33, 286
Chronicles of the Society of American Bacteriologists, 46, 230, 232-33
Cohen, Elisabeth P. (330, 508), 286, 304
Cohen, Gary H., 368
Cohen, Harold (236), 270
Cohen, Seymour S. (180, 184, 214, 232), 259-60, 265, 269
Cole, W. (CM22), 293, 340
College of New York Infirmary, 56
College of Physicians of Philadelphia, 10, 51, 371
Collins, Mary K., 370
comma bacillus, 38, 191
Committee on Materials for Visual Instruction in Microbiology, 84, 88, 209, 256
Commonwealth of Pennsylvania, 87, 209
Compendium on Bacteriology Including Animal Parasites, A (Pitfield), 53, 68, 202, 231
complement fixation test, 249, 269
computerization, xv, 116, 125-26, 215,

292
Conaron, Julie, 160, 383
Conn, Herbert W., 46, 49, 61
Conner, Robert (302), 281
Cook, Margaret, 139, 383
Cooper, Billy H. (405, CM17, CM26), 292, 294, 298, 339-40
Cooper Medical Center, 138-39, 152, 185, 219, 239-42, 244
Copeland, James (CM5), 103, 288, 338
Coplin, W. M. L., 38-39, 43, 67, 194, 196-97, 199, 202
Coriell, Lewis L. (344), 262, 288
Cornell University, 366
Corner, George W., 21, 37
Corning Clinical Laboratories, 152, 244
Covert, Scott V. (196), 263
coxsackie, 120, 268, 279, 283, 285
Coy, Nettie H. (245), 271
Coyle, Marie B. (595), 159, 375
Coyne, Veronica (350), 290
Creamer, Alan A. (246), 272
Creech, H. J. (216), 266
Creeden, Linda, 131, 139, 383
Critchley, Ian A., 159, 352, 355
Croce, Carlo, 154, 219
Crouch, Ruth W. (174), 259
Crowell, Richard L. (288, 428, 544, 674), 107, 121-22, 129, 131, 139, 153-54, 157, 215, 237, 239, 242-44, 307, 347-48, 350, 354, 375-76
Crumb, Cretyl (187, 195), 261-62
Cundy, Kenneth R. (614, CM14, CM18, CM21, CM23, CM30), 130, 144, 153, 160, 243, 290, 292-93, 295, 312, 339-40, 347, 350, 354, 375
Curtiss, Roy, III (531), 363, 373
Cystic Fibrosis, 309, 316
Czarnetzky, E. J. (119, 124, 127, 132), 246, 248-49

D

DaCosta, Jacob M., 38-39, 191, 195
Daneo-Moore, Lolita, 170
Dannenberg, Arthur M. (234, 309), 269, 282
Darrah, Trish, 365
DaSilva, Alexandre J. (675), 316
Davis, Harriett, 262
Davis, John C. (233, 242, 262), 269, 271, 275
Davis, Kimberly (702), 318
DeCourcy, Samuel (297), 154, 281
Defendi, Vittorio (268, 307, 338), 282
Deforest, Adamadia (331, 518, 607), 151, 286, 311, 380
DeLamater, Edward D. (222, 227), 267-68
Delaware Hospital, 95, 186
Delette, B. (CM14), 290, 339
dengue, 106, 272
dental bacteriology, 36-37, 189
deRivas, Domaso, 72-73
dermatomycosis, 275
Derrick, Melva J. (235), 270
diabetes, 300
diagnostic microbiology, 96
Diarrhea, 258, 302
diarrheal disease, 141, 302, 360
Dickinson College, 17
DiCuollo, C. John (391), 297
Dienes, Louis L. (203), 264
Dietz, Catherine C. (183, 186), 260
Diller, Irene Corey (216), 266
diphtheria (*see also C. diphtheriae*), 12, 18, 21, 43-45, 49, 62, 196, 229, 246, 253, 271, 277, 299, 372
diphtheria antitoxin, 12, 21, 43, 45, 62
diphtheria bacillus, 49, 246
diphtheria diagnosis, 44, 196
disinfectants, 85, 96, 260
disk susceptibility testing, 96, 273
Distinguished Branch Member

Lecture Series. *See under*
Eastern Pennsylvania Branch
of the American Society for
Microbiology (ASM)
Dixon, Samuel Gibson, 15, 21, 23-24, 31, 191, 203, 229
DNA fingerprinting, 317
Dole, Marjorie, 153
Doms, Robert W. (603), 367
Dozois, T. F. (146, 155), 252, 254
Drake, Miles E. (222), 267
Drexel Institute, 75, 106
Drexel University College of Medicine, 165, 171-73, 179, 186, 367-70
Drutz, David J. (516, 532), 306
Dubos, Rene (355), 123, 378
DuPont Laboratories, 135, 165
Dutcher, Ray (326), 285
dysentery, 257, 273

E

Eagle, Harry (143, 237, 321), 251, 377
Earley, Elizabeth (259, 275), 274, 277
Eastern Pennsylvania Branch of
the American Society for
Microbiology (ASM)
and the Annual Industry Sponsored
Guest Lecture, 109-12, 123-24, 281-82, 284, 288
basic science symposia of, 3, 116, 120-21, 126, 129, 131-32, 140, 154, 173, 183, 215, 224, 342-43, 345-46, 355
Branch Archives of, 1, 4, 71, 77, 92, 187, 207, 210, 233, 245, 342, 345-46, 364, 372
branch meeting milestones
500th, 135, 137, 142, 218, 379
600th, 146, 151, 221, 380
branch meetings summaries (by decade), 85, 94-96, 105, 107, 118-20, 135-36, 146-48, 155, 158-59, 164, 166
clinical microbiology section of, 101-5, 107, 111-12, 115-16, 118-19, 123-27, 132, 173, 182-83, 213, 215, 220, 224, 235, 245, 337-38
committees of
education, 134, 143-44, 148, 153, 156-57, 163, 174, 177, 184, 224
executive, 4, 74, 115, 126, 133, 153, 159, 168, 170, 174, 176-78, 183, 215, 223, 225, 374
program, 77, 83, 89, 93-94, 101, 115, 208-9
symposium, 113, 115, 123, 170, 343, 345
and the Distinguished Branch
Member Lecture Series, 146, 150-51, 158, 183, 220
emeritus members of, 139-40, 142-43, 153-56
founding of, 71-73
and the gap of the regular branch
and clinical microbiology, 3-4, 113-17, 123-29, 131-32, 134, 145, 162, 172-73, 182-83, 185, 213, 215
and hosting the general meetings of
ASM, 73-74, 84, 101-2, 108, 116, 125, 162-63, 184, 201, 224
and the November Symposiums of, 117, 124, 126-32, 139-42, 144, 154-60, 168-74, 183, 342
origin and early years of, 73-77
and the Philadelphia Annual
Infection and Immunity Forums, 3, 146, 150, 155-60, 168-74, 177, 183-84, 219, 235, 362
regular branch of, 64, 78, 104, 112-15, 117, 123-29, 134, 145, 162, 172, 182-83, 185, 213, 215, 223, 337-39
and the Stuart Mudd Lectures, 116,

130-33, 139-44, 146, 150-51, 153-58, 175, 183, 216-17, 220, 297-300, 302-9, 371-72, 374
  student chapter of, 3, 150, 157-59, 171, 173-74, 177, 184, 221, 362, 364
Eastern State Penitentiary, 50, 63
Easton Hospital, 81
EB virus, 106, 120, 148
E. coli, 96, 106, 120, 147, 165, 275, 279, 282, 285, 289, 291-92, 298, 308, 373
ecology, 5, 120, 294
Egbert, Seneca, 31, 43, 191
Ehrich, William E. (176, 255), 259, 273
Ehrlichia sp., 148
Eichberg, Jorg, 157, 351, 355
Eiman, John A. (140, 167, 179), 250, 257, 259
Eisenberg, George (254), 273
Eisenstein, Toby K. (345, 363, 592), 121, 138, 140, 142, 153, 155, 157, 171, 217, 238-39, 241-42, 244, 289, 349-51, 354-55, 366-67
electron microscopy, 85-86, 95, 106, 139, 154, 187, 219, 260, 271, 300
Ellis, Alice (160), 255
emerging infections, 137, 144, 166, 171, 317, 344-45, 350, 353
Emini, Emilio A. (674), 172, 376
Emmart, Emily E. (176), 259
Emmons, Ellen (239), 270
Engle, Claire (300), 281
Englehardt, John F. (569), 309
Engley, Frank B., Jr. (180, 198), 260, 263
Ensinger, Maria (365), 125, 295
Ensminger, Marcia (364), 294
Enteric Bacteria, 123, 265, 307, 314, 359
Enterobacteriaceae, 128, 136, 296, 299, 359
enterococci, 165, 264, 317

enterotropic virus, 258
epidemiology, 99, 127, 129, 131, 136, 141, 148, 158, 269, 274, 281, 286, 288, 305-6, 344, 359-60
Episcopal Hospital, 152, 167, 241-42
Erf, L. A. (164), 256
Ericson, Lance, 127
Ernst, H. C., 61
Ernst, Paul, 31, 193
E. R. Squibb Laboratories, 128
Essentials of Bacteriology (Ball), 46, 50, 193, 230
Evangelista, Alan T. (696), 138-41, 152-53, 155, 158, 219, 238-39, 241-44, 349, 351, 354-55
Evans, G. (CM6), 288, 338
Eveleigh, Douglas, 163
Ewing, William (375, 425), 296
Eyring, Henry B. (168), 257
Ezekiel, David (293), 280

F

Facklam, Richard R. (383), 297
Fagan, Raymond (244, 261), 271, 275
Fang, Ferric C. (630), 313
Farber, Miriam B. (228), 269
Fare, Louis (338), 287
Faries Academy, 51
Favorite, Grant O., 91
Favour, Cutting B. (223), 268
Fearing, Mary (240), 271
Feinberg, Robert (212), 265
Feldman, William (263), 352-53, 355
Feldman-Rosen, Anna, xiii, 143, 152, 160, 167, 173, 241, 352-53, 355, 383
Felton, Harriet M. (162, 179, 195), 256, 259, 262
Felton, Lloyd D. (122), 247
Female Medical College of Pennsylvania, 10, 39
Fernandes, Prabhavathi (CM31), 295

Fertig, J. (121), 247
Field, A. Kirk (340), 288
Fighting Foes Too Small to See (McFarland), 68, 204
filtration, 60, 248, 262
Finegold, S. M. (CM15, CM23), 291, 293, 339-40
Finger, Irving (256), 274
Finkelstein, Richard (511), 305, 373
Fisher, Margaret (517), 305
Fishman, M. (330), 286
Fitzpatrick, Florence K. (163, 193, 213, 224), 262, 265, 268
Flexner, Abraham, 41
Flexner, Simon, 31-32, 38, 44, 49, 55, 63, 192, 199-201
Flick, J. A. (267), 275
Flick, Lawrence Francis, 15, 19-20, 23-25, 62, 66-67, 194, 201, 208, 229
Flippin, Harrison (264), 275
Flora and Fauna within Living Animals, A (Leidy), 16, 28
Flosdorf, Earl W. (119, 123, 146, 155, 162), 92, 246-47, 252, 254, 256
Forbes, Martin (273), 276
Force, E. E. (291), 280
Formad, Henry F., 15, 18-19, 23-25, 30-31, 54, 190, 192, 194, 229
Foster, C. (193), 262
Foster, J. W. (168), 257
Fowler, Elizabeth H. (178, 191. 224, 250, 389), 97, 107, 241, 259, 261, 268, 273, 297
Fowler, Haraold W. (167), 257
Fowler, Russell H. (140, 157, 179), 250, 255, 259
Fox, Herbert, 72
Fox Chase Cancer Center, 106, 119, 135, 139, 186, 242
Fraenkel, Carl, 50, 52
Fraimow, Henry (548), 156, 307, 351, 354
Fraimow, W. (CM10), 290, 339

Frank, Hilmer (282), 278
Frankel, Jack (247, 275), 277
Franklin, Benjamin, 10
Franklin Medical College, 16, 54
Franklin Plaza Hotel, 140-41, 349
Free, B. (CM6), 288, 338
Free, Elizabeth (277), 121, 241, 277
Freedman, Henry (310), 283
Free Hospital Association for Poor Consumptives, 20
frei test, 250
Fresh-Water Rhizopods of North America (Leidy), 16, 29, 189
Friedlander's bacterium, 252
Friedman, Herman (259, 281, 294, 353, 360, 363, CM11, CM24, CM25), 103-4, 107, 113, 121, 124-29, 131, 140, 173, 214, 223, 237, 290-91, 293-94, 339-40, 346-49, 354
Fritz, Mary Ann (312), 283, 295
Fukuyama, T. T. (278), 277
fungal chemotherapy, 99, 274
fungal infection, 156, 313, 317, 361
fungi, 31, 125, 130, 190, 266, 317
Furcolow, Michael L. (154), 254

G

Gaby, William L. (219, 277, 281, 283), 267, 277-78
Galan, Jorge, 365
Garcia, Lynne S. (575, 635), 168
Garreston Hospital, 67. *See also* Temple University
gas chromatography, 292, 294
Gaylor, Steven D., 309
Genco, Caroline, 169, 313, 376
Genetics, 5, 121, 131-32, 272, 274, 299, 302, 313, 317, 343, 348, 372
Genital Tract Infections, 155, 166, 172, 302, 345, 353, 361
genomics, 148, 160, 162, 165, 187,

316-17, 344, 352
Georgetown College, 20
Gerety, Robert (490), 303
Gerhard, William Wood, 27-28
German Hospital, xv, 45, 50, 57. *See also* Lankenau Hospital
Gettysburg College, 54
Ghannoum, Mahmoud A. (629), 312
Ghriskey, Albert A., 34, 49, 51, 56-57, 63, 193-94, 231
"Bacteria of Wounds and Skin Stitches, The," 51
Gibbs, Clarence (426), 299
Gibier, Paul, 45
Gibson, E. V. (155), 254
Giger, Olarae, 144, 152, 166-67, 169, 222, 238, 240-44, 350, 352, 354-55
Gilden, R. (319), 284
Gilead, Z. (319), 284
Gillespie, W. G., 199
Gillespie, W. J., 43-44, 196
Gilligan, Peter H. (513, 682), 305
Gilmore, Michael S. (695), 317
Ginsberg, Harold S. (288, 319), 107, 213, 237, 239, 279, 284
Ginsburg, D. (209), 265
Ginsburg, H. S. (357), 292
Ginsburg, Isaac (295), 280
Giovenella, Albert, 139, 383
Girardi, Anthony J. (222, 247, 319), 267, 272, 284
GlaxoSmithKline (GSK) Pharmaceuticals, 165
Glenn, John T. (149), 253
Gliocladium, 263
Goddard, Paul, 15
Goldfine, Howard (357), 292, 365
Goldman, Donal (530), 306
Golenbock, Douglas T., 365, 368
Gollub, S. (295), 280
gonorrhea, 85, 124, 254, 258, 282, 293, 340, 363

Goode, William (136), 250
Goodgal, S. H. (357), 292
Goodner, Kenneth (221, 244), 67, 97, 111, 210, 212, 237, 267, 271
Gots, Joseph S. (198, 215, 224, 225, 232, 248, 256, 268, 287, 300, 359, 389, 421), 107, 123, 131-32, 143, 212, 237, 239, 242, 263, 268-69, 272, 276, 279, 281, 297, 371-72
Goulet, Normand (291), 280
Graduate Hospital, 75, 81, 86, 186, 243
Graham, A. F. (293), 280
Grainger, T. H. (155), 254
gram-positive cocci, 147, 165, 252
Granoff, Allan (502), 373
Green, Judith (318), 284
Greenbaum, S. S. (151), 253
Greene, Arthur, 154
Greene, Joyce (331), 286
Greenhalgh, P. (CM19), 292, 340
Greenspan, George (391), 297
Gregory, James E. (CM7), 288
Griffin, Charles (290), 279
Griffin, Diane, 364
Grosebeck, Marjorie E. (252), 273
Groupe, Vincent (274), 277
Guiteras, Juan, 30-32, 37-38, 191-93, 195, 198-200
Gunn, John, 368
Gunter, J. H. (151), 253
Gunther, Cora B. (132, 145, 180, 229), 249, 252, 260, 269
Gutekunst, Richard R. (CM7, CM18, CM25, CM28, CM32), 113, 127, 288, 292, 294-95, 339-40, 342, 346-47, 354
Gwynedd-Mercy College, 167, 242

H

H1N1 influenza, 184
Haemophilus influenzae, 106, 123, 147, 290, 339, 369

Haemophilus pertussis, 252, 259, 264, 266, 270
Haemophilus sp., 106, 123, 147
Haff, Richard (340), 288
Hahnemann Medical Center, 242-43
Hahnemann Medical College, 10, 26, 40, 86, 95, 103, 105, 107-8, 119, 121-22, 135, 138-39, 147, 157, 185, 239-44
Hahnemann University, 137, 150, 153, 155, 167, 243, 362-63, 365, 380
Hall, Smith, 146, 148-50, 155, 183, 219, 307
Hamel, Mark, 150
Hampil, Betty Lee (141, 200, 221, 238), 251, 264, 267, 270
Hanna, Nabil, 142, 349-50, 354
Hansen, Sharon L. (547), 307
Harris, Susanna (195, 200, 228, 240, 255, 276, 294, 325, 334), 262, 264, 268-69, 271, 273, 277, 280, 285, 287
Harris, T. N. (176, 219, 228, 240, 255, 276, 294, 295, 311, 325, 334), 259, 267-69, 271, 273, 277, 280, 283, 285, 287
Harrison, James A. (178, 191, 224, 246, 250), 86, 97, 122, 211, 237, 239, 242-43, 259, 261, 268, 273
Hartman, Philip E. (233, 471), 269, 302, 372
Hartshorne, Henry, 29, 188
Haskin, Allen L. (391), 297
Hatfield, Charles J., 67
Havas, H. Francis (284, 303), 273, 278
Havens, W. Paul, Jr. (147, 255), 252, 273
Haverford College, 106, 186
Hayflick, Leonard (280, 323, 481), 141, 278, 285, 303, 372
Haymann, Hans (298), 281
Hechemy, K. (357), 292
Hegh, C. (CM28), 295, 340

Heifets, Leonid B. (572), 309
Heineman, Herbert (CM9), 156, 289
Heliobacter pyloris, 148
Henderson, Howard J. (120, 222, 258), 246, 267, 274
Henle, Gertrude (167, 174, 191, 195, 200, 222, 326, 359, 410), 124, 257-58, 262, 264, 267, 285, 293
Henle, Werner (152, 157, 158, 167, 174, 191, 195, 200, 214, 222, 272, 326, 342, 359, 410), 124, 253, 255, 257-58, 262, 264-65, 267, 276, 285, 288, 293, 299
Henry Phipps Institute for the Study, Prevention and Treatment of Tuberculosis, 20, 25, 52-54, 62-63, 65-66, 73, 75, 81, 84-85, 179, 185, 201-2, 205, 232, 238, 240
hepatitis, 85, 120, 136, 148, 166, 168, 259, 267, 290, 298-301, 303, 344-45, 349, 352, 372
herpes, 106, 120, 148, 293
Hewell, Barbara (154), 254
Hickman, Somia Perdow, 365
High, Robert H. (192), 295, 369
Hildreth, E. A. (333), 286
Hill, Gale (448), 378
Hiller, J. (216), 266
Hilton Hotel, 129-32, 347-48
Hinks, Eileen T., 138, 141-42, 144, 241, 349-50, 354
Hipp, Sally (470), 302
histology, 11, 24, 50-51, 63, 227
History of the Eastern Pennsylvania Branch of the ASM, A, 137
Hitch, Lola S., 74
Hitchens, Arthur Parker (135, 138), 71-73, 75, 79, 86, 201-4, 207-8, 211, 242, 250
H. K. Mulford Company. *See* Sharp and Dohme Pharmaceuticals
Hoban, R. T. (164), 256
Hobby, Gladys L. (264), 154, 275

Hodges, John (174), 258
Hodinka, Richard L. (579), 172, 353, 355
Hogan, R. B. (306), 282
Holiday Inn, 113, 342, 346
Holy Redeemer Hospital, 138, 152, 185, 219, 239-41, 244
Homeopathic Medical College of Pennsylvania, 40
Hoogerheide, J. C. (146), 252
hookworm, 55
Hopkins, W. J. (250), 273
Horner, William Edmonds, 15, 27
Hornick, Richard (473), 302
Horton, R. E. (273), 276
hospital infections, 99, 277
Hospital of the University of Pennsylvania, 28, 118, 123-24, 139, 168, 222, 306
Hottle, G. A. (121), 247
Houser, Enoch (297), 281
H. parapertussis, 256
H. pertussis, 85, 99, 254, 256, 262, 274
Hraba, Tomas (346), 289
Huang, N. (CM4, CM14), 288, 290, 338-39
Huff, J. W. (220), 267
Hughes, David (299), 364
human immunodeficiency virus (HIV), 311, 313, 318, 367
Hummeler, Klaus (227, 290, 334), 268, 279, 287
Hunt, Andrew D., Jr. (215), 266
Hunt, W. R. (121, 236), 247, 270
Hunter, Christopher A., 366
Huntoon, F. M., 75-76
Hutchinson, Wesley G. (165, 218, 352, 365), 86, 97, 125, 142, 241, 256, 266, 291, 295
Hwang, M. S. (139), 250
hygiene, 11-12, 21, 27-29, 33-34, 38, 46, 49, 57-58, 67, 149, 188, 191, 194

Hygiene of Transmissible Diseases, The (Abbott), 46, 50, 200, 230
Hygienisches Institut, 53

I

Ichelson, Rose (131), 249
Iglewski, Barbara Hotham (486, 577), 366, 373
immunoglobulins, 120, 343, 346
immunology, 4, 76, 85, 95-97, 99, 106, 111, 119, 130, 135, 147, 165, 173, 186-87, 206, 223-24
Immunology Club of Philadelphia, 111
Incubator, 84, 91
Indiana Normal College, 55
industrial microbiology, xvii
Infection Control, 12, 120, 130-31, 136, 142, 144, 343-44, 347, 350, 360
influenza (*see also* swine influenza), 60, 85, 96, 148, 165, 184, 203, 225, 247, 251, 253, 255, 257-58, 261-63, 275-77, 279-80
Ingalls, T. H. (327), 286
Ingham, J. L. (121), 247
Institute for Cancer Research, 97, 125, 139, 243
Institute of Allergy and Infectious Diseases, 364, 378
Institute of Hygiene, 21, 32-33
Institute of Infectious Diseases (Berlin), 22
Institute of the History of Medicine, 228, 230, 232, 380
interferon, 106-7, 187, 286, 288, 293, 301, 377
intestinal, 85, 147, 165, 298
intestinal flora, 98, 259, 276
intestinal parasites, 172, 361
Iralu, Vichazelu (CM17, CM26, CM29), 132, 292, 294-95, 339-40,

348, 354
Irr, Joseph D. (498, 509), 304
irradiation therapy, 256, 258, 261, 280
Isaacs, Stuart N., 367
Isberg, Ralph, 364, 368
Isenberg, Henry D. (375, CM27), 294, 296, 340
Isquith, Alan (314, 318, 324), 283-85
Israel, Harold L. (156), 254

J

Jackson, Sterling (271), 276
Jackson, W. (CM10), 290, 339
Jacobs, William R., 364
Jakubowitch, R. A. (CM5), 288, 338
Jarvis, William (533), 306
Jasewicz, L. (252), 273
Jeans Hospital, 97
Jefferson Alumni Hall, 118-19, 124-32, 135, 147, 164
Jefferson Hospital, 54, 75-78, 83, 103, 207, 238, 242
Jefferson Medical College, 10, 19, 23-24, 42-43, 54-57, 66-67, 75, 77-78, 85-86, 121, 123-24, 185, 194-97, 201-2, 209-10, 238-42
Jensen, Erling M. (247, 291), 272, 280
Jewish Hospital of Philadelphia, 85, 186
Johns Hopkins Hospital, 33, 51, 61
Johns Hopkins University, 29, 35, 44, 49, 81, 189, 227-28, 364, 366, 370
Johnson, Caroline C. (666), 171, 315, 376
Johnson, Frank H. (146, 168), 252, 257
Johnson, Wyatt, 61
Johnston, Richard (CM4), 155, 288, 338, 362
Joiner, Keith, 364
Jones, Charles A. (135, 139, 143), 249-51
Jones, H. W. (164), 256

Jones, J. H. (193), 262
Jordan, Edwin O., 46
Joseph Leidy (Warren, L.), 17, 228
Journal of Bacteriology, 61, 77, 79, 82, 90, 182, 203, 206, 211, 245
Journal of Clinical Microbiology, 117
Juneja, Vijay (567), 308
Jungkind, Donald J. (480, 554), 139, 142-44, 152-54, 156, 158, 166-68, 172, 174, 222, 238, 240, 242-44, 350-55, 383

K

Kalle, G. P. (287, 300), 279, 281
Kalter, Seymour S. (183), 260
Kandel, Judy (567), 308
Kaper, James, 369
Kaplan, Abram (267), 275
Kaplan, Albert (275, 293), 277, 280
Kapral, Frank (281, 317, 492), 284, 372
Karafin, L., 287, 338
Kast, Clara (123), 247
Kaye, Donald (370, 371), 296
Kazal, Henry, 154
Kearns, Wallace (168), 257
Kelner, Albert N. (186), 261
Kelser, Raymond A. (197, 221), 263, 267
Kemper, Walter, 22
Kereluk, Karl (282), 278
Ketchum, Karen A. (615), 312
Kiefer, Raymond H., 309
Kiley, Patricia J. (690), 317
Kimball, A. C. (146), 252
Kimmelman, Leonard (187), 261
Kincov, J. (121), 247
Kings College, 21
Kinyoun, J. J., 45, 201-2
Kirby, W. W., 291
Klebsiella sp., 165
Kleger, Bruce (607, 623), 138-39, 142, 144, 151, 153, 155, 167, 242-43,

311-12, 350-51, 354, 375, 380
Klein, Emanuel, 21
Klein, Morton (183, 187, 192, 200, 215, 229, 233, 244, 247, 259, 269, 275), 97, 131, 142, 156, 212, 237, 239, 260-63, 269, 271-72, 274, 276-77
Kligman, Albert M. (217, 229, 271), 266, 269, 276
Klinman, Norman (318, 320, 357), 284, 292
Kneass, Samuel Stryker, 31, 34, 36, 41, 49, 51, 56-57, 63, 193-94, 196, 198, 231
Knerr, W. P. (121), 247
Knight, Ralph A. (332, CM4, CM10), 103, 122, 142, 156, 220, 242-43, 286, 288, 290, 338-39
Knight, Simon, 172-74, 368-70
Koch, Robert, 12, 14, 20-22, 25-26, 30, 40, 42, 49-50, 53, 62, 191, 200, 202, 227-28, 309
Koch Institute, 40, 200
Kocholaty, Walter (158, 167), 255, 257
Koft, Bernard E. (215, 241), 266, 271
Koh, Won Young (224), 268
Kohler, Robert (549), 149-50
Kolmer, John A. (123, 160, 196, 239), 68, 73, 203, 247, 255, 263, 270
Practical Text-book of Infection, Immunity and Specific Therapy, 68, 203
Konzelmann, F., 76
Koprowski, Hilary (230, 442), 98, 140, 269, 372
Korean hemorrhagic fever, 299
Korman Suites Hotel, 156
Kozub, W. (CM6), 288, 338
Kraemer, Paul (312), 283
Kral, Frank (265), 275
Kravis, Lillian P. (218, 227), 266, 268
Krebs, Fred C., 367
Kreidler, William A., 67, 86, 209-10, 237-38

Kretschman, Georganne, 127
Kritchevsky, David (289), 279
Kruse, Connie, 155
Kundsin, Ruth (388), 297
Kunkel, L. O. (125), 248
Kutzler, Michele, 370
Kyle, D. Brandon, 39, 195

L

Laboratory Manual Project, 156
Lackman, David B. (119, 124, 132, 145, 148), 246-49, 252
lactose-fermenting organism, 250
Lafayette College, 81
laminar air flow, 288
Landau, Burton J. (558), 122, 243, 308
Landis, H. R. M., 67
Landy, Maurice (310), 283
Langner, Paul H. (140), 250
Lankenau Hospital, 51, 85, 95, 186
Laplace, Ernest, 15, 20-21, 23-25, 40, 54, 62, 192, 198, 204, 229
Laplace's solution, 21, 263
Larkin, Edward (326), 285
laryngitis, 259
La Salle College, 119, 186
Lattimer, Gary L. (402), 298
Lauderbach Academy, 51
Lautenbach, Ebbing, 172, 353, 355
Lawrence, H. Sherwood (348), 378
Lea, Henry C., 32
Leberman, Paul R. (170, 211, 217), 258, 265-66
Leboy, P. S. (302), 281
Lecce, James G. (245), 271-72
Lederle, 95
Lee, Henry F. (184, 190), 260-61
Lee, Margaret P. (184), 260
Leffingwell, J., 31, 192
Legionella, 136, 301, 305, 367
Legionnaire's disease, 120
Lehigh Valley Clinical Microbiology

Group, 134
Lehrer, Robert, 363
Leidy, Joseph, 2, 15-17, 23, 26, 28-30, 164, 188-89, 191, 193, 228-29
   Flora and Fauna within Living Animals, A, 16, 28
   Fresh-Water Rhizopods of North America, 16, 29, 189
Leidy, Joseph Jr., 73
Leishmania major, 368
Lenhart, Nancy (297), 281
Lennette, David A., 131, 348
leprosy, 246
Leptospira, 252, 272
leptospirosis, 252, 272
Lerman, Stephen (351), 291
Lesher, Eilene (289), 279
Lessner, James M. (236), 270
Lessons and Laboratory Exercises in Bacteriology (Smith, A.), 55
"Lesson Taught by the Epidemic at Plymouth Concerning Typhoid Fever, The" (Shakespeare), 17
Leszynski, Werner (159), 255
leukemia, 106, 111, 246, 282, 285, 290, 293
Lev, M. (273), 276
Levine, Arnold (324), 285
Levine, Myron (466), 302
Levine, Stanley (297), 281
Levinson, H. S. (242), 271
Levitt, Neil (324), 285
Levy, Richard (258), 274
Levy, Stuart B., 168, 313, 375
Lewis, Paul A., 67, 73
Lichstein, Herman C. (285), 108
Lief, Florence (296), 280
Lincoln, Clarence W., 45
Lincoln, I. (179), 259
Linna, T. Juhani (CM20), 131, 292, 340, 348, 354
Lippincott, J. B., 37, 228, 230, 232
Lister, Joseph, 13, 189

Listeria, 147, 306, 364-65, 368
listeriosis, 136
Listgarten, Max (427), 300
Liu, O. C. (291), 280
Liu, Zhi, 369
Live, Israel (144, 162, 265), 142, 251, 256, 275
Livermore, David M. (692), 173, 317
Llinas, Manuel (702), 318
Lockwood, John S. (137, 141), 250-51
Long, Carole A., 363
Long, Esmond R. (120, 149), 67, 228, 246, 253
Lorber, Bennett (504, 601, 650), 159, 170, 314, 376
Louis, Pierre A. C., 27
Love, W. G. (121, 144), 247, 251
Loveland, R. P. (205), 264
Lowery, Barbara, 156
Lucke, Baldwin (123), 247
Ludlum, S. de W. (193), 262
Lukens, Marguerite (118, 135, 143, 160), 246, 249, 251, 255
luminous bacteria, 252, 259
Luria, Salvador (322), 110, 285
Lurie, Max B. (120, 127, 142, 149, 156, 163, 173, 194, 234, 258), 248, 251, 254, 256, 258, 262, 269, 274
lyme disease, 136
Lymphopathia venereum, 80, 250
Lynch, Elsa R. (239), 270
Lynch, Frank, 73
Lynch, Helen M. (137), 250
Lynn, Raymond J. (245), 272
Lytle, Jean (240), 271

M

MacGregor, R. R. (CM24), 293, 340
MacLean, Estelle (168), 257
MacLeod, Colin (260), 274
MacNeal, Ward J. (130, 168), 257
Magasanick, Boris (308), 377

Mahoney, J. F. (153), 254
Main Line Clinical Laboratories, 166-67, 185, 222, 240, 243-44
malaria, 30, 42, 49, 125, 190, 295, 300, 302, 304, 312, 318, 340, 363, 366, 368, 370
Mandle, Robert J. (297), 121-22, 142, 157, 216, 220, 237, 239, 242, 244, 281, 371
Manko, M. (CM3), 287, 338
Manor Junior College, 138, 244
Manson, Lionel A. (267, 289), 275, 279
Mantua Academy, 21
Manuel, Sharron L., 369
Maragliano Institute, 53
Mark, George, III (351, 358, 359), 124, 291, 293
Marples, R. R. (CM26), 294, 340
Marriott Hotel, 123-24, 126-29, 131, 139, 346-47
Martin, Christopher (418), 299
Masachussetts Institute of Technology (MIT), 110
Maulitz, Russell C. (497), 14, 189, 227-28
Maurer, P. H. (334), 287
McCarthy, Daniel, 67
McCaughan, J. A. (262), 275
McCellan, George, 38
McCracken, Mary R. (165), 256
McCrea, J. F. (251), 273
McFarland, Joseph, xvii, 2, 14, 31-32, 38-41, 43-46, 48-49, 51, 53, 55-57, 63-66, 193, 195-96, 198-201, 228, 230-32
  "Beginning of Bacteriology in Philadelphia, The," 14, 48, 207, 228, 230, 232
  Fighting Foes Too Small to See, 68, 204
  Pathogenic Bacteria, The, 46, 52, 198, 231
McGarrity, Gerard (344, CM33), 288, 296, 341

McGowan, Karin L., 170, 353, 355
McGuinness, Aims C. (124, 162, 195, 221), 248, 256, 262, 267
McHale, Barbara, 158, 167, 242
McKenna, John (294), 280
McKinney, John D., 367
McKitrick, John C. (454), 103, 121-22, 138, 217, 238-39, 242, 244
McLean, Richard (284), 278
MCP-Hahnemann University, 152-53, 167, 169, 185, 221, 240, 243, 365-66
Medical College of Pennsylvania, 10, 26, 39-40, 52, 63, 65, 122, 135, 137-39, 142, 147, 157, 169, 199-200, 242-43, 379
Medical College of the State of South Carolina, 53, 56
Medical Corps of the U.S. Army, 52, 65. *See also* U.S. Army
Medical Microbiology, 67, 99, 120, 128, 173, 224, 269, 274, 276, 291, 343, 347
Medical Officers Training Corps, 52, 65
Medico-Chirurgical College of Philadelphia, 21, 24, 26, 40, 52-54, 56-57, 62-63, 65, 198
Mekalanos, John, 155, 363
Melvin, Mae (CM29), 295
meningitis, 28, 247, 257, 309, 367
Menzel, Arthur E. O. (168), 257
Mercantile College, 21
Mercer Memorial, 51
Mercersburg College, 54
Merck Sharpe & Dohme Laboratories, 98-99, 105, 108, 119, 135, 139, 177-78, 185. *See also* Sharp & Dohme Laboratories
Methodist Hospital, 127
MetPath Laboratories, 152, 185, 220, 239-41
Meyer, H. M. (327), 286

Michalka, Jack (351), 291
Micklin, Eugene (CM10), 154, 169, 290, 339
microbial diversity, 309, 311
microbial ecology, 5, 299, 311
Microbiological Club, 71-74, 181, 203, 233. *See also* Eastern Pennsylvania Branch of the American Society for Microbiology (ASM)
microbiology, xv-xvii, 1-5, 101-7, 110-20, 122-32, 134-37, 140-66, 168-74, 176-84, 212-16, 218-24, 290-92, 337-40, 342-45, 359-61, 378-81
Micrococcus pasteuri, 18, 30, 190
microscopy, 11, 15, 27, 39, 50, 55, 57-58, 63, 85-86, 95, 106-7, 112, 139, 154, 196, 260
Mikroorganismen der Mundhöhle, Die (Miller, W.), 36, 192
Miley, G. P. (164), 256
Millar, A. F. (121), 247
Miller, A. Katherine (192, 198, 203, 225), 262-64, 268
Miller, C. S. (220), 267
Miller, Gail (347), 290
Miller, I. George, Jr. (540), 373
Miller, Linda A., xvii, 137-38, 152, 158, 172-73, 219, 233, 238-39, 241, 244, 351, 353, 355, 379
Miller, Ruth Emma (127, 211, 235), 86, 97-98, 137, 142, 154, 169, 211, 222, 237, 239, 241-43, 248, 265, 270, 379
Miller, Virginia L., 365
Miller, Willoughby Dayton, 36, 192
  Mikroorganismen der Mundhöhle, Die, 36, 192
Millman, Irving (350, 396, 441), 139-40, 155, 242-43, 290, 298, 349, 354
Milton S. Hershey Medical Center, 123
Minnefor, Anthony (468), 302

Miovic, Margaret (345), 289
Moat, Albert G. (239, 259, 270, 314, 352, 389), 103-4, 107, 122-23, 127, 170, 213, 222, 237, 239, 242-43, 270, 274, 276, 291, 297, 383
Mobley, Harry L. T. (608), 311
Modlin, Robert, 364
Mohan, R. Ram (246, 298), 272
Molavi, John (488), 303
Molecular Diagnosis, 144, 148, 156, 158, 166, 170-71, 315
Moller, Goran (353), 291
monoclonal antibody, 302, 309
Moon, Robert (324), 285
Moore, Dan H. (390), 297
Morahan, Page Smith (460), 243, 302
Morgan, Isabel (124), 248
Morgan, Shannon L., 368
Morman, Edward, 137, 379-80
Morse, Stephen A., 367
Mortensen, Joel E., 155, 158, 351, 355, 362
Morton, Harry E. (119, 127, 141, 144, 150, 157, 163, 165, 167, 170, 180, 186, 211, 217, 245, 477, 549, 585 CM7, CM11, CM18), 76-78, 83, 86-87, 102-3, 137, 142-43, 146, 148-51, 206-7, 218-19, 237-38, 241-42, 255-58, 265-66, 374-75, 379
Mosser, David M. (612), 160, 352, 355, 369
Mudd, Stuart (119, 124, 145, 148, 155, 165, 179, 182, 186, 190, 205, 213, 221, 222, 233, 242, 258, 262, 269, 283, 297), 76, 86, 128-29, 132, 164, 183, 205, 208, 237-38, 246, 252, 267, 274-76, 278, 281, 371-72
Mulford, H. K., 44-45
Munder, R. (178), 259
Munich Institute, 49
Munoz, J. (227, 235, 238, 249), 268,

270, 273
mycobacteria, 106-7, 113, 126, 130, 136, 147, 165, 267, 274, 290, 309, 339, 359-60, 362
mycobacterial disease, 266, 308, 364
mycology, 85, 95-97, 106, 119, 128-29, 135-36, 139-40, 147, 153, 165-66, 168, 170, 172, 186, 344-45, 359-61
mycoplasmas, 96, 107, 111-12, 119-20, 136, 147, 187, 283, 285, 288, 297, 305, 338
Myrvik, Quentin N. (386), 297
myxomatosis, 275

N

Nachamkin, Irving (617), 152-53, 155, 162, 166-67, 169, 171, 221, 238, 240, 242-44, 352-53, 355, 362
Nair, Meera, 370
Natale, Phyllis (259), 274
Nathanson, Neal (439), 300
National Association for the Study and Prevention of TB, 20, 62
National Drug Company, 50, 65, 77, 81, 205, 243
National Institute of Health, 81, 378
natural sciences, 15-16, 21, 27-29, 189
Nazerian, Keyvan (326), 285
Neefe, John R. (177), 259
Nelson, John B. (125), 248
Nelson, Waldo E. (154, 156), 254
neomycin, 96, 265
Neter, Erwin (349, CM13), 290
Neuman, Robert (286), 279
Neumann College, 143-44
neurological disease, 302
New York Times, 19, 229
Niagara University, 50, 63
Nicholas, Leslie (384), 297
Nichols, Warren (307), 282
Nickel, Lois (312), 283

Nicolas, L. (CM20), 292, 340
Nightingale, F. F. (283), 278
nitrofurans, 96, 268
Normark, Staffan J., 363
Norris, George W., 67
Norris, Robert, 143
North, Leon L. (180), 260
Northrop, John H. (125), 248
Nosocomial Infections, 125, 142, 148, 295, 304, 306, 317, 340, 360
Nowotny, Alois (310), 283
Nussenzweig, Victor, 363

O

Obold, Walter L. (171), 76, 258
O'Brien, Alison D. (559, 660), 164, 224, 373
O'Donnell, Edward (339), 287
O'Donnell, Judith A. (582), 309
Ogburn, C. A. (325), 285
Ohio State University, 368
O'Kane, Daniel (278), 277
Okono, C. (121), 247
O'Leary, D. K. (159), 255
ophthalmia, 251
Opie, Eugene L., 67, 75, 205, 237-38
opportunistic infections, 148, 364
oral flora, 96, 273
"Original Minutes of the Society of American Bacteriologists, The" (Roos, Christopher G.), 79, 249
Origin of Species, The, 15
Osborne, William, 154
Oskay, John J. (245), 271
Osler, William, xv, 30, 42, 51, 190
Ossman, E. (CM24), 293, 340
Ottolenghi, Abramo (295), 280
ozone hole, 314

P

Pacis, Macello (168), 257

Pagano, Joseph F. (391), 121-22, 138, 169, 216, 222, 238-39, 241-44, 297
Painter, Heather J., 368
Pakpo ur, Nazzy, 368
Palmer, Carroll E. (154), 254
Pamer, Eric G., 365
Pancoast, William, 40, 192
Panos, C. (323), 285
Pappenheimer, Alwin M. (152, 411), 253, 372
Paracolon Bacillus, 262
parasitic diseases, 142, 299, 360
parasitology, 16, 23, 55, 63, 85, 96, 119-20, 132, 154, 158, 160, 221, 298, 344, 348, 359-61
paratyphoid infection, 247
Park, J. T. (273), 276
Park, William H., 43, 45
Parke, Davis and Company, 52, 65
Parker, Ernestine R. (168), 257
parotitis, 262
Parr, Erma I. (142, 149), 251, 253
parvovirus, 136, 301
Pasculle, A. William (443), 301
Pasternak, V. Z. (211), 265
Pasteur, Louis, 12, 18, 21-22, 40, 192, 229
Pasteur Institute, 31, 45, 51-53, 65, 193
Paterson, Yvonne, 365
Pathogenic Bacteria, The (McFarland), 46, 52, 198, 231
Pathogens, 120, 126, 136, 140, 142, 297, 307, 317, 343-44, 347, 349, 360, 364, 367-68
Pathological Society of Philadelphia, 79, 89
pathology, 1, 11-12, 15, 17, 24, 27, 29-30, 49, 51, 54, 57-58, 64, 71, 188-89, 201-3, 227-29
Paucker, Kurt (251, 275), 273, 277
Paul-Vermitsky, John, 368
Pearce, Edward, 367

Pearson, Leonard, 45, 53
Pease, Herbert D., 41, 43-44, 196, 198-99
Peckham, Adelaide Ward, 22, 34, 39, 49, 52, 56-57, 63, 194-95, 200, 231
Pedersen, Kai C. (133), 249
Pelouze, P. S. (153), 254
pelvic inflammatory disease, 136
penatin, 257
Penicillin, 85, 96, 259-60, 262-64, 268, 270, 273, 276
Penicillium, 257, 309
Pennell, Robert (190, 218), 266
Pennsylvania College of Podiatric Medicine, 121, 137, 153, 242-43, 380
Pennsylvania Department of Health, 106, 113, 121, 138-39, 149, 151, 153, 170, 179, 186, 210-11, 214, 239, 242, 359, 380
Pennsylvania Department of Health Laboratories, 75, 81, 113, 121, 138, 149, 170, 214, 239, 241, 244, 359
Pennsylvania Hospital, 10, 26, 28, 38, 41-42, 45, 102-3, 118-19, 124-25, 153, 186, 242, 383
Pennsylvania Medical Journal, 51
Pennsylvania Medical Society, 87, 209
Pennsylvania Society for the Prevention of Tuberculosis, 19, 194
Penn Tower Hotel, 156-58, 363-64
Peoples, Don M. (245), 272
Pepper, D. Sergeant (124), 248
Pepper, William, Jr., xv, 32, 35, 193
Perez, J. E. (200), 263
Persing, David, 158, 351, 355
Petermann, M. L. (152), 253
Peterson, Arthur (282), 278
Pett, Donald (351), 291
Pettenkoffer, Max, 49

Pettit, Horace (119, 124, 483), 246, 248
Pfizer Inc., 135, 178
Phalle, M. de Saint (209), 265
pharmaceutical industry, 9, 44, 174, 177-78, 228
phase microscopy, 260, 268
Philadelphia Academy of Natural Sciences, 15
Philadelphia Almshouse. *See under* almshouse
Philadelphia Annual Infection and Immunity Forum. *See under* Eastern Pennsylvania Branch of the American Society for Microbiology (ASM), 150, 155-60, 168-74, 184, 219, 235, 362
Philadelphia Bug Club. *See* Microbiological Club
Philadelphia Chapter of the Society of American Bacteriologists, 1
Philadelphia Civic Center, 125
Philadelphia College of Osteopathic Medicine, 10, 137, 139, 153, 242-43, 379-81
Philadelphia College of Pharmacy and Science, 10
Philadelphia Convention Center, 163
Philadelphia County Medical Society, 76, 78, 83, 94, 206, 212
Philadelphia Dental College, 67. *See also* Temple University
Philadelphia Ledger, 19, 229
Philadelphia Naval Hospital, 85, 186
Philadelphia Polyclinic and College for Graduates in Medicine, 26, 41, 51, 198
Philadelphia Public Health Department, 26, 43, 55, 63, 119, 121, 138, 196, 199, 217, 240, 242
Philadelphia Public Health Laboratory, 72
Philadelphia State Hospital, 127
Philadelphia VA Medical Center, 166-67, 185, 224, 240, 244
Phipps, Henry, 20, 66
Physiological Society of Philadelphia, 79
Pickard, Allan (364), 294
Pidcoe, Vern (CM9, CM11), 128, 155, 289-90, 339, 347, 354
Pierce, Nathaniel (CM31), 295
Pijper, Adrianus (180, 205), 264
Pillsbury, D. M. (209), 265
Pinto, Carl (340), 288
Pioneer Microbiologists of America (Clark, P.), 11, 227-28, 230
Pitfield, Robert Lucas, 34, 41, 45, 49, 52, 56-57, 63, 68, 198, 202, 231
Compendium on Bacteriology Including Animal Parasites, A, 53, 68, 202, 231
Pizer, L. I. (314), 283
plague, xv, 22, 164, 224, 276, 297, 369
platinum culture transfer wire loop, 53
Platsoucas, Chris, 156, 220
Plenum Press, 140-42, 155-56, 345, 349-51
pleuropneumonia-like organisms (PPLO) (*see also* mycoplasmas), 95-96, 98-99, 106, 187, 264-66, 271-72, 274, 278, 283
Plotkin, Stanley A. (327, 469, CM3), 286-87, 302, 338
pneumococcal vaccine, 153, 172, 223
pneumococci, 52, 65, 74, 106, 147, 165, 172, 223, 263, 294, 301
pneumonia, 74, 151, 288, 301, 318, 381
Polevitzky, Katherine (148, 241), 252, 271
poliomyelitis, 85, 96, 247, 253, 262, 272, 274-75
polio vaccine, 99, 280
poliovirus, 106, 283, 285
Pollikoff, Ralph (234), 269
Polyoma Virus, 282
Porges, Nandor (243, 252), 271, 273

Porphyromonas gingivali, 165, 376
Portney, Daniel (534), 155, 306, 362, 364
Poupard, James A. (543, 607), 122, 127, 137-41, 143, 150-53, 155, 163-64, 171, 217, 229, 233, 238-39, 242-44, 349, 353-55, 379-81
Poussin, Mathilde A., 368
Practical Text-book of Infection, Immunity and Specific Therapy (Kolmer), 68, 203
Preer, John R. (204, 276, 302), 264, 277, 281
Price, Alison H. (174, 202), 258, 264
Prier, James E. (286, 333, 358, 364, 607, 658, CM3, CM9, CM22), xvi-xvii, 103, 111, 113, 118, 121, 124-29, 131, 137, 151, 214, 286-87, 293-94, 346-48, 354, 375-76
Principles of Bacteriology, The (Abbott), 46, 193
Principles of Hygiene, The (Bergey), 50
promin, 256
Protez Pharmaceuticals, 165, 186
protozoa, 16, 28, 109, 188, 261, 281
protozology, 16, 29, 189
Pseudomonas sp., 96, 136, 277, 298, 303, 366
pseudo rabies. *See* rabies
psittacosis-lymphogranuloma venereum group, 267, 269

Q

Quest Diagnostics Inc, 153, 167, 243
Quinn, J. (CM5), 288, 338
Quivik, Frederick, 150

R

rabies, 12, 96, 98, 106, 136, 140, 252, 269, 275, 286, 290, 298, 300, 369, 372

Rabin, Harvey (259), 274
Rabinowitsch, Lydia, 15, 21-22, 25, 35, 39-40, 197, 199-200, 229
Raefler, Janet (156), 254
Rajan, T. V., 366
Rake, Adrian (293), 280
Rake, Geoffrey (168), 257
Rakoff, A. E. (137), 250
Ranck, E. V., 45
Randall, Eileen L. (303, 440, CM2, CM6, CM12, CM15, CM17, CM21, CM23, CM27, CM30, CM31, CM32), 103, 112, 121, 123, 154, 218, 241, 282, 287-88, 290-95, 300, 338-40
Rao, N. U. (238), 270
rapid diagnosis, 124, 136, 140, 142, 168, 178, 252, 274, 284, 294, 314, 340, 344, 350, 360-61
Ratcliff, H. L. (187), 261
Rauscher, Frank (274), 277
Ravdin, I. S. (124), 248
Raven, Clara (147), 252
Ravenel, Mazyck Porcher, 34-35, 38, 43, 49, 53, 56-57, 63, 66, 194-97, 199, 201, 229, 231
Rawlings, W. B. (121), 247
Rawnsley, H. M., 123, 346, 354
Ray, Verne (432), 300
Rebell, G. (209), 265
Redowitz, Edward (140), 250
Reed, Walter, 17
Reeves, H. C. (314), 283
Reich, P. R. (323), 285
Reichel, John, 72-73, 75-76
Reimann, Hobart A. (126, 136, 147, 154, 174, 185, 202), 248, 250, 252, 254, 258, 260, 264
Reiner, Steven L. (628), 312
Reinhold, John G. (202), 264
Reisman, David, 32, 199
replicate plating, 273
Report on Cholera in Europe and India

(Shakespeare), 17, 42, 192
Research Institute of Cutaneous
   Medicine, 75, 81
respiratory syncytial virus (RSV), 148,
   165
respiratory tract infections, 107, 112,
   141, 157, 250-51, 272, 275, 287-
   88, 303, 305, 309, 338, 360-61,
   364
Rest, Richard F., 141, 150, 152-53, 155-
   58, 167, 170-72, 217, 221, 238,
   240, 243-44, 362-64, 366-67
retroviruses, 136, 148, 165, 302
Reyniers, James A. (210), 265
rheumatic fever, 249-50, 268
Rhodes, Russel (262), 275
Rich, Marvin (307, 326), 282
rickettsial disease, 107, 113, 251, 264,
   289
Rights, F. L. (121), 247
Rigney, Elizabeth (171), 258
Riley, Eleanor, 368
Riley, Randolph (CM9), 289
Robb, Hunter, 51
   "Bacteria of Wounds and Skin
      Stitches, The," 51
Robert Hare Laboratory, 36
Robinson, George, 72
Rocco, Loretta, 127
Rockborn, Gunnar (265), 275
Rockefeller Institute for Medical
   Research, 80-81, 98, 109, 377
Rocky Mountain Spotted Fever, 256,
   269
Rogers, M. L. (268), 276
Rogers, Thomas J., 366
Rolling Hill Hospital, 138, 241
Ronkin, R. R. (302), 281
Rooney, Tom, 158
Roos, Christopher G. (134, 146), 75,
   78-79, 92, 109, 207-8, 213, 237-
   38, 249, 252, 365
   "Original Minutes of the Society of
American Bacteriologists, The,"
   79, 249
Roos, David, 365
Rosanoff, Eugene (225), 268
Rose, B., 76
Rose, Elizabeth Kirk (221), 267
Rose, Milton J. (151), 253
Rose, Noel (451), 301, 372
Rose, S. Brandt (127, 158, 202), 248,
   255, 264
Rosebury, Theodor (160), 255
Rosen, George, 19, 228
Rosenberg, Martin (520), 305
Rosenberger, Randle C. (127, 174), 3,
   39, 49, 54, 56-57, 64, 66-67, 72-
   73, 75-78, 199, 201-2, 204, 206,
   210, 231, 237-38
rotavirus vaccine, 176
Rothstein, E. L. (267), 275
Rous sarcoma, 96
Rowan, Regina (377, 389), 296-97
Rowe, Wallace (335), 378
Roy, Anirban, 369
Royal College of Physicians, 10
Roy and Diana Vagelos Laboratories,
   149
rubella, 105-7, 113, 214, 286, 342-43,
   346
Rubenstein, Arthur, 164, 224
Rubin, Ben A. (288, 325, 340, 408),
   279, 285, 287-88, 298, 338
Rule, Anna M. (123), 247
Rush Hospital for Consumption and
   Related Diseases, 20
Russell, David G., 366
Rutgers University, 143, 168, 222

S

Sage, Dorothy N. (173, 179), 258-59
Saint Christopher's Hospital, 135, 147,
   186
Saint Joseph's Hospital, 16

Saint Joseph's University, 147, 158, 186
Salgame, Padmini, 159-60, 168, 364-66
Salinger, Julius, 38, 191
Sall, T. (283), 278
Sallman, Bennett (206), 264
Salmonella, 106, 120, 147, 165, 263, 273-74, 289, 306, 363, 370, 373
Samaritan Hospital, 67. *See also* Temple University
Sanghvi, Viraj (702), 318
Sangree, Ernest Brewster, 40, 49, 54, 56-57, 63, 231
Santer, Melvin (287), 279
Santer, Ursula (263), 275
Sappington, S. W., 73
sarcoma. *See* Rous sarcoma
Satterlee Military Hospital, 16
Satz, Jay E. (318, 438, CM20), 113, 123-24, 126-31, 284, 292, 340, 342, 346-48, 354, 383
Sawitz, William G. (229), 269
Saylor, Robert M. (142, 149), 251, 253
Sbarra, Anthony J. (397), 298
Schaechter, Moselio (461), 140, 372
Schaedler, Russell W. (394, CM15), 172, 223, 291, 297, 339
Schatz, Albert (242, 246, 271), 271-72, 276
Schatz, Vivian (242), 271
Scheidy, Samuel F. (246), 272
Schering-Plough Corp., 178
Scherp, H. W. (123), 247
Schlottman, Dorothy W. (225), 268
Schoble, Otto, 72
Schorr, Sonia E. (215, 233), 266, 269
Schreck, Kenneth (277, 297, 317), 277, 284
Schuchardt, Lee F. (235, 238, 249), 270, 273
Schultz, Mark P. (140), 250
Schwartzman, Robert (420), 299
Scolnick, Edward (459), 302
Scott, Dwight McNair (263, 289, 303), 275, 282
Scott, Elvyn G. (CM12, CM15, CM18, CM30, 254), 127, 273, 290-92, 295, 339
Scott, Joseph P. (144, 147, 164), 251-52, 256
Scott, Philip A., 365
Scott, T. F. McNair (150, 195), 253, 262
Sedgwick, W. T., 61
Segall, Stanley (282), 278
Seibert, Florence B. (120, 133, 149, 156, 176), 92, 246, 249, 253-54, 259
Seibles, T. S. (250), 273
septicemia, 18, 30, 190, 259
serology, 85, 95, 106, 187
Serratia sp., 96, 266, 273, 275, 285
Sevag, Manasseh G. (132, 145, 152, 155, 196, 198, 206, 211, 215, 225, 235, 242), 111, 249, 252-54, 263-66, 268, 270-71
severe acute respiratory syndrome (SARS), 165, 314
sewage disposal, 60
sexually transmitted diseases (STDs), 131, 140, 166, 172, 345, 353, 360
Shadomy, S. (CM17), 292, 339
Shaffer, F. W. (195), 262
Shakespeare, Edward Orem, 15, 17-18, 23-25, 42-43, 181, 190-92, 201, 229
  "Lesson Taught by the Epidemic at Plymouth Concerning Typhoid Fever, The," 17
  Report on Cholera in Europe and India, 17, 42, 192
Shapiro, L. (CM22), 293, 340
Sharp & Dohme Laboratories, 26, 44, 52-53, 57, 63, 65, 72-75, 78-79, 81, 85-86, 95, 185-86, 195, 207-8, 211, 238
Sharp and Dohme Pharmaceuticals,

26, 44, 52-53, 57, 63, 65, 195
Sharpless, G. (216), 266
Shaw, Dorothy R. (123, 136), 247, 250
Shay, Donald, 137, 143, 380
Shelburne, Myrtle (155), 254
Shenker, Bruce (489), 303
Sher, Alan, 364
Sheraton Hotel, 110
Shlaes, David M. (559), 311
Shockman, Gerald D. (283, 298), 160, 221, 278, 281
Shramm, J. R. (138), 250
Sigel, M. Michael (195, 203, 218, 222, 227, 229 234, 356), 262, 264, 266-69
Silberman, Ronald (283), 278
Singer, Arthur G. (191), 261
Siu, R. G. H. (215), 265
Skeggs, Helen R. (220, 232), 142, 267
skin diseases, 141, 159-60, 255, 275, 360-61
skin tests, 256, 299
Sloan, Bruce (331), 286
slow virus, 120, 136, 165, 299-300
Smalley, David (524), 305
smallpox, 106, 120, 148, 280, 287, 367
Smibert, Robert (280), 278
Smith, Allen John, 31, 49, 54, 56-57, 63, 192, 202, 231
  Lessons and Laboratory Exercises in Bacteriology, 55
Smith, Andrew G. (213), 265
Smith, Dale (487), 303
Smith, George H., 72
Smith, Harry (243, 404, CM31), 143, 271, 295, 298, 383
Smith, J. Bruce (436), 300
Smith, Louis DeSpain (145), 252
Smith, L. W. (216), 266
Smith, Paul F. (211, 217, 245, 259, 280), 265-66, 274, 278
Smith, Theobald, 61, 128
Smith, William Elliott (190, 199), 261,

263
Smith Kline & French Animal Health Farm, 124
Smith Kline & French Award for Excellence, 122-23
Smith Kline & French Pharmaceuticals, 59, 95, 99, 105, 112, 116, 119, 121-22, 135, 137-38, 143, 152, 160, 177, 185, 239-44
SmithKline Beecham Pharmaceuticals, 139, 147, 150, 152, 159-60, 380-81
Smith Kline Diagnostics, 121-22, 241
Smolens, Joseph (132, 152, 179), 91, 249, 253, 259
Smythe, Dorothy (284), 278
Sobernheim, George, 52
Society of American Bacteriologists (SAB). *See* American Society for Microbiology (ASM)
Solowey, Mathilde (198), 263
Somerson, N. L. (323), 285
Somia Perdow Hickman, 365
Somkuti, George (407), 298
Sommer, Harriet E. (158), 255
Spanish American War, 18
Sparling, Philip Frederick (568), 309, 373
Spaulding, Earle H. (136, 156, 186, 219, 224, 229, 238, 254, 264, 277, 352, 365, 595, CM1, CM15, CM24), 86, 104, 125-26, 131, 139, 157, 209-10, 220, 237-38, 254, 269-70, 291, 338-40, 347, 354, 375
Specter, Steven, 130-31, 348, 354, 383
spirochetes, 255, 268
Spitzer, Judy A. (303, 311), 282-83
Spizizen, John (196, 220, 238), 263, 267, 270
Spongiform Encephalopathies, 303
Sprague, J. M. (186), 261

Springer, Georg F. (254, 267, 273), 273, 275-76
Sputnik, 111
Squib, 95
staining, 31, 41, 53-55, 192, 198, 265
Stanley, Wendell M., 99
staphylococci, 85, 96, 106, 136, 147, 165, 257, 281, 311
State Live Stock Sanitary Bureau of Pennsylvania, 53
Stavitsky, Abram (184), 260
Steers, Edward (196, 215), 263, 266
Stelos, Peter (320), 284
Stempen, Henry (218), 143, 168, 222, 266
Stempin, Harry, 131, 216
Stengel, Alfred, 36
sterilization, 249
Sterling Pharmaceuticals, 135, 186
Stern, Kurt G. (152), 253
Sternberg, George Miller, 18, 30, 190, 229
Stevens, Kingsley (255), 273
Stewart, Alonzo H., 39, 44, 49, 55-57, 63, 196-97, 199, 211, 231
Stewart, Gordon (341), 288, 378
Stieritz, Donald D., 122, 138-39, 141, 169-70, 217, 238-39, 241-42, 244, 349, 352-55
stilbamidine, 96, 271
Stinebring, Warren (247, 274), 272
Stipcevich, Jane, 127
Stock, Neale (326), 285
Stockton, J. R. (238), 270
Stokes, Joseph, Jr. (123, 124, 136, 141, 158, 168, 222), 247-48, 250-51, 255, 257, 267
Stoller, J. (CM19), 292, 340
stomatitis, 253
Strauss, Robert R., 132, 348, 354
Strawinski, R. J. (173, 178), 258-59
streptococci, 52, 65, 80, 85, 96, 106, 120, 136, 147, 165, 246-50, 253-54, 260, 280, 282, 297
streptomycin, 85, 96, 260-62, 265-66, 268, 270, 272
Strumia, Max M. (119, 126), 246
Stuart Mudd Lectures. *See under* Eastern Pennsylvania Branch of the American Society for Microbiology (ASM)
Stubbs, Evan L. (142, 144), 75-76, 92, 142, 154, 206, 218, 237-38, 251
St. Vincent's College, 19
SugarLoaf Conference Center, 160, 168, 365
sulfa drugs, 85, 96
Summers, William C., 164
Supina, Walter (354), 292
Sussdorf, Dieter H. (294), 280
Sutter, V. L. (CM23), 293, 340
SV40 (Simian virus 40), 106, 120, 284, 287, 378
Swanson, Michele S., 367
Swarthmore College, 16, 165
Sweeney, Frank J. (261, 317, CM2), 274, 284, 338
Swenson, R. M. (CM24), 293, 340
swine influenza, 85, 163, 251, 256, 298
symbiosis, 253
syphilis, 85, 249, 251, 255, 261, 270
Systemic Mycosis, 123, 305, 359
Szybalski, Waclaw (248), 272

T

Tabachnick, Joseph (236), 270
Taichman, Norton (489), 155, 303, 362
Takeya, Kinji (258), 274
Talbot, George (476), 303
Tarleton, Rick L., 367
Tashman, Sylvia (215), 266
Taylor, Inez (168), 257
Taylor, Mark (318), 284
Teller, David (287), 279
Teller, Ida (192), 262

Temple University Conference Center, 128-30
Temple University School of Medicine, 65-67, 85-86, 97, 103-5, 108-13, 121-22, 124-25, 128-30, 138-39, 153-54, 165-68, 210-12, 220-21, 223, 238-44, 365-68
Tenover, Fred C. (523, 609), 305, 311
terramycin, 96, 266
test. *See* individual tests
tetanus, 45, 205, 249
Thiele, Elizabeth H. (202, 218), 264, 266
Thomas, Charles C., 113, 228, 345-46
Thomas A. Scott Fellow in Hygiene, 50, 53, 194-96
Thomas Jefferson Hospital, 54, 76, 207
Thomas Jefferson University, 118-19, 122-24, 127-28, 130, 135, 139-40, 142, 147, 150-55, 157, 164-67, 171-72, 219-23, 230-32, 240-44, 362
Thompson, Harriet C. W. (363), 294
Thompson, John H., Jr. (409, 423, 542), 307
Thompson, Kenneth, 131, 348, 354
Thomson, Richard B., Jr. (549), 149-50, 307
Thorat, Swati, 370
Thornsberry, Clyde (605), 159, 375
Tiedje, James M. (687), 316
Tilghman, Shirley Caldwell (437), 300
Tilton, Richard (403), 298
Tint, H. (296), 280
Tiselius, Arne (133), 249
tissue culture, 96, 187, 272, 280
Toennies, Gerrit (268), 276
Tolchin, Sidney (236), 270
Topics in Clinical Microbiology audio set, 116-18, 125, 215
Topping, Norman (233), 269

toxic-shock syndrome, 136, 303
traveler's illnesses, 313
Trelawny, Gilbert (246), 272
trichina larvae, 16
Tripodi, Daniel (324), 285
Truant, Alan, 143, 383
tubercular meningitis, 28
tuberculin, 12, 45, 249, 253-54, 268, 272
tuberculosis, 12, 18-22, 24-25, 53-54, 62-63, 66-67, 85, 194, 200-202, 232, 248, 251, 253-54, 258-59, 267-68, 282
Tudor, John (558, 567), 308
Tuft, Louis (132), 249
Tufts University, 364, 368
Tulane University, 21
tularemia, 85, 251-52
Tumilowicz, Joseph (326), 285
tumors, 110, 148, 266, 273, 276, 278, 284, 293, 296
typhoid fever, 17, 28, 60, 247, 249-50, 256, 260, 262, 265, 302, 368
typhus fever, 28, 260, 262, 265
Tyson, James, 14, 29-31, 189, 191-92
Tytell, Alfred (286), 279

U

Umbreit, Gerald (354), 292
Umbreit, W. W. (292), 280
Unger, J. B. (291), 280
University Hospital, 30, 35, 190, 193
University of Berne, 22
University of California, 99, 363, 370
University of Chicago, 46, 89
University of Connecticut, 366
University of Delaware, 126-27, 369
University of Heidelberg, 18
University of Louisiana, 20, 23-24
University of Maryland, 49, 56, 369, 380
University of Massachusetts, 368
University of Michigan, 21, 367

University of Minnesota, 108
University of Missouri, 53
University of Pennsylvania, 14-20, 23, 83-86, 94-99, 121-25, 137-39, 149-53, 165-74, 183-85, 192-93, 205-8, 221-22, 238-44, 353, 363-64, 366-67
  Department of Comparative Pathology and Parasitology, 55, 63, 202
  Laboratory of Hygiene, 21-22, 26-27, 30, 32-34, 37, 43, 49-53, 57-58, 64, 74-75, 149, 162-64, 183-84, 193-98, 200-201, 224
  School of Dental Medicine, 26, 29, 36, 179, 189, 199, 238, 369
  School of Veterinary Medicine, 26, 30, 37, 179, 190-91, 199-200, 238, 365, 368-69
  William Pepper Laboratory of Clinical Medicine, 26, 35-36, 51, 63, 179, 196, 229
University of Pennsylvania Faculty Club, 137, 149-50, 218-19, 379
University of Rochester, 366
University of South Florida, 173, 224
University of Texas, 55, 63, 122
University of the South, 53
University of Virginia, 44
University of Wisconsin, 53, 63
University of Zurich, 21
University Park Press, 124, 126-31, 345-48
Ureaplasma, 302
urinary tract infection (UTI), 107, 111, 283, 338
U.S. Army, 17, 52, 65, 73, 87-89, 93, 182, 207, 209, 265
Utz, John (380), 296

V

Vaccines, 45, 85, 95-96, 106, 119, 125, 135-36, 140, 147-48, 157, 165, 300-302, 305-6, 344, 349, 363
vaginosis, 313
Vaidya, Akhil, 314, 366
Valenta, Joseph (391), 297
Vali, N. A. (283), 278
Valley Forge General Hospital, 106, 186
Vanderbilt University, 54, 63
van Furth, Ralph (570), 309
Van Slyke, C. J. (153), 254
Vaughan, Victor, 17
Velicer, Leland (331), 286
venereal disease, 80, 120, 124, 136, 147, 258, 292, 340
Verwey, Willard F. (163,173, 178, 179, 192, 202, 203, 206, 225, 235, 238,240,257), 84, 86, 99, 211, 237-38, 242, 256, 258-59, 262, 264, 268, 270-71, 274
Verwey, William, 86, 211, 237-38
veterinary medicine, xvii, 26, 30, 33, 37-38, 45, 49, 53, 63-64, 66, 75, 190-91, 198-201, 271-72, 365, 368-69
Vibrio, 16, 147, 165, 275, 295, 369
viola, 339
virology, 2, 4, 85, 95-97, 103, 106, 119-20, 131, 135-36, 147-48, 151, 165, 186, 299, 305, 360
Virus Detection, 148, 154, 344, 350
in vitro assays, 298
Vogt, Agnes Beebe (145, 149), 252-53

W

Wagman, Gerard (391), 297
Wagner, Robert R. (251, 328), 273, 378
Wagner Free Institute of Science, 16
Waksman, Byron (329, 472), 286, 302
Waksman, Selman A. (161), 84, 88, 209
Walker, L. (179), 259
Wallis, A. D. (188), 261

Walsh, Joseph, 67
Walsh, Lori, 143, 155-56, 160, 351, 354, 383
Wang, Zi-Xuan, 174, 353, 355
Warren, George H. (178), 103, 107-8, 131, 214, 237, 239, 242-43, 259
Warren, Leonard, 17, 228
Joseph Leidy, 17, 228
Washington University, 363, 365
Wasserman, Aaron E. (236, 250), 266, 270
water filtration system, 60
Waters, J. R. (CM27), 294, 340
W. B. Saunders and Company, 46, 50, 68, 193, 198, 200, 230-31
Weaver, R. E. (CM21), 293, 340
Webster, Guy (606), 159, 311
Weed, L. L. (220), 267
Wegner, W. (314), 283
Weinstein, Marvin (391), 297
Weintraub, S. E. (121), 247
Weiser, Jeffrey N. (661), 364
Weiss, Alison (539), 306
Weiss, Charles (186, 236, 262), 260, 270, 275
Weiss, Susan R., 366
Weissman, S. M. (323), 285
Welch, William H., 35, 43-44, 49
Wells, Dorothy (145, 154), 252, 254
Wells, William F. (131, 142, 145, 160, 171, 187, 195), 92, 249, 251-52, 255, 258, 261-62
Weltman, Andre (623), 167, 375
Wesleyan University, 46
West, Margaret King (236, 240), 270-71
West Jersey Hospital, 127
West Nile virus, 165, 314
West Philadelphia Academy, 51
Wheelock, Frederick (360), 293
White, Benjamin, 71
White, C. Y., 72-74, 76
White, Phillip R. (168), 257

Whitman, William B., 164, 224
whooping cough, 85, 256, 259
Wigdahl, Brian (686), 316, 367
Wiktor, Tadeusz (333, 408), 286
Wilcke, Burton W., Jr. (614), 160, 375
Wilhelm, J. M. (357), 292
Willard, Cecelia (195), 262
Willett, Norman P. (389, 558, 567, 594, 642), 121, 131, 157, 169, 216, 237, 239, 241, 244, 308, 313, 375-76
William Pepper Laboratory of Clinical Medicine, 26, 35, 179, 193, 196
Williams & Wilkins Publishing Company, 117-18, 214-15
Williams, Ned B. (204, 241, 243, 250, 271), 264, 271, 273, 276
Willits, C. O. (282), 278
Will's Eye Hospital, 135, 186
Wilmer, Dorothy (203), 264
Wilmington Medical Center, 127
Wilson, D. W. (220), 267
Winterscheid, Loren C. (222), 267
Wintersteiner, O. (168), 257
Wise, Robert (252, 261, 264), 273-75
Wistar, Isaac, 43
Wistar and Horner Museum of Anatomy, 43
Wistar Institute of Anatomy and Biology, 10, 26, 43, 105, 119, 135, 179, 185, 195
Witlin, Bernard, 154
Wohl, Michael G. (202), 264
Wolfe, Francis D. (139), 250
Wolfe, Ralph S. (361, 419), 294, 299, 378
Wolman, I. L. (123), 247
Woman's Medical College of Pennsylvania, 22, 25-26, 39, 52, 54, 56, 63, 65-66, 75, 81, 86, 97, 106, 109, 119, 199-200
Wong, Samuel (286), 279
Wood, Thomas (256), 274

Woodward, C. R., Jr. (168), 257
Woodward, George, 43
Woolley, D. W. (189), 261
World Affairs Council of Philadelphia, 158
World War I, 60-61
World War II, xvi, 61, 83-84, 88, 93, 167, 182, 209, 221
Wright, James Homer, 34, 194
Wright, Lemuel D. (197, 220, 241), 263, 267, 271
Wuerthele-Caspe, V. (216), 266
Wyeth Laboratories, 95, 99, 103, 105, 107-8, 119, 165, 177-78, 185-86, 214, 240, 242-43

Y

Yale University, 364-65
yeast, 125, 158, 265, 270, 274, 276, 281, 294, 340, 361
yellow fever, 2, 30, 191
York Academy, 54
Young, Clarence L. (689), 317
Young, Frank (379), 296
Young, Viola Mae (CM14, CM19), 290, 292, 339-40
Yuk, Ming H., 366
Yungbluth, Margaret M., 160, 352, 355

Z

Zablotney, Sharon L. (558), 308
Zajac, Ihor (283, 312), 278
Zappasodi, Peter (156, 234, 258), 254, 269, 274
Zavala, Fidel P., 370
Zeldow, B. J. (250), 273
Zellat, Joseph (275), 277
Zilinskas, Raymond A. (585), 158, 375
Zillessen, F. O. (121), 247
Zinder, Norton D. (315), 284, 377
Zintis, Harold A. (209), 265
Zittle, Charles A. (137, 152), 250, 253
Zi-Xuan Wang, 174, 353, 355
Zubrzycki, Leonard, 104, 107, 110-11, 122, 241-42, 383

www.ingramcontent.com/pod-product-compliance
Lightning Source LLC
Chambersburg PA
CBHW031814170526
45157CB00001B/53